Sustainable Utilization of Nanoparticles and Nanofluids in Engineering Applications

Sampath Boopathi
Mechanical Engineering, Muthayammal Engineering College, India

J. Paulo Davim
University of Aveiro, Portugal

A volume in the Advances in Chemical and Materials Engineering (ACME) Book Series

IGI Global
PUBLISHER of TIMELY KNOWLEDGE

Published in the United States of America by
IGI Global
Engineering Science Reference (an imprint of IGI Global)
701 E. Chocolate Avenue
Hershey PA, USA 17033
Tel: 717-533-8845
Fax: 717-533-8661
E-mail: cust@igi-global.com
Web site: http://www.igi-global.com

Library of Congress Cataloging-in-Publication Data

Names: Boopathi, Sampath, 1980- editor. | Davim, J. Paulo, editor.
Title: Sustainable utilization of nanoparticles and nanofluids in engineering applications / edited by Sampath Boopathi and J. Paulo Davim.
Description: Hershey, PA : Engineering Science Reference, [2023] | Includes bibliographical references and index. | Summary: "Sustainable Utilization of Nanoparticles and Nanofluids in Engineering Applications will aim to provide the latest research and significant potential to apply nanomaterials and nanofluids in various engineering applications. It will be identified as a reference guide because it provides real-life problems with feasible potential systems, models, and examples related to the application, synthesis, innovations, and properties of advanced nanomaterials and nanofluids. Covering key topics include such as sustainable development, utilization, and innovation, this reference work is ideal for industry professionals, material scientists, manufacturing industry people, Material selectors, Design engineers, Machine instructors, administrators, researchers, academicians, under and postgraduates, scholars, instructors, and students, etc"-- Provided by publisher.
Identifiers: LCCN 2023012738 (print) | LCCN 2023012739 (ebook) | ISBN 9781668491355 (h/c) | ISBN 9781668491362 (s/c) | ISBN 9781668491379 (eISBN)
Subjects: LCSH: Nanoparticles. | Nanofluids. | Sustainable development.
Classification: LCC TA418.9.N35 S887 2023 (print) | LCC TA418.9.N35 (ebook) | DDC 620/.50286--dc23/eng/20230727
LC record available at https://lccn.loc.gov/2023012738
LC ebook record available at https://lccn.loc.gov/2023012739

This book is published in the IGI Global book series Advances in Chemical and Materials Engineering (ACME) (ISSN: 2327-5448; eISSN: 2327-5456)

British Cataloguing in Publication Data
A Cataloguing in Publication record for this book is available from the British Library.

All work contributed to this book is new, previously-unpublished material.
The views expressed in this book are those of the authors, but not necessarily of the publisher.

For electronic access to this publication, please contact: eresources@igi-global.com.

Advances in Chemical and Materials Engineering (ACME) Book Series

ISSN:2327-5448
EISSN:2327-5456

Editor-in-Chief: J. Paulo Davim, University of Aveiro, Portugal

MISSION

The cross disciplinary approach of chemical and materials engineering is rapidly growing as it applies to the study of educational, scientific and industrial research activities by solving complex chemical problems using computational techniques and statistical methods.

The **Advances in Chemical and Materials Engineering (ACME) Book Series** provides research on the recent advances throughout computational and statistical methods of analysis and modeling. This series brings together collaboration between chemists, engineers, statisticians, and computer scientists and offers a wealth of knowledge and useful tools to academics, practitioners, and professionals through high quality publications.

COVERAGE

- Nanomaterials
- Coatings and surface treatments
- Materials to Renewable Energies
- Wear of Materials
- Fatigue and Creep
- Computational methods
- Sustainable Materials
- Ceramics and glasses
- Ductility and Crack-Resistance
- Thermo-Chemical Treatments

IGI Global is currently accepting manuscripts for publication within this series. To submit a proposal for a volume in this series, please contact our Acquisition Editors at Acquisitions@igi-global.com or visit: http://www.igi-global.com/publish/.

The Advances in Chemical and Materials Engineering (ACME) Book Series (ISSN 2327-5448) is published by IGI Global, 701 E. Chocolate Avenue, Hershey, PA 17033-1240, USA, www.igi-global.com. This series is composed of titles available for purchase individually; each title is edited to be contextually exclusive from any other title within the series. For pricing and ordering information please visit http://www.igi-global.com/book-series/advances-chemical-materials-engineering/73687. Postmaster: Send all address changes to above address.

Titles in this Series

For a list of additional titles in this series, please visit:
http://www.igi-global.com/book-series/advances-chemical-materials-engineering/73687

Handbook of Research on Advanced Functional Materials for Orthopedic Applications
R. Ranjith (SNS College of Technology, India) and J. Paulo Davim (University of Aveiro, Portugal)
Engineering Science Reference • copyright 2023 • 400pp • H/C (ISBN: 9781668474129) • US $325.00 (our price)

Production, Properties, and Applications of Engineered Cementitious Composites
S. Praveenkumar (PSG College of Technology, India) and J. Paulo Davim (University of Aveiro, Portugal)
Engineering Science Reference • copyright 2023 • 300pp • H/C (ISBN: 9781668481820) • US $250.00 (our price)

Innovative Multifunctional Nanomaterial for Photocatalysis, Sensing, and Imaging
Azad Kumar (Siddharth University, India)
Engineering Science Reference • copyright 2023 • 300pp • H/C (ISBN: 9781668487433) • US $270.00 (our price)

Modeling, Characterization, and Processing of Smart Materials
Ajay . (Mechanical Engineering Department, School of Engineering and Technology, JECRC University, India) Parveen . (Mechanical Engineering Department, Rawal Institute of Engineering and Technology, India) Ashish Kumar Srivastava (G.L. Bajaj Institute of Technology and Management, India) and Vikas Goyat (Mechanical Engineering Department, SRM Institute of Science and Technology, India)
Engineering Science Reference • copyright 2023 • 320pp • H/C (ISBN: 9781668492246) • US $270.00 (our price)

For an entire list of titles in this series, please visit:
http://www.igi-global.com/book-series/advances-chemical-materials-engineering/73687

IGI Global
PUBLISHER of TIMELY KNOWLEDGE

701 East Chocolate Avenue, Hershey, PA 17033, USA
Tel: 717-533-8845 x100 • Fax: 717-533-8661
E-Mail: cust@igi-global.com • www.igi-global.com

Table of Contents

Detailed Table of Contents

Nanoparticles, which are particles with dimensions on the nanoscale, have a wide range of applications across many different fields. Nanoparticles are used in many fields, such as electronics, medicine, cosmetics, manufacturing, and materials science. They have unique properties such as a high surface area-to-volume ratio, high reactivity, and the ability to interact with biological systems. In this chapter, the applications and effects of nanoparticles in conventional—Lathe, Grinding, Milling, and drilling processes—and unconventional machining processes—electro-chemical machining, electro-discharge, ultrasonic, abrasive waster-jet machining—processes are illustrated. Special applications of nanoparticles in medicine, agriculture, food, construction, biotechnology, and IC engines have also been discussed. The limitations of using nanoparticles in engineering applications have also been explained.

Lightweight materials and their alloys are increasingly used in automobiles and aircraft due to their high specific strength and strength and weight ratio. The characteristics

Nanomaterials reduce biodegradable pollutants before promoting standard levels. Thus, nanomaterials could efficiently and sustainably treat environmental contaminants. However, additional research is needed to determine the destiny of environment remediation nanomaterials. This review covers biological and plant-based bioremediation nanotechnologies. Nanomaterials reduce waste and harmful material degradation costs. Nanomaterials/nanoparticles immediately catalyse waste and toxic material breakdown, which is hazardous to microorganisms, and enable microorganisms degrade waste and toxic materials more efficiently and sustainably.

Nanotechnology is a cutting-edge scientific discipline that focuses on manipulating matter at the nanoscale scale to create new and useful materials, gadgets, and systems (1–100 nm). One of the most rapidly expanding, and commercially viable niches in the cosmetics industry is the use of micellar nanoparticles in these products. The improved qualities of the nano-sized particles in terms of colour, transparency, solubility, etc., are largely responsible for the extensive impact of nanotechnology in the cosmetics sectors. Cosmetics use a wide variety of nano materials, including nanosomes, liposomes, fullerenes, solid lipid nanoparticles, etc. Recent worries about the potential dangers of nano cosmetics have prompted the cosmetics industry to restrict the use of nanotechnology in cosmetics and push for stricter regulations requiring a thorough safety review of the products before they hit store shelves. This chapter looks at the latest developments in nano-sized materials for use in cosmetics in an effort to address these problems despite the not-so-obvious hidden hazards.

Increased energy consumption is causing a constant rise in global temperature above pre-industrial levels, with harmful gas and radiation emissions causing catastrophic climatic conditions. To address this issue, it is essential to plan thermal solutions that are continuing and extremely efficient. Because of its exceptional qualities, scientists are doing research on diverse nanofluids to develop routine in thermal applications. has attracted substantial interest in recent years. However, because they are more reactive in solvents, it raises safety and health problems. Due to the presence of powerful chemicals, other undesirable consequences such as rusting of equipment and non-biodegradable by products develop. The current study provides a thorough overview of the many methods used to create green nanofluids utilizing a variety of natural extracts. Green nanoparticles are a good choice for improving the efficiency of thermal systems due to their distinctive shape, optical characteristics, stability, large surface area, low toxicity, and improved thermo-physical properties.

Advancement of nanotechnology leads to the creation of innovation in nanostructures which are characterized by their explorative utilization in a variety of applications in biosciences. Nanofluid technology is an interdisciplinary field, where nanoscience, nanotechnology, and thermal engineering have enhanced over the past decades. The diagnosis and therapeutic aspects of diseases are found to be improved very efficiently in drug delivery platform by using nanofluids. Nanofluids contain novel properties which are potentially employed in several applications like biotechnology, national security and defense, food and agriculture, information technology, aerospace, plastics and textile industries, energy and environment, cosmetics, and health and medicine. This chapter discusses the future perspectives and applications of nanofluids.

Nanoparticles and nanofluids are dynamic area of research due to their broad applications in various emerging field such as medicine, energy, and environmental science. Recent advancements in this field include the development of new synthesis techniques for creating nanoparticles with specific properties, important uses of nanoparticles in drug delivery systems in medical field, and the exploration of using nanofluids as coolants in thermal management systems. However, there are also challenges that need to be addressed in this field. One of the main challenges is understanding the behavior and properties of nanoparticles in different environments, as they can be affected by factors such as size, shape, and surface chemistry. Additionally, there are concerns about the possible health and environmental hazards associated with the use of nanoparticles, and more research is needed to understand and mitigate these risks.

Preface

Nanotechnology has revolutionized industries by harnessing the unique properties of nanoparticles and nanofluids, which offer applications in engineering processes, materials science, medicine, cosmetics, and etc., *Sustainable Utilization of Nanoparticles and Nanofluids in Engineering Applications* explores the wide-ranging impact of nanoparticles and nanofluids across different sectors. Nanotechnology has emerged as a powerful scientific discipline with potential to revolutionize industries. Nanoparticles and nanofluids have been used in engineering, medicine, cosmetics, environmental science, and more.

The book explores the wide-ranging impact of these nanoscale entities and provides a comprehensive analysis of their sustainable use in different fields. Nanoparticles possess remarkable properties such as a high surface area-to-volume ratio, high reactivity, and the ability to interact with biological systems. Their applications extend to conventional manufacturing processes such as lathe, grinding, milling, and drilling, as well as unconventional machining processes like electro-chemical machining, electro-discharge, ultrasonic machining, and abrasive water-jet machining.

This book delves into the effects of nanoparticles in these processes, shedding light on their advantages, limitations, and potential for enhancing engineering applications. The book travels the special applications of nanoparticles in fields such as medicine, agriculture, food, construction, biotechnology, and internal combustion engines. It focuses on their role in improving concrete quality, reducing environmental pollutants, enhancing cosmetic products, advancing thermal systems, and transforming healthcare. Readers gain insights into the potential benefits of integrating nanoparticles and nanofluids into various domains.

The target audience of this book is researchers, engineers, scientists, professionals, and students interested in understanding the applications and advancements in nanotechnology. It caters to individuals seeking to explore the innovative utilization of nanoparticles and nanofluids and their implications in engineering processes, materials science, healthcare, cosmetics, and environmental science.

Additionally, it serves as a valuable resource for policymakers and regulators involved in establishing guidelines and regulations for the safe and sustainable use of nanotechnology. This book provides a comprehensive analysis of sustainable utilization in engineering applications, bridging the gap between research and practical implementation. It aims to inspire further research, innovation, and responsible use of nanoparticles and nanofluids by presenting a wide range of applications, discussing current limitations, and exploring future directions.

Chapter 1 delves into the applications of nanoparticles in conventional and unconventional manufacturing processes. It highlights their effects on processes such as lathe, grinding, milling, drilling, electro-chemical machining, electro-discharge, ultrasonic machining, and abrasive water-jet machining. Furthermore, special applications of nanoparticles in medicine, agriculture, food, construction, biotechnology, and IC engines are discussed, along with the limitations encountered in engineering applications.

Chapter 2 focuses on the influences of nanoparticles in the friction stir welding process, particularly in joining aluminum alloys. As lightweight materials gain prominence in automotive and aerospace industries, traditional joining processes face challenges. This chapter provides an extensive review of recent research on nanoparticle-influenced friction stir welding, addressing advancements, applications, and unaddressed issues in joining aluminum alloys.

Chapter 3 explores the impact of nanomaterials on enhancing concrete quality. By utilizing nanotechnology, researchers have discovered methods to prevent or reduce concrete deterioration, improve compressive strength, and reduce permeability. The chapter also examines the role of nano-silica particles in cement hydration, rheology, and strength development.

In Chapter 4, the prospects for ecological advantages through nanoparticle characterization and bioremediation are explored. Nanomaterials have shown potential in reducing biodegradable pollutants and promoting sustainable treatment of environmental contaminants. This chapter discusses biological and plant-based bioremediation nanotechnologies and their ability to catalyze waste breakdown and enhance microbial degradation.

Chapter 5 focuses on the applications of nanoparticles in the cosmetic field. Nanotechnology has had a profound impact on the cosmetics industry, with micellar nanoparticles playing a crucial role. The chapter examines various nano materials used in cosmetics, recent developments, and the need for safety regulations to address potential hazards.

Chapter 6 delves deeper into nanotechnology's role in cosmetics and cosmeceuticals. Nanoparticles offer innovative solutions in formulation and delivery systems, improving product efficacy and consumer satisfaction. However, concerns about nanotoxicity necessitate careful consideration of health hazards. This chapter discusses nanotechnology approaches in delivering cosmetics and cosmeceuticals, as well as their benefits, safety, and future directions.

Chapter 7 explores the integration of the Internet of Things (IoT) and Artificial Intelligence (AI) with nano-materials research. The chapter presents a system design and architecture that utilizes AI, IoT, and ML-based solutions for data collection, analysis, and decision-making in the nano-materials industry. The incorporation of robust security measures ensures data privacy and enhances productivity and quality.

Chapter 8 examines the applications of nanofluids in various engineering fields. Nanofluids, fluid suspensions of nanoparticles, find use as coolants, lubricants, additives in coatings and adhesives, and even printing inks. The chapter explores their role in enhancing machining accuracy, reducing tool wear, improving drug delivery, and increasing the visibility of tissues and organs.

Chapter 9 focuses specifically on the applications of nanofluids, or nano-cutting fluids, in advanced machining processes. By improving lubrication and cooling properties, nanofluids enhance surface finish, reduce tool wear, and increase material removal rates. Their thermal conductivity, lubrication properties, and ability to reduce friction make them valuable in various cutting processes.

Chapter 10 addresses the innovative study of green nanofluids in thermal systems. As energy consumption and global temperature rise, efficient thermal solutions become crucial. The chapter explores the use of green nanoparticles to improve the efficiency of thermal systems, highlighting their distinctive properties and potential health and environmental implications.

Chapter 11 discusses the future perspectives and applications of nanofluids in healthcare. Nanofluid technology offers promising advancements in drug delivery, biotechnology, and various industries. The chapter explores the potential of nanofluids in improving diagnosis, therapy, and other biomedical applications.

Finally, Chapter 12 presents the advancements and future challenges in nanoparticle and nanofluid research. Recent developments, such as new synthesis techniques and applications in drug delivery and thermal management systems, are highlighted. However, challenges remain, including understanding the behavior of nanoparticles in different environments and addressing health and environmental concerns.

The book *Sustainable Utilization of Nanoparticles and Nanofluids in Engineering Applications* provides a comprehensive analysis of their applications and sustainable use across various industries. It explores the unique properties and potential of nanoparticles and nanofluids, and highlights their special applications in medicine, agriculture, food, construction, biotechnology, and internal combustion engines. The book explores the potential of nanomaterials in concrete quality, reducing environmental pollutants, and catalyzing bioremediation processes. It also examines the role of nanoparticles in the cosmetic industry, emphasizing the need for regulation and safety assessment. The proposed system design and architecture for data collection, analysis, and decision-making in the nano-materials industry presents a blueprint for improved quality, efficiency, and waste reduction. Nanofluids are used in thermal systems, cutting processes, and green thermal applications to improve heat transfer, lubrication, and material removal rates. This book provides valuable knowledge for researchers, engineers, scientists, professionals, students, and policymakers.

The book *Sustainable Utilization of Nanoparticles and Nanofluids in Engineering Applications* encourages sustainable engineering practices, improved material performance, enhanced healthcare solutions, and eco-friendly applications.

Sampath Boopathi
Mechanical Engineering, Muthayammal Engineering College, India

J Paulo Davim
University of Aveiro, Portugal

Acknowledgment

The Almighty has given us the strength, wisdom, and perseverance to complete this book chapter on the sustainable utilization of nanoparticles and nanofluids in engineering applications.

IGI Global's Publisher, Editor-in-Chief, and reviewers provided valuable contributions and support during the publication process, which helped shape the content of this book chapter.

The authors of the book chapters in "Sustainable Utilization of Nanoparticles and Nanofluids in Engineering Applications" have contributed to the overall content and made the publication a comprehensive resource.

Thank you to all those who have contributed to the success of this book chapter, as their support, guidance, and inspiration have been instrumental in my professional growth.

Dr. Sampath Boopathi is grateful to his father, Mr. V. Sampath, and mother, Mrs. S. Banumathi, for their love, encouragement, and support in all his endeavors. He also acknowledges the moral support and understanding provided by his wife, Mrs. B. Sangeetha, and daughters, B. Prathiksha and B. Subhiksha, for their patience, understanding, and encouragement during the challenging phases of completing this work.

The content of this acknowledgment section is intended to express gratitude and appreciation to those who contributed to the creation and completion of the book chapter. It adheres to IGI Global guidelines.

Thank you for your support and dedication to the project.

Sincerely,

Sampath Boopathi
Mechanical Engineering, Muthayammal Engineering College, India

J Paulo Davim
University of Aveiro, Portugal

Chapter 1
Applications of Nanoparticles in Various Manufacturing Processes

Sampath Boopathi
Mythayammal Engineering College, India

J. Paulo Davim
https://orcid.org/0000-0002-5659-3111
University of Aveiro, Portugal

ABSTRACT

Nanoparticles, which are particles with dimensions on the nanoscale, have a wide range of applications across many different fields. Nanoparticles are used in many fields, such as electronics, medicine, cosmetics, manufacturing, and materials science. They have unique properties such as a high surface area-to-volume ratio, high reactivity, and the ability to interact with biological systems. In this chapter, the applications and effects of nanoparticles in conventional—Lathe, Grinding, Milling, and drilling processes—and unconventional machining processes—electro-chemical machining, electro-discharge, ultrasonic, abrasive waster-jet machining—processes are illustrated. Special applications of nanoparticles in medicine, agriculture, food, construction, biotechnology, and IC engines have also been discussed. The limitations of using nanoparticles in engineering applications have also been explained.

INTRODUCTION

Nanoparticles are particles that have a size range of 1-100 nanometers. They exhibit

DOI: 10.4018/978-1-6684-9135-5.ch001

unique and useful properties due to their small size, such as high surface area-to-volume ratio, quantum confinement, and surface effects. These properties make nanoparticles highly desirable for a wide range of applications in various fields such as electronics, medicine, energy, and materials science. Nanoparticles can be made from a variety of materials such as metals, metal oxides, polymers, and ceramics. They can be synthesized using various techniques such as chemical methods, physical methods, and biological methods. The unique properties of nanoparticles have led to a wide range of potential applications, such as targeted drug delivery in medicine, high-performance coatings in materials science, and high-efficiency solar cells in energy. However, nanoparticles also pose potential risks, such as toxicity and environmental impact, which must be carefully considered and addressed. Nanoparticles have enormous potential in many different fields due to their unique properties, and their use is expected to continue to grow in the future as scientists and engineers continue to explore their applications and capabilities(Mohanraj & Chen, 2006).

Nanoparticles and Nano powders are both materials with particle sizes in the nanometre range, but there are some differences between them. Nanoparticles are typically individual particles with a size range of 1-100 nanometres in at least one dimension. They have a high surface area-to-volume ratio and can exhibit unique properties due to their small size. Nanoparticles can be synthesized using various methods such as chemical, physical, and biological methods. Nano powders, on the other hand, are collections of nanoparticles that can be composed of one or more materials. Nano powders can be produced using methods such as gas-phase synthesis, wet chemical synthesis, and solid-state reaction. Nano powders can have a wide range of particle sizes and shapes and may exhibit different properties compared to their constituent nanoparticles. Overall, while nanoparticles and Nano powders both have small particle sizes, they can have different properties and applications depending on how they are synthesized and used(Welch & Compton, 2006).

Nanoparticles have a wide range of applications due to their unique properties such as small size, high surface area-to-volume ratio, and high reactivity (Figure 1). Here are some of the applications of nanoparticles in different fields(Biswas & Wu, 2005; Mohanraj & Chen, 2006):

- Electronics: Nanoparticles such as quantum dots are used in electronic displays and solar cells.
- Energy: Nanoparticles are used in the production of fuel cells, batteries, and solar panels.
- Medicine: Nanoparticles are used in drug delivery, imaging, diagnostics, and tissue engineering.

Figure 1. Applications of nano-particles in engineering fields

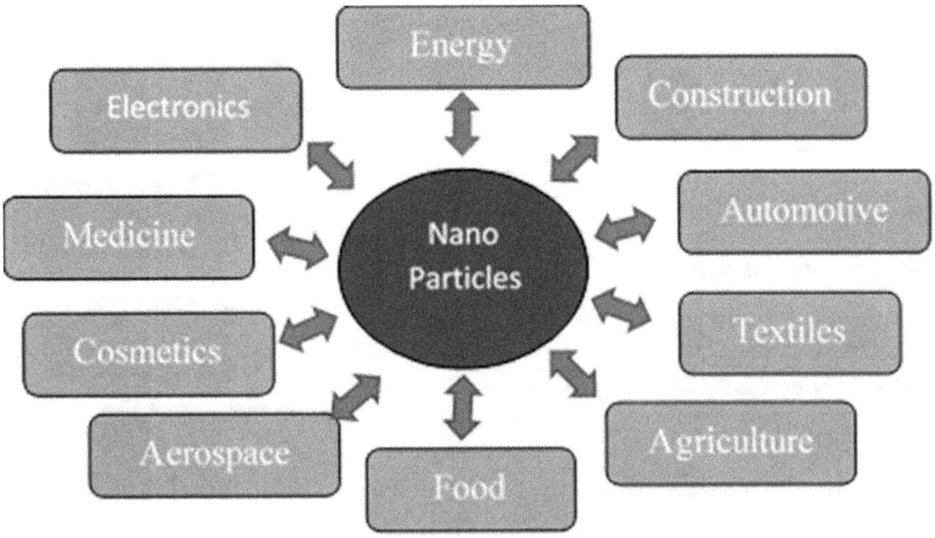

- Cosmetics: Nanoparticles such as titanium dioxide and zinc oxide are used in sunscreen and other cosmetic products.
- Food and agriculture: Nanoparticles are used in food packaging to improve shelf life and reduce contamination. They are also used in agriculture for crop protection and nutrient delivery.
- Environmental remediation: Nanoparticles are used in the removal of pollutants from air and water.
- Construction: Nanoparticles are used in the production of concrete and other construction materials to improve strength and durability.
- Textiles: Nanoparticles are used in fabrics to improve properties such as water repellence, stain resistance, and durability.
- Automotive: Nanoparticles are used in the production of automotive parts to improve strength, durability, and performance.
- Aerospace: Nanoparticles are used in the production of aerospace materials to improve strength, reduce weight, and enhance performance.

Overall, nanoparticles have a wide range of applications across different fields due to their unique properties, and their use is expected to grow in the coming years as more applications are discovered and developed.

Nanoparticles have emerged as a significant area of interest in various machining processes due to their unique properties and potential benefits (Figure 2). In this review,

Figure 2. Application of nano-particles in the various manufacturing processes

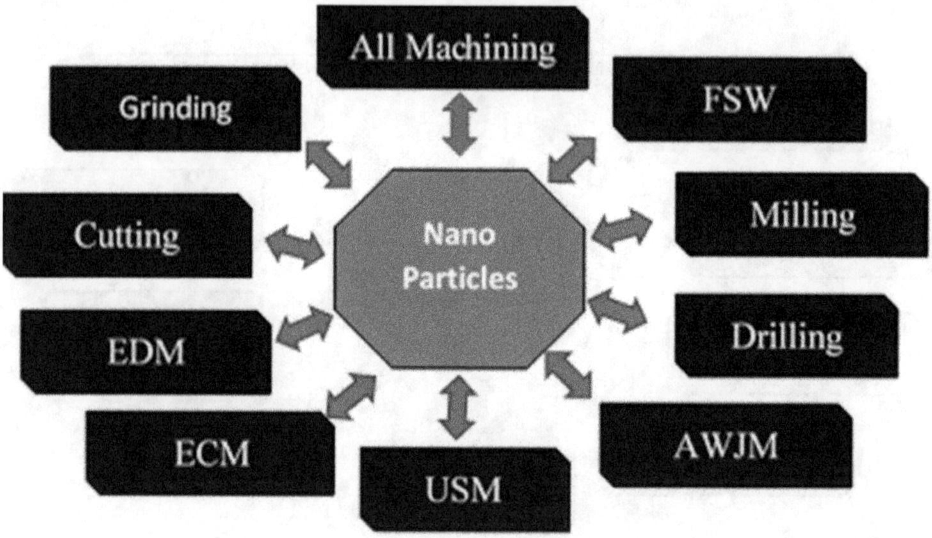

we will discuss the role of nanoparticles in different machining processes(Hasan, 2015; Rajput, 2015).

- **Grinding:** Nanoparticles are used as a grinding aid to improve the grinding efficiency and reduce the energy consumption. Nanoparticles of different materials such as alumina, silica, and diamond have been used as grinding aids in various applications.
- **Cutting:** Nanoparticles have been used as cutting fluids to reduce the cutting forces, improve the surface finish, and increase the tool life. Nanoparticles of titanium dioxide, alumina, and graphite have been used as cutting fluids in various applications.
- **Electrochemical Machining:** Nanoparticles have been used as additives in the electrolyte to improve the surface finish and reduce the machining time. Nanoparticles of diamond, silicon carbide, and alumina have been used as additives in the electrolyte in various applications.
- **Abrasive Jet Machining:** Nanoparticles have been used as abrasive particles to improve the cutting performance and reduce the surface roughness. Nanoparticles of alumina, silicon carbide, and diamond have been used as abrasive particles in various applications.
- **Ultrasonic Machining:** Nanoparticles have been used as abrasive particles to improve the machining performance and reduce the surface roughness.

Nanoparticles of alumina, silicon carbide, and diamond have been used as abrasive particles in various applications.

In conclusion, the use of nanoparticles in various machining processes has shown significant improvements in machining efficiency, surface finish, and tool life. However, further research is needed to fully understand the role of nanoparticles in these processes and to optimize their use for different applications(Indira & Lakshmi, 2010; Sharma et al., 2015). The literature suggests that the addition of nanoparticles to grinding fluids can significantly improve the grinding performance, reduce wheel wear, enhance cooling, reduce surface damage, and improve the machinability of difficult-to-grind materials. However, further research is needed to fully understand the underlying mechanisms and optimize the use of nanoparticles for specific grinding applications.

APPLICATIONS OF NANO-PARTICLES IN UNCONVENTIONAL MACHINING PROCESSES

ECM Process

The addition of nanoparticles to the electrolyte in Electrochemical Machining (ECM) has shown potential for improving the machining performance and surface quality. Here are some of the key findings from the literature on the effects of nanoparticles in the ECM process(Esfe et al., 2020; Sharma et al., 2015):

- **Improved Surface Finish:** Addition of nanoparticles to the electrolyte in ECM has been shown to improve the surface finish of the machined part. This is due to the high specific surface area of nanoparticles, which enhances the mass transfer and improves the dissolution of the workpiece material.
- **Reduced Machining Time:** Nanoparticles of materials such as diamond, silicon carbide, and alumina have been used as additives in the electrolyte to reduce the machining time in ECM. This is due to the increased electrochemical reaction rate and improved mass transfer provided by the nanoparticles.
- **Improved Machinability of Difficult-to-Machine Materials:** Nanoparticles have been shown to improve the machinability of difficult-to-machine materials such as titanium alloys, nickel-based superalloys, and ceramics. This is due to the increased dissolution rate and reduced tool wear provided by the nanoparticles.
- **Reduced Electrode Wear:** Nanoparticles of materials such as diamond and silicon carbide have been used as additives in the electrolyte to reduce the

electrode wear in ECM. This is due to the increased lubrication and reduced friction provided by the nanoparticles.

- **Enhanced Mechanical Properties:** Addition of nanoparticles to the electrolyte in ECM has been shown to enhance the mechanical properties of the machined part. This is due to the improved surface quality and reduced residual stress provided by the nanoparticles.
- **Optimization of Particle Size:** The particle size of the nanoparticles added to the electrolyte is a critical factor in the ECM process. The optimum particle size depends on the material being machined and the machining conditions. Therefore, optimizing the particle size of the nanoparticles can improve the material removal rate, surface finish, and dimensional accuracy of the machined parts.
- **Optimization of Particle Concentration:** The concentration of nanoparticles in the electrolyte is another important factor in the ECM process. The optimum concentration depends on the material being machined, the particle size, and the machining conditions. Therefore, optimizing the particle concentration can improve the material removal rate, surface finish, and tool life in the ECM process.
- **Optimization of Machining Parameters:** Machining parameters such as applied voltage, current density, electrolyte flow rate, and temperature can significantly affect the ECM process. Therefore, optimizing these parameters can improve the material removal rate, surface finish, and dimensional accuracy of the machined parts.
- **Optimization of Electrolyte Composition:** The composition of the electrolyte can also affect the ECM process. Addition of various additives such as surfactants, stabilizers, and buffering agents can improve the stability, conductivity, and pH of the electrolyte, leading to improved ECM performance.
- **Optimization of Electrode Design:** The electrode design can also affect the ECM process. The shape, size, and material of the electrode can affect the distribution of the electric field and the material removal rate. Therefore, optimizing the electrode design can improve the ECM performance.

In conclusion, the addition of nanoparticles to the electrolyte in ECM can significantly improve the machining performance, but further optimization techniques can be used to enhance the ECM process. Optimization of particle size, particle concentration, machining parameters, electrolyte composition, and electrode design can all improve the material removal rate, surface finish, and dimensional accuracy of the machined parts in the ECM process.

EDM Process

Electrical discharge machining (EDM) is a non-traditional machining process that uses electrical sparks to erode material from a workpiece. The addition of nanoparticles to the dielectric fluid used in the EDM process has been shown to have several effects on the machining performance. Here are some of the key findings from the literature on the effects of nanoparticles in the EDM process(Ishfaq et al., 2022):

- **Improved Material Removal Rate:** Addition of nanoparticles such as SiC, TiO2, and Al2O3 to the dielectric fluid has been shown to improve the material removal rate in the EDM process. This is due to the abrasive and cutting effects of the nanoparticles, which help to erode the workpiece material at a faster rate(Boopathi, 2022a; Boopathi & Sivakumar, 2016; Janardhana et al., 2023; Sampath & Myilsamy, 2021).
- **Reduced Electrode Wear:** Nanoparticles have been shown to reduce the electrode wear in the EDM process. This is due to the reduced adhesion and friction between the electrode and the workpiece, which helps to reduce the wear rate of the electrode.
- **Improved Surface Finish:** Addition of nanoparticles to the dielectric fluid has been shown to improve the surface finish of the machined parts in the EDM process. This is due to the abrasive and polishing effects of the nanoparticles, which help to remove the surface roughness and improve the surface quality.
- **Improved Accuracy:** Nanoparticles have been shown to improve the machining accuracy in the EDM process. This is due to the reduced thermal expansion and improved thermal conductivity provided by the nanoparticles, which help to minimize the dimensional deviations of the machined parts.
- **Reduced Machining Time:** Nanoparticles have been shown to reduce the machining time in the EDM process. This is due to the improved material removal rate and reduced electrode wear, which result in faster and more efficient machining.

In conclusion, the addition of nanoparticles to the dielectric fluid in the EDM process can significantly improve the machining performance. Nanoparticles can improve the material removal rate, reduce electrode wear, improve surface finish, improve accuracy, and reduce machining time in the EDM process. However, the optimum concentration and size of nanoparticles depend on the material being machined and the EDM conditions, and further research is needed to fully understand the underlying mechanisms and optimize the use of nanoparticles for specific EDM applications.

Optimizing the use of nanoparticles in the electrical discharge machining (EDM) process involves determining the optimum concentration, size, and type of nanoparticles to use for a given application. Here are some of the optimization techniques used in the literature to optimize the effects of nanoparticles in the EDM process(Boopathi et al., 2021; Boopathi & Myilsamy, 2021; Myilsamy et al., 2021; Sampath et al., 2021):

- **Optimization of Particle Concentration:** The concentration of nanoparticles in the dielectric fluid is a critical factor in the EDM process. The optimum concentration depends on the material being machined and the EDM conditions. Too high a concentration can lead to clogging and reduced machining performance, while too low a concentration may not have a significant effect on the machining performance. Therefore, optimizing the particle concentration can improve the material removal rate, reduce electrode wear, and improve surface finish in the EDM process(Boopathi, 2022c, 2022a; Boopathi & Sivakumar, 2012a, 2012b, 2014; Myilsamy & Sampath, 2021).
- **Optimization of Particle Size:** The particle size of the nanoparticles added to the dielectric fluid is another important factor in the EDM process. The optimum size depends on the material being machined and the EDM conditions. Smaller nanoparticles can provide better surface finish and accuracy, but may have a lower material removal rate. Larger nanoparticles can provide higher material removal rate, but may have a negative impact on surface finish and accuracy. Therefore, optimizing the particle size can improve the machining performance and achieve the desired surface finish and accuracy.
- **Optimization of Particle Type:** The type of nanoparticles added to the dielectric fluid is also an important factor in the EDM process. Different types of nanoparticles have different properties and can affect the machining performance differently. For example, SiC nanoparticles can provide a high material removal rate, while TiO2 nanoparticles can provide a good surface finish. Therefore, optimizing the type of nanoparticles can improve the machining performance and achieve the desired machining characteristics.
- **Optimization of EDM Parameters:** The EDM parameters such as discharge current, pulse duration, pulse frequency, and gap voltage can significantly affect the EDM process. Therefore, optimizing these parameters in conjunction with the nanoparticles can improve the machining performance and achieve the desired machining characteristics(Boopathi, 2022b, 2022e; Boopathi & Sivakumar, 2013; Dass james & Boopathi, 2016; Kannan et al., 2022; Yupapin et al., 2023).

- **Optimization of Dielectric Fluid:** The dielectric fluid used in the EDM process can also affect the machining performance. Addition of various additives such as surfactants, stabilizers, and buffering agents can improve the stability, conductivity, and pH of the dielectric fluid, leading to improved EDM performance(Boopathi & Sivakumar, 2012c, 2016; Gunasekaran et al., 2022; Myilsamy & Boopathi, 2017; Sampath & Myilsamy, 2021).

In conclusion, optimization of the use of nanoparticles in the EDM process involves determining the optimum concentration, size, and type of nanoparticles to use for a given application, and optimizing the EDM parameters and dielectric fluid accordingly. Optimization of particle concentration, particle size, particle type, EDM parameters, and dielectric fluid can all improve the material removal rate, reduce electrode wear, and improve surface finish and accuracy in the EDM process.

Abrasive-Jet Machining Process

Abrasive jet machining (AJM) is a non-traditional machining process that uses a high-velocity jet of abrasive particles to remove material from a workpiece. The addition of nanoparticles to the abrasive particles used in the AJM process has been shown to have several effects on the machining performance. Here are some of the key findings from the literature on the effects of nanoparticles in the AJM process(Nargundkar et al., 2023):

- **Improved Material Removal Rate:** Addition of nanoparticles such as SiC, Al2O3, and TiO2 to the abrasive particles used in the AJM process has been shown to improve the material removal rate. This is due to the abrasive and cutting effects of the nanoparticles, which help to erode the workpiece material at a faster rate.
- **Reduced Surface Roughness:** Nanoparticles have been shown to reduce the surface roughness of the machined parts in the AJM process. This is due to the polishing and smoothing effects of the nanoparticles, which help to remove the surface irregularities and improve the surface quality.
- **Improved Surface Integrity:** Nanoparticles have been shown to improve the surface integrity of the machined parts in the AJM process. This is due to the reduced damage and deformation caused by the abrasive particles, which helps to minimize the residual stresses and improve the surface properties.
- **Reduced Tool Wear:** Nanoparticles have been shown to reduce the tool wear in the AJM process. This is due to the reduced adhesion and friction between the abrasive particles and the tool, which helps to reduce the wear rate of the tool.

- **Improved Accuracy:** Nanoparticles have been shown to improve the machining accuracy in the AJM process. This is due to the reduced tool wear and improved surface finish, which result in better dimensional control and accuracy of the machined parts.

In conclusion, the addition of nanoparticles to the abrasive particles used in the AJM process can significantly improve the machining performance. Nanoparticles can improve the material removal rate, reduce surface roughness, improve surface integrity, reduce tool wear, and improve accuracy in the AJM process. However, the optimum concentration and size of nanoparticles depend on the material being machined and the AJM conditions, and further research is needed to fully understand the underlying mechanisms and optimize the use of nanoparticles for specific AJM applications.

Abrasive Water-Jet Machining Process

Abrasive waterjet machining (AWJM) is a non-traditional machining process that uses a high-velocity jet of abrasive particles suspended in water to remove material from a workpiece. The addition of nanoparticles to the abrasive particles used in the AWJM process has been shown to have several effects on the machining performance. Here are some of the key findings from the literature on the effects of nanoparticles in the AWJM process(Biswas & Wu, 2005):

- **Improved Material Removal Rate:** Addition of nanoparticles such as SiC, Al2O3, and TiO2 to the abrasive particles used in the AWJM process has been shown to improve the material removal rate. This is due to the abrasive and cutting effects of the nanoparticles, which help to erode the workpiece material at a faster rate.
- **Reduced Kerf Width:** Nanoparticles have been shown to reduce the kerf width of the machined parts in the AWJM process. This is due to the smaller size of the nanoparticles, which helps to create a finer abrasive stream and improve the cutting accuracy.
- **Improved Surface Quality:** Nanoparticles have been shown to improve the surface quality of the machined parts in the AWJM process. This is due to the polishing and smoothing effects of the nanoparticles, which help to remove the surface irregularities and improve the surface finish.
- **Reduced Tool Wear:** Nanoparticles have been shown to reduce the tool wear in the AWJM process. This is due to the reduced adhesion and friction between the abrasive particles and the tool, which helps to reduce the wear rate of the tool.

- **Reduced Heat-Affected Zone:** Nanoparticles have been shown to reduce the heat-affected zone (HAZ) of the machined parts in the AWJM process. This is due to the reduced thermal energy and localized heating caused by the smaller and more evenly distributed abrasive particles.

In conclusion, the addition of nanoparticles to the abrasive particles used in the AWJM process can significantly improve the machining performance. Nanoparticles can improve the material removal rate, reduce kerf width, improve surface quality, reduce tool wear, and reduce the heat-affected zone in the AWJM process. However, the optimum concentration and size of nanoparticles depend on the material being machined and the AWJM conditions, and further research is needed to fully understand the underlying mechanisms and optimize the use of nanoparticles for specific AWJM applications.

Ultrasonic Machining Process

Ultrasonic machining (USM) is a non-traditional machining process that uses ultrasonic vibrations to remove material from a workpiece. The addition of nanoparticles to the abrasive slurry used in the USM process has been shown to have several effects on the machining performance. Here are some of the key findings from the literature on the effects of nanoparticles in the USM process(Katna et al., 2020; Patil-Tekale et al., 2021):

- **Improved Material Removal Rate:** Addition of nanoparticles such as SiC, Al2O3, and TiO2 to the abrasive slurry used in the USM process has been shown to improve the material removal rate. This is due to the abrasive and cutting effects of the nanoparticles, which help to erode the workpiece material at a faster rate.
- **Improved Surface Quality:** Nanoparticles have been shown to improve the surface quality of the machined parts in the USM process. This is due to the polishing and smoothing effects of the nanoparticles, which help to remove the surface irregularities and improve the surface finish.
- **Reduced Tool Wear:** Nanoparticles have been shown to reduce the tool wear in the USM process. This is due to the reduced adhesion and friction between the abrasive particles and the tool, which helps to reduce the wear rate of the tool.
- **Reduced Recast Layer:** Nanoparticles have been shown to reduce the recast layer of the machined parts in the USM process. This is due to the reduced thermal energy and localized heating caused by the smaller and more evenly

distributed abrasive particles, which helps to minimize the formation of recast layers.

- **Reduced Tool Clogging:** Nanoparticles have been shown to reduce the tool clogging in the USM process. This is due to the improved dispersion and flow properties of the abrasive slurry, which helps to prevent the accumulation and clogging of the abrasive particles in the tool.

In conclusion, the addition of nanoparticles to the abrasive slurry used in the USM process can significantly improve the machining performance. Nanoparticles can improve the material removal rate, surface quality, reduce tool wear, reduce recast layer, and reduce tool clogging in the USM process. However, the optimum concentration and size of nanoparticles depend on the material being machined and the USM conditions, and further research is needed to fully understand the underlying mechanisms and optimize the use of nanoparticles for specific USM applications.

WEDM Process

Wire electrical discharge machining (WEDM) is a non-traditional machining process that uses a thin wire electrode to cut and shape a workpiece by discharging electrical sparks. The addition of nanoparticles to the dielectric fluid used in the WEDM process has been shown to have several effects on the machining performance. Here are some of the key findings from the literature on the effects of nanoparticles in the WEDM process(Boopathi et al., 2021; Sampath et al., 2021):

- **Improved Surface Quality:** Addition of nanoparticles such as SiC, Al2O3, and TiO2 to the dielectric fluid used in the WEDM process has been shown to improve the surface quality of the machined parts. This is due to the polishing and smoothing effects of the nanoparticles, which help to remove the surface irregularities and improve the surface finish.
- **Reduced Kerf Width:** Nanoparticles have been shown to reduce the kerf width of the machined parts in the WEDM process. This is due to the smaller size of the nanoparticles, which helps to create a finer discharge channel and improve the cutting accuracy.
- **Improved Material Removal Rate:** Nanoparticles have been shown to improve the material removal rate of the WEDM process. This is due to the abrasive and cutting effects of the nanoparticles, which help to erode the workpiece material at a faster rate.
- **Reduced Tool Wear:** Nanoparticles have been shown to reduce the tool wear in the WEDM process. This is due to the reduced adhesion and friction

between the tool and the workpiece, which helps to reduce the wear rate of the tool.

- **Reduced Recast Layer:** Nanoparticles have been shown to reduce the recast layer of the machined parts in the WEDM process. This is due to the reduced thermal energy and localized heating caused by the smaller and more evenly distributed discharge channels, which helps to minimize the formation of recast layers.

In conclusion, the addition of nanoparticles to the dielectric fluid used in the WEDM process can significantly improve the machining performance. Nanoparticles can improve the surface quality, reduce kerf width, improve material removal rate, reduce tool wear, and reduce recast layer in the WEDM process. However, the optimum concentration and size of nanoparticles depend on the material being machined and the WEDM conditions, and further research is needed to fully understand the underlying mechanisms and optimize the use of nanoparticles for specific WEDM applications.

EFFECTS OF NANO-PARTICLES IN CONVENTIONAL MACHINING PROCESSES

Lathe Machining Process

Lathe machining is a traditional machining process that involves rotating the workpiece against a stationary cutting tool to remove material and shape the part. The addition of nanoparticles to the cutting fluid used in the lathe process has been studied to evaluate the effects on the machining performance. Here are some of the key findings from the literature on the effects of nanoparticles in the lathe process(Peña-Parás et al., 2023):

- **Improved Cutting Performance:** Addition of nanoparticles such as Al2O3, SiO2, and TiO2 to the cutting fluid used in the lathe process has been shown to improve the cutting performance. This is due to the abrasive and lubricating effects of the nanoparticles, which help to reduce tool wear and improve the surface quality of the machined parts.
- **Improved Surface Quality:** Nanoparticles have been shown to improve the surface quality of the machined parts in the lathe process. This is due to the polishing and smoothing effects of the nanoparticles, which help to remove the surface irregularities and improve the surface finish.

- **Reduced Tool Wear:** Nanoparticles have been shown to reduce the tool wear in the lathe process. This is due to the reduced adhesion and friction between the tool and the workpiece, which helps to reduce the wear rate of the tool.
- **Improved Lubrication:** Nanoparticles have been shown to improve the lubrication of the cutting fluid used in the lathe process. This is due to the ability of the nanoparticles to form a protective film on the tool and workpiece surfaces, which helps to reduce friction and wear.
- **Reduced Cutting Force:** Nanoparticles have been shown to reduce the cutting force in the lathe process. This is due to the reduced friction between the tool and the workpiece, which helps to reduce the cutting force required to remove material.

In conclusion, the addition of nanoparticles to the cutting fluid used in the lathe process can improve the machining performance. Nanoparticles can improve the cutting performance, surface quality, reduce tool wear, improve lubrication, and reduce cutting force in the lathe process. However, the optimum concentration and size of nanoparticles depend on the material being machined and the lathe conditions, and further research is needed to fully understand the underlying mechanisms and optimize the use of nanoparticles for specific lathe applications.

Grinding Process

The addition of nanoparticles to the grinding process has been shown to have several effects on the process and the resulting ground surface. Here are some key findings from the literature(Abdolmohammadi et al., 2016; Gupta et al., 2021):

- **Improved grinding efficiency:** The addition of nanoparticles as grinding aids has been shown to improve the grinding efficiency by reducing the energy consumption and increasing the material removal rate. For example, the addition of alumina nanoparticles to a grinding wheel has been shown to increase the material removal rate by up to 50%.
- **Reduced surface roughness:** The use of nanoparticles as grinding aids has also been shown to reduce the surface roughness of the ground surface. This is thought to be due to the increased lubrication and reduced friction provided by the nanoparticles.
- **Improved surface integrity: The** addition of nanoparticles to the grinding process has also been shown to improve the surface integrity of the ground surface. This is due to the reduced thermal damage and residual stresses caused by the grinding process.

- **Improved tool life:** The addition of nanoparticles to the grinding process can also improve the tool life by reducing the wear and damage to the grinding wheel. For example, the addition of diamond nanoparticles to a grinding wheel has been shown to increase the tool life by up to 50%.
- **Particle dispersion:** The dispersion of the nanoparticles in the grinding fluid is an important factor in achieving the desired effects. The use of surfactants and sonication can improve the dispersion and distribution of the nanoparticles in the grinding fluid.
- **Improved Grinding Performance:** Addition of nanoparticles to grinding fluids has been shown to improve the grinding performance in terms of reducing grinding forces, increasing material removal rate, and improving surface quality. This is due to the high specific surface area and superior mechanical properties of nanoparticles.
- **Reduced Wheel Wear:** Nanoparticles of materials such as diamond and silicon carbide have been used as grinding aids and have been shown to reduce the wear of the grinding wheel. This is due to the increased lubrication provided by the nanoparticles, which reduces the friction and heat generated during the grinding process.
- **Enhanced Cooling:** Nanoparticles have also been used as cooling agents in the grinding process. The high thermal conductivity and specific heat of nanoparticles allow them to absorb and dissipate heat more efficiently than conventional coolants, resulting in reduced temperatures and improved cooling of the workpiece.
- **Reduced Surface Damage:** Addition of nanoparticles to grinding fluids has been shown to reduce the surface damage and subsurface deformation of the workpiece. This is due to the lubricating and cushioning effect of the nanoparticles, which reduces the contact pressure between the grinding wheel and the workpiece.
- **Improved Machinability of Difficult-to-Grind Materials:** Nanoparticles have been shown to improve the machinability of difficult-to-grind materials such as ceramics, composites, and hardened steels. This is due to the increased lubrication and reduced heat generation provided by the nanoparticles, which reduce the risk of surface cracking and workpiece damage.

Overall, the addition of nanoparticles to the grinding process has shown promising results in improving the efficiency, surface finish, and tool life of the process. However, further research is needed to optimize the use of nanoparticles for different materials and applications.

Milling Process

Nanoparticles have been studied as additives in the milling process, which is a machining operation that involves rotating a cutting tool with multiple cutting edges to remove material from the workpiece surface. The addition of nanoparticles to the cutting fluid used in the milling process has been shown to have several effects on the machining performance. Here are some of the key findings from the literature on the effects of nanoparticles in the milling process(Esfe et al., 2020; Sharma et al., 2015):

- **Improved Cutting Performance:** The addition of nanoparticles such as Al2O3, TiO2, SiO2, and CNTs to the cutting fluid used in the milling process has been shown to improve the cutting performance. This is due to the abrasive and lubricating effects of the nanoparticles, which help to reduce tool wear and improve the surface quality of the machined parts.
- **Improved Surface Quality:** Nanoparticles have been shown to improve the surface quality of the machined parts in the milling process. This is due to the polishing and smoothing effects of the nanoparticles, which help to remove surface irregularities and improve the surface finish.
- **Reduced Tool Wear:** Nanoparticles have been shown to reduce the tool wear in the milling process. This is due to the reduced adhesion and friction between the tool and the workpiece, which helps to reduce the wear rate of the tool.
- **Improved Lubrication:** Nanoparticles have been shown to improve the lubrication of the cutting fluid used in the milling process. This is due to the ability of the nanoparticles to form a protective film on the tool and workpiece surfaces, which helps to reduce friction and wear.
- **Reduced Cutting Force:** Nanoparticles have been shown to reduce the cutting force in the milling process. This is due to the reduced friction between the tool and the workpiece, which helps to reduce the cutting force required to remove material.

In conclusion, the addition of nanoparticles to the cutting fluid used in the milling process can improve the machining performance. Nanoparticles can improve the cutting performance, surface quality, reduce tool wear, improve lubrication, and reduce cutting force in the milling process. However, the optimum concentration and size of nanoparticles depend on the material being machined and the milling conditions, and further research is needed to fully understand the underlying mechanisms and optimize the use of nanoparticles for specific milling applications.

Drilling Process

Nanoparticles can be used in the drilling process to improve various aspects of the machining performance. Here are some specific uses of nanoparticles in the drilling process(Cheraghian, 2021):

- **Lubrication:** Nanoparticles can act as solid lubricants, reducing the friction between the drill bit and the workpiece. This can reduce the tool wear and improve the surface quality of the machined hole.
- **Heat transfer:** Nanoparticles can increase the heat transfer capacity of the drilling fluid, reducing the heat generated during the drilling process. This can help to prevent thermal damage to the workpiece and the drill bit, and improve the drilling performance.
- **Abrasion:** Nanoparticles can act as abrasives, helping to remove burrs and improve the surface finish of the machined hole.
- **Chemical stability:** Nanoparticles can increase the chemical stability of the drilling fluid, preventing the formation of corrosive compounds that can damage the workpiece or the drill bit.
- **Reducing cutting forces:** Nanoparticles can reduce the cutting forces required to remove material during the drilling process. This can reduce the wear on the drill bit and improve the drilling performance.

Overall, the use of nanoparticles in the drilling process can help to improve the machining performance, reduce tool wear, and improve the surface quality of the machined hole. However, the optimum concentration and size of nanoparticles depend on the material being drilled and the drilling conditions, and further research is needed to fully understand the underlying mechanisms and optimize the use of nanoparticles for specific drilling applications.

APPLICATION OF NANO-PARTICLES IN SPECIAL PURPOSES

Cosmetics

Nanoparticles are used in cosmetics due to their unique physical and chemical properties, which can enhance the performance of cosmetic products. Some of the most common uses of nanoparticles in cosmetics include(Esfe et al., 2020; Sharma et al., 2015):

- **UV protection:** Nanoparticles such as titanium dioxide and zinc oxide are commonly used in sunscreens and other cosmetic products to provide protection from ultraviolet (UV) radiation. These nanoparticles are effective at reflecting and absorbing UV radiation, preventing it from penetrating the skin.
- **Improved texture:** Nanoparticles such as silica, mica, and talc are used in cosmetic products to improve their texture and feel. These nanoparticles can help to smooth out the surface of the skin, reduce the appearance of fine lines and wrinkles, and provide a silky smooth feel.
- **Enhanced color:** Nanoparticles such as iron oxides and titanium dioxide are used in cosmetic products to provide color. These nanoparticles can be finely tuned to produce a wide range of colors, making them useful for a variety of cosmetic applications.
- **Improved delivery:** Nanoparticles can be used to improve the delivery of active ingredients in cosmetic products. By encapsulating active ingredients in nanoparticles, they can be more easily absorbed by the skin and targeted to specific areas, improving their effectiveness.
- **Antimicrobial properties:** Nanoparticles such as silver and copper have antimicrobial properties and can be used in cosmetic products to prevent the growth of bacteria and other microorganisms. This can help to improve the shelf life of cosmetic products and prevent skin infections.

While the use of nanoparticles in cosmetics has many benefits, there are also concerns about their safety. Some studies suggest that nanoparticles may be able to penetrate the skin and enter the bloodstream, potentially causing harm to the body. As a result, the safety of nanoparticles in cosmetics is closely monitored, and regulations are in place to ensure their safe use.

Painting Process

Nanoparticles are increasingly being used in painting applications due to their unique physical and chemical properties, which can improve the performance of paints and coatings. Some of the most common uses of nanoparticles in painting include(Biswas & Wu, 2005; Hasan, 2015; Indira & Lakshmi, 2010):

- **UV protection:** Nanoparticles such as titanium dioxide and zinc oxide are commonly used in paints and coatings to provide UV protection. These nanoparticles are effective at reflecting and absorbing UV radiation, preventing it from damaging the underlying surface.

- **Improved durability:** Nanoparticles can be used to improve the durability of paints and coatings. For example, nanoparticles such as silica and alumina can be added to paint formulations to improve their hardness, scratch resistance, and abrasion resistance.
- **Improved adhesion:** Nanoparticles can improve the adhesion of paints and coatings to the underlying surface. By modifying the surface chemistry of nanoparticles, they can create a stronger bond between the paint and the substrate, reducing the risk of peeling and flaking.
- **Enhanced color:** Nanoparticles such as iron oxides and titanium dioxide can be used to provide vibrant, long-lasting colors in paints and coatings. These nanoparticles can be finely tuned to produce a wide range of colors, making them useful for a variety of painting applications.
- **Self-cleaning properties:** Nanoparticles such as titanium dioxide and zinc oxide can be used to create self-cleaning coatings. These nanoparticles react with UV radiation to produce highly reactive oxygen species, which can break down dirt and other organic matter on the surface of the coating.

While the use of nanoparticles in painting has many benefits, there are also concerns about their safety. Some studies suggest that nanoparticles may be released into the environment during the painting process, potentially causing harm to human health and the environment. As a result, the safety of nanoparticles in painting is closely monitored, and regulations are in place to ensure their safe use.

Friction Stir Welding

Friction Stir Welding (FSW) is a solid-state welding process that involves the use of a rotating tool to generate frictional heat and plastic deformation, which results in a welded joint. Nanoparticles can be added to the weld material or tool to improve the welding process and enhance the properties of the resulting weld. Some of the applications of nanoparticles in FSW include(Boopathi, 2022d; Boopathi et al., 2017, 2022; Raja et al., 2020; Trojovský et al., 2023; Vimalraj & Kah, 2021):

- **Improved mechanical properties:** The addition of nanoparticles such as carbon nanotubes, alumina, and titanium dioxide can significantly improve the mechanical properties of the weld, including strength, ductility, and toughness.
- **Enhanced wear resistance:** The addition of nanoparticles such as silicon carbide and diamond can improve the wear resistance of the weld, making it more suitable for high-wear applications.

- **Reduced defects:** The addition of nanoparticles can also help to reduce defects in the weld, such as porosity and cracking. This is because the nanoparticles can act as nucleation sites for the formation of a uniform and defect-free microstructure.
- **Reduced thermal degradation:** The addition of nanoparticles can also help to reduce the thermal degradation of the weld material during the welding process. This is because the nanoparticles can act as heat sinks, absorbing the excess heat generated during the welding process and preventing the material from overheating.

Overall, the use of nanoparticles in FSW has the potential to improve the welding process and enhance the properties of the resulting weld. However, the optimal amount and type of nanoparticles will depend on the specific application and materials being welded, and further research is needed to fully understand the effects of nanoparticles on the FSW process.

Medicine

Nanoparticles have shown potential applications in medicine due to their unique properties such as small size, high surface area-to-volume ratio, and ability to penetrate biological barriers. Here are some of the way's nanoparticles are being used in medicine(Shafiq et al., 2021):

- **Drug delivery:** Nanoparticles can be used to deliver drugs to specific cells or tissues in the body. This is because nanoparticles can be engineered to have specific surface properties, enabling them to target specific cells or tissues. Additionally, nanoparticles can protect drugs from degradation and improve their solubility, thereby increasing their effectiveness.
- **Imaging:** Nanoparticles can be used as contrast agents in medical imaging, such as magnetic resonance imaging (MRI) and computed tomography (CT). This is because nanoparticles can be engineered to have magnetic or radioactive properties, enabling them to be detected in the body.
- **Diagnostics:** Nanoparticles can also be used in diagnostics, such as in the detection of biomarkers for disease. This is because nanoparticles can be engineered to have specific properties that enable them to bind to specific biomolecules, allowing for their detection in biological samples.
- **Tissue engineering:** Nanoparticles can be used in tissue engineering to create scaffolds that support the growth of new tissues. This is because nanoparticles can be engineered to have specific surface properties that enable them to interact with cells and promote their growth.

- Overall, the use of nanoparticles in medicine has the potential to revolutionize drug delivery, imaging, diagnostics, and tissue engineering. However, more research is needed to fully understand the long-term effects of nanoparticles in the body and to optimize their use in medical applications.

IC Engines

Nanoparticles have shown potential in improving the performance of internal combustion engines (ICEs). Here are some of the ways nanoparticles can be used in IC engines(Kittelson, 1998; Murari et al., 2023):

- **Fuel additives:** Nanoparticles can be added to fuel to improve combustion efficiency and reduce emissions. For example, adding nanoparticles such as cerium oxide to diesel fuel has been shown to reduce particulate matter emissions and improve fuel efficiency.
- **Lubricants:** Nanoparticles can be added to engine lubricants to improve wear resistance and reduce friction. For example, adding nanoparticles such as tungsten disulphide to engine oil has been shown to improve fuel efficiency and reduce engine wear.
- **Coatings:** Nanoparticle coatings can be applied to engine components such as pistons and cylinder walls to improve wear resistance and reduce friction. For example, applying coatings of nanoparticles such as diamond or carbon nanotubes to engine components has been shown to improve engine performance and reduce emissions.
- **Nanofluids:** Nanoparticles can be added to engine coolant to improve heat transfer and reduce engine temperature. For example, adding nanoparticles such as aluminum oxide to engine coolant has been shown to improve engine performance and reduce emissions.

Overall, the use of nanoparticles in IC engines has the potential to improve engine performance, reduce emissions, and increase fuel efficiency. However, more research is needed to fully understand the effects of nanoparticles on engine components and to optimize their use in IC engines.

Bio-Medical Applications

Nanoparticles have shown promising potential in various biomedical applications due to their unique properties such as high surface area-to-volume ratio, size, and reactivity. Here are some of the biomedical applications of nanoparticles(Jones et al., 2016; Shafiq et al., 2021):

- **Drug delivery:** Nanoparticles can be used to deliver drugs to targeted cells or tissues, improving drug efficacy and reducing side effects. Nanoparticles can be designed to release drugs in a controlled manner, leading to sustained drug release and improved therapeutic outcomes.
- **Imaging:** Nanoparticles can be used as contrast agents in medical imaging techniques such as magnetic resonance imaging (MRI), computed tomography (CT), and positron emission tomography (PET). Nanoparticles can also be used to label cells and track them in vivo, enabling the study of disease progression and treatment efficacy.
- **Tissue engineering:** Nanoparticles can be incorporated into scaffolds to improve tissue regeneration and promote wound healing. Nanoparticles can also be used to stimulate cell growth and differentiation, leading to improved tissue formation.
- **Diagnostics:** Nanoparticles can be used in biosensors for the detection of biomolecules such as proteins and DNA. Nanoparticles can also be used to detect diseases such as cancer, infectious diseases, and neurological disorders, enabling early diagnosis and treatment.
- **Cancer therapy:** Nanoparticles can be used to target cancer cells and deliver drugs directly to them, improving drug efficacy and reducing side effects. Nanoparticles can also be used in photodynamic therapy, which uses light to activate nanoparticles that selectively destroy cancer cells.
- **Antibacterial agents:** Nanoparticles such as silver nanoparticles have shown promising antibacterial properties, making them potential candidates for the development of new antibacterial agents.

Overall, nanoparticles have the potential to revolutionize the field of biomedicine by improving drug delivery, imaging, tissue engineering, diagnostics, cancer therapy, and antibacterial agents. However, more research is needed to fully understand the effects of nanoparticles on biological systems and to optimize their use in biomedical applications.

Concrete-Civil Applications

Nanoparticles can be added to concrete to improve its mechanical and physical properties. Here are some of the way's nanoparticles are used in concretes(Huseien et al., 2022):

- **Improving strength:** The addition of nanoparticles such as silica, titanium dioxide, and carbon nanotubes can improve the strength of concrete by enhancing the hydration process and reducing the porosity of the concrete.

- **Enhancing durability:** Nanoparticles can also enhance the durability of concrete by reducing the permeability of the concrete, which prevents the ingress of water, chemicals, and other deleterious substances.
- **Improving workability:** Nanoparticles can be used to improve the workability of concrete by reducing the amount of water needed to achieve a desired consistency.
- **Reducing shrinkage:** The addition of nanoparticles can also reduce the shrinkage of concrete, which can lead to cracking and reduced durability.
- **Enhancing thermal properties:** Nanoparticles can improve the thermal properties of concrete by enhancing its ability to absorb and release heat, making it more suitable for use in passive solar buildings and other energy-efficient applications.
- **Reducing environmental impact:** The use of nanoparticles in concrete can also reduce its environmental impact by reducing the amount of cement needed, which in turn reduces the amount of carbon dioxide emissions associated with concrete production.

Agricultural Field

Nanoparticles have a wide range of potential applications in agriculture, from improving crop yields to enhancing plant growth and nutrition. Here are some ways nanoparticles are being used in agriculture(Katna et al., 2020)(Boopathi et al., 2023; Domakonda et al., 2023; Kumara et al., 2023; S. et al., 2022; Samikannu et al., 2023; Vanitha et al., 2023; Vennila et al., 2023):

- **Fertilizers:** Nanoparticles can be added to fertilizers to improve their efficiency and reduce nutrient losses. For example, nanoparticles of iron, zinc, and copper can be added to fertilizers to increase their uptake by plants.
- **Pesticides:** Nanoparticles can also be used to develop more effective and environmentally friendly pesticides. For example, nanoparticles of silver and copper can be used to create coatings that protect plants from pests and diseases.
- **Soil remediation:** Nanoparticles can be used to remediate contaminated soils by removing heavy metals and other pollutants.
- **Crop protection:** Nanoparticles can be used to protect crops from environmental stresses such as drought and salinity. For example, nanoparticles of titanium dioxide can be used to enhance the photosynthesis and water-use efficiency of plants.

- **Seed treatment:** Nanoparticles can also be used to enhance the germination and growth of seeds. For example, nanoparticles of silicon and zinc oxide can be used to improve seedling growth and root development.
- **Food preservation:** Nanoparticles can be used to develop food packaging materials that are more effective at preserving food freshness and preventing spoilage.

Food Technology

Nanoparticles are increasingly being used in the food industry for a variety of applications, such as improving food texture, enhancing flavor, and increasing shelf life. However, the use of nanoparticles in food has raised concerns about their safety and potential health risks(Das et al., 2009).

Here are some examples of how nanoparticles are being used in food:

- **Food packaging:** Nanoparticles are being used in food packaging to create barriers that protect food from oxygen, moisture, and other environmental factors. This can help to extend the shelf life of food and reduce food waste.
- **Food additives:** Nanoparticles are being used as food additives to enhance flavor, color, and texture. For example, nanoparticles of titanium dioxide are commonly used as a whitening agent in products such as toothpaste and candy.
- **Food supplements:** Nanoparticles are being used in food supplements to increase their bioavailability and effectiveness. For example, nanoparticles of iron and zinc can be added to food supplements to improve their absorption by the body.
- **Food processing:** Nanoparticles are being used in food processing to improve the efficiency and effectiveness of processes such as mixing, emulsifying, and homogenizing.

LIMITATIONS

While nanoparticles have many advantages and potential applications, they also have several limitations that must be considered:

- **Toxicity:** Some nanoparticles can be toxic to humans and the environment. It is important to understand the potential risks associated with nanoparticles and take appropriate precautions to prevent exposure.

- **Cost:** The production and use of nanoparticles can be expensive, which can limit their practical applications.
- **Manufacturing challenges:** The production of nanoparticles can be difficult and requires specialized equipment and processes.
- **Stability:** Nanoparticles can be unstable and prone to agglomeration, which can affect their properties and performance.
- **Regulatory challenges:** The use of nanoparticles is subject to regulatory scrutiny and there may be restrictions on their use in certain applications.
- **Lack of standardization:** There is currently a lack of standardization in the characterization and measurement of nanoparticles, which can make it difficult to compare results across different studies.
- **Environmental impact:** The impact of nanoparticles on the environment is not yet fully understood, and more research is needed to determine their potential effects on ecosystems.

SUMMARY

Nanoparticles are used in many fields, such as electronics, medicine, cosmetics, manufacturing, and materials science. They have unique properties such as a high surface area-to-volume ratio, high reactivity, and the ability to interact with biological systems. This chapter discusses the applications and effects of nanoparticles in conventional and unconventional machining processes, as well as special applications in medicine, agriculture, food, construction, biotechnology, and IC engines. The limitations of using nanoparticles in engineering applications are also discussed.

REFERENCES

Abdolmohammadi, S., Ghiasi, R., & Ahmadzadeh-Vatani, S. (2016). A highly efficient CuI nanoparticles-catalyzed synthesis of tetrahydrochromenediones and dihydropyrano [c] chromenediones under grinding. *Zeitschrift für Naturforschung. Section B. A Journal of Chemical Sciences*, *71*(7), 777–782. doi:10.1515/znb-2015-0195

Biswas, P., & Wu, C.-Y. (2005). Nanoparticles and the environment. *Journal of the Air \& Waste Management Association, 55*(6), 708–746.

Boopathi, S. (2022a). An Extensive Review on Sustainable Developments of Dry and Near-Dry Electrical Discharge Machining Processes. *Journal of Manufacturing Science and Engineering*, *144*(5), 50801. doi:10.1115/1.4052527

Boopathi, S. (2022b). An investigation on gas emission concentration and relative emission rate of the near-dry wire-cut electrical discharge machining process. *Environmental Science and Pollution Research International*, *29*(57), 86237–86246. doi:10.100711356-021-17658-1 PMID:34837614

Boopathi, S. (2022c). Cryogenically treated and untreated stainless steel grade 317 in sustainable wire electrical discharge machining process: A comparative study. *Environmental Science and Pollution Research International*, 1–10. doi:10.100711356-022-22843-x PMID:36057706

Boopathi, S. (2022d). Experimental investigation and multi-objective optimization of cryogenic Friction-stir-welding of AA2014 and AZ31B alloys using MOORA technique. *Materials Today. Communications*, *33*, 104937. doi:10.1016/j.mtcomm.2022.104937

Boopathi, S. (2022e). Performance Improvement of Eco-Friendly Near-Dry Wire-Cut Electrical Discharge Machining Process Using Coconut Oil-Mist Dielectric Fluid. *Journal of Advanced Manufacturing Systems*, 1–20. doi:10.1142/S0219686723500178

Boopathi, S., Haribalaji, V., Mageswari, M., & Asif, M. M. (2022). Influences of Boron Carbide Particles on the Wear Rate and Tensile Strength of Aa2014 Surface Composite Fabricated By Friction-Stir Processing. *Materiali in Tehnologije*, *56*(3), 263–270. doi:10.17222/mit.2022.409

Boopathi, S., Kumaresan, A., Manohar, N., & Krishna Moorthi, R. (2017). Review on Effect of Process Parameters-Friction Stir Welding Process. [IRJET]. *International Research Journal of Engineering and Technology*, *4*(7), 272–278.

Boopathi, S., Lewise, K. A. S., Subbiah, R., & Sivaraman, G. (2021). Near-dry wire-cut electrical discharge machining process using water-air-mist dielectric fluid: An experimental study. *Materials Today: Proceedings*, *49*(5), 1885–1890. doi:10.1016/j.matpr.2021.08.077

Boopathi, S., & Myilsamy, S. (2021). Material removal rate and surface roughness study on Near-dry wire electrical discharge Machining process. *Materials Today: Proceedings*, *45*(9), 8149–8156. doi:10.1016/j.matpr.2021.02.267

Boopathi, S., Siva Kumar, P. K., & Meena, R. S. J., S. I., P., S. K., & Sudhakar, M. (2023). Sustainable Developments of Modern Soil-Less Agro-Cultivation Systems. In Human Agro-Energy Optimization for Business and Industry (pp. 69–87). IGI Global. doi:10.4018/978-1-6684-4118-3.ch004

Boopathi, S., & Sivakumar, K. (2012a). Experimental Analysis of Eco-friendly Near-dry Wire Electrical Discharge Machining Process. *Archives des Sciences*, *65*(10), 334–346.

Boopathi, S., & Sivakumar, K. (2012b). Experimental comparative study of near-dry wire-cut electrical discharge machining (WEDM). *European Journal of Scientific Research*, *75*(4), 472–481.

Boopathi, S., & Sivakumar, K. (2012c). Parametric Study of Dry WEDM Using Taguchi Method. *International Journal of Engineering Research and Development*, *2*(4), 63–68. www.ijerd.com

Boopathi, S., & Sivakumar, K. (2013). Experimental investigation and parameter optimization of near-dry wire-cut electrical discharge machining using multi-objective evolutionary algorithm. *International Journal of Advanced Manufacturing Technology*, *67*(9–12), 2639–2655. doi:10.100700170-012-4680-4

Boopathi, S., & Sivakumar, K. (2014). Study of water assisted dry wire-cut electrical discharge machining. *Indian Journal of Engineering and Materials Sciences*, *21*(1), 75–82.

Boopathi, S., & Sivakumar, K. (2016). Optimal parameter prediction of oxygen-mist near-dry Wire-cut EDM. *International Journal of Manufacturing Technology and Management*, *30*(3–4), 164–178. doi:10.1504/IJMTM.2016.077812

Cheraghian, G. (2021). Nanoparticles in drilling fluid: A review of the state-of-the-art. *Journal of Materials Research and Technology*, *13*, 737–753. doi:10.1016/j.jmrt.2021.04.089

Das, M., Saxena, N., & Dwivedi, P. D. (2009). Emerging trends of nanoparticles application in food technology: Safety paradigms. *Nanotoxicology*, *3*(1), 10–18. doi:10.1080/17435390802504237

Dass james, A., & Boopathi, S. (2016). Experimental Study of Eco-friendly Wire-Cut Electrical Discharge Machining Processes. *International Journal of Innovative Research in Science, Engineering and Technology*, *5*(1), 666–675.

Domakonda, V. K., Farooq, S., Chinthamreddy, S., Puviarasi, R., Sudhakar, M., & Boopathi, S. (2023). Sustainable Developments of Hybrid Floating Solar Power Plants. In *Human Agro-Energy Optimization for Business and Industry* (pp. 148–167). IGI Global. doi:10.4018/978-1-6684-4118-3.ch008

Esfe, M. H., Bahiraei, M., & Mir, A. (2020). Application of conventional and hybrid nanofluids in different machining processes: A critical review. *Advances in Colloid and Interface Science*, *282*, 102199. doi:10.1016/j.cis.2020.102199 PMID:32679398

Gunasekaran, K., Boopathi, S., & Sureshkumar, M. (2022). Analysis of a Cryogenically Cooled Near-Dry Wedm Process Using Different Dielectrics. *Materiali in Tehnologije*, *56*(2), 179–186. doi:10.17222/mit.2022.397

Gupta, M. K., Khan, A. M., Song, Q., Liu, Z., Khalid, Q. S., Jamil, M., Kuntoğlu, M., Usca, Ü. A., Sarıkaya, M., & Pimenov, D. Y. (2021). A review on conventional and advanced minimum quantity lubrication approaches on performance measures of grinding process. *International Journal of Advanced Manufacturing Technology*, *117*(3-4), 729–750. doi:10.100700170-021-07785-x

Hasan, S. (2015). A review on nanoparticles: Their synthesis and types. *Research Journal of Recent Sciences*, *2277*, 2502.

Huseien, G. F., Khalid, N. H. A., & Mirza, J. (2022). *Nanotechnology for smart concrete*. CRC Press.

Indira, T. K., & Lakshmi, P. K. (2010). Magnetic nanoparticles—A review. [IJPSN]. *International Journal of Pharmaceutical Sciences and Nanotechnology*, *3*(3), 1035–1042. doi:10.37285/ijpsn.2010.3.3.1

Ishfaq, K., Maqsood, M. A., Anwar, S., Harris, M., Alfaify, A., & Zia, A. W. (2022). EDM of Ti6Al4V under nano-graphene mixed dielectric: A detailed roughness analysis. *International Journal of Advanced Manufacturing Technology*, *120*(11–12), 7375–7388. doi:10.100700170-022-09207-y

Janardhana, K., Anushkannan, N. K., Dinakaran, K. P., Puse, R. K., & Boopathi, S. (2023). *Experimental Investigation on Microhardness, Surface Roughness, and White Layer Thickness of Dry EDM*. Engineering Research Express. doi:10.1088/2631-8695/acce8f

Jones, D. E., Ghandehari, H., & Facelli, J. C. (2016). A review of the applications of data mining and machine learning for the prediction of biomedical properties of nanoparticles. *Computer Methods and Programs in Biomedicine*, *132*, 93–103. doi:10.1016/j.cmpb.2016.04.025 PMID:27282231

Kannan, E., Trabelsi, Y., Boopathi, S., & Alagesan, S. (2022). Influences of cryogenically treated work material on near-dry wire-cut electrical discharge machining process. *Surface Topography : Metrology and Properties*, *10*(1), 15027. doi:10.1088/2051-672X/ac53e1

Katna, R., Agrawal, N., & Suhaib, M. (2020). A Short Review on Machining with Ultrasonic MQL Method. *IOP Conference Series. Materials Science and Engineering*, *802*(1), 12011. doi:10.1088/1757-899X/802/1/012011

Kittelson, D. B. (1998). Engines and nanoparticles: A review. *Journal of Aerosol Science*, *29*(5–6), 575–588. doi:10.1016/S0021-8502(97)10037-4

Kumara, V., Mohanaprakash, T. A., Fairooz, S., Jamal, K., Babu, T., & B., S. (2023). Experimental Study on a Reliable Smart Hydroponics System. In *Human Agro-Energy Optimization for Business and Industry* (pp. 27–45). IGI Global. doi:10.4018/978-1-6684-4118-3.ch002

Mohanraj, V. J., & Chen, Y. (2006). Nanoparticles-a review. *Tropical Journal of Pharmaceutical Research*, *5*(1), 561–573.

Murari, G., Nahak, B., & Pratap, T. (2023). Hybrid surface modification for improved tribological performance of IC engine components--a review. *Proceedings of the Institution of Mechanical Engineers, Part E: Journal of Process Mechanical Engineering*, 09544089221150718.

Myilsamy, S., & Boopathi, S. (2017). Grey Relational Optimization of Powder Mixed Near-Dry Wire Cut Electrical Discharge Machining of Inconel 718 Alloy. *Asian Journal of Research in Social Sciences and Humanities*, *7*(3), 18. doi:10.5958/2249-7315.2017.00157.5

Myilsamy, S., Boopathi, S., & Yuvaraj, D. (2021). A study on cryogenically treated molybdenum wire electrode. *Materials Today: Proceedings*, *45*(9), 8130–8135. doi:10.1016/j.matpr.2021.02.049

Myilsamy, S., & Sampath, B. (2021). Experimental comparison of near-dry and cryogenically cooled near-dry machining in wire-cut electrical discharge machining processes. *Surface Topography : Metrology and Properties*, *9*(3), 35015. doi:10.1088/2051-672X/ac15e0

Nargundkar, A., Gulia, V., & Khan, A. (2023). Nano-Abrasives Assisted Abrasive Water Jet Machining of Bio-Composites—An Experimental and Optimization Approach. *Journal of Advanced Manufacturing Systems*, 1–24. doi:10.1142/S0219686723500373

Patil-Tekale, R. A., Gadekar, A., Gadhade, Y., Parakh, L., Balaji, R., & Selokar, A. (2021). Recent Investigation on Ultrasonic Machining of Aluminum Metal Matrix Composite. *Intelligent Manufacturing and Energy Sustainability: Proceedings of ICIMES 2020*, (pp. 619–636). IEEE.

Peña-Parás, L., Maldonado-Cortés, D., Rodríguez-Villalobos, M., Sauceda, A., Aguilar, I. J., Villanueva, J. A., & Cárdenas, P. A. (2023). Nanoparticle lubricant additives applied in a CNC lathe ball screw component for improving the quality of machined workpieces: A case study. *Wear, 523*, 204752. doi:10.1016/j.wear.2023.204752

Raja, S., Muhamad, M. R., Jamaludin, M. F., & Yusof, F. (2020). A review on nanomaterials reinforcement in friction stir welding. *Journal of Materials Research and Technology*, *9*(6), 16459–16487. doi:10.1016/j.jmrt.2020.11.072

Rajput, N. (2015). Methods of preparation of nanoparticles-a review. *International Journal of Advances in Engineering \& Technology, 7*(6), 1806.

S., P. K., Sampath, B., R., S. K., Babu, B. H., & N., A. (2022). Hydroponics, Aeroponics, and Aquaponics Technologies in Modern Agricultural Cultivation. In *Trends, Paradigms, and Advances in Mechatronics Engineering* (pp. 223–241). IGI Global. doi:10.4018/978-1-6684-5887-7.ch012

Samikannu, R., Koshariya, A. K., Poornima, E., Ramesh, S., Kumar, A., & Boopathi, S. (2023). Sustainable Development in Modern Aquaponics Cultivation Systems Using IoT Technologies. In *Human Agro-Energy Optimization for Business and Industry* (pp. 105–127). IGI Global. doi:10.4018/978-1-6684-4118-3.ch006

Sampath, B., & Myilsamy, S. (2021). Experimental investigation of a cryogenically cooled oxygen-mist near-dry wire-cut electrical discharge machining process. *Strojniski Vestnik. Jixie Gongcheng Xuebao*, *67*(6), 322–330. doi:10.5545v-jme.2021.7161

Sampath, B., Sureshkumar, M., Yuvaraj, T., & Velmurugan, D. (2021). Experimental investigations on eco-friendly helium-mist near-dry wire-cut edm of m2-hss material. *Materials Research Proceedings*, *19*, 175–180. doi:10.21741/9781644901618-22

Shafiq, T., Uzair, M., Iqbal, M. J., Zafar, M., Hussain, S. J., & Shah, S. A. A. (2021). Green Synthesis of Metallic Nanoparticles and Their Potential in Bio-Medical Applications. *Nano Biomedicine \& Engineering, 13*(2).

Sharma, A. K., Tiwari, A. K., & Dixit, A. R. (2015). Mechanism of nanoparticles functioning and effects in machining processes: A review. *Materials Today: Proceedings*, *2*(4–5), 3539–3544. doi:10.1016/j.matpr.2015.07.331

Trojovský, P., Dhasarathan, V., & Boopathi, S. (2023). Experimental investigations on cryogenic friction-stir welding of similar ZE42 magnesium alloys. *Alexandria Engineering Journal*, *66*(1), 1–14. doi:10.1016/j.aej.2022.12.007

Vanitha, S. K. R., & Boopathi, S. (2023). Artificial Intelligence Techniques in Water Purification and Utilization. In *Human Agro-Energy Optimization for Business and Industry* (pp. 202–218). IGI Global., doi:10.4018/978-1-6684-4118-3.ch010

Vennila, T., Karuna, M. S., Srivastava, B. K., Venugopal, J., Surakasi, R., & B., S. (2023). New Strategies in Treatment and Enzymatic Processes. In *Human Agro-Energy Optimization for Business and Industry* (pp. 219–240). IGI Global. doi:10.4018/978-1-6684-4118-3.ch011

Vimalraj, C., & Kah, P. (2021). Experimental review on friction stir welding of aluminium alloys with nanoparticles. *Metals*, *11*(3), 390. doi:10.3390/met11030390

Welch, C. M., & Compton, R. G. (2006). The use of nanoparticles in electroanalysis: A review. *Analytical and Bioanalytical Chemistry*, *384*(3), 601–619. doi:10.100700216-005-0230-3 PMID:16402180

Yupapin, P., Trabelsi, Y., Nattappan, A., & Boopathi, S. (2023). Performance Improvement of Wire-Cut Electrical Discharge Machining Process Using Cryogenically Treated Super-Conductive State of Monel-K500 Alloy. *Iranian Journal of Science and Technology. Transaction of Mechanical Engineering*, *47*(1), 267–283. doi:10.100740997-022-00513-0

Chapter 2
Influences of Nanoparticles in Friction Stir Welding Processes

S. Suresh
https://orcid.org/0000-0001-8840-0786
Erode Sengunthar Engineering College, India

D. Velmurugan
https://orcid.org/0000-0002-5392-3461
Muthayammal Engineering College, India

J. Balaji
Erode Senguthar Engineering College, India

Elango Natarajan
https://orcid.org/0000-0003-1215-0789
UCSI University, Kuala Lumpur, Malaysia

P. Suresh
Muthayammal Engineering College, India

S. Rajesh
Knowledge Institute of Technology, India

ABSTRACT

Lightweight materials and their alloys are increasingly used in automobiles and aircraft due to their high specific strength and strength and weight ratio. The characteristics of aluminium materials and their alloys make applying traditional mechanical joining and welding processes difficult. Recently much research has been conducted to obtain reliable and high mechanical strength joints on aluminium alloys using various methods of joining process. This chapter provides a detailed review of recent research on nanoparticle-influenced friction stir welding on different aluminium alloys. The latest advancements and application of nanoparticle reinforcements in the joining behaviour of aluminium alloys are also systematically described. Finally, several unaddressed issues and future development in joining aluminium alloys with the addition of nanoparticles are included.

DOI: 10.4018/978-1-6684-9135-5.ch002

1. INTRODUCTION

Lightweight alloys such as Aluminium, Magnesium, Titanium, and Copper are potential engineering alloys from research and industrial point of view. These alloys have become more significant in transport manufacturing, including aircraft, cars, heavy trucks, trains, ships, and defense products (Daehn, 2014; Danish et al., 2018; Mayyas et al., 2016)(Ravikumar et al., 2020). These alloys are appealing for steel structures in some engineering applications due to their characteristics, such as a high strength-to-weight ratio, effective heat conductivity, and excellent damping capacity (Sukkasamy & Gopal P. M., 2022)(Madesh et al., 2022).

Methods to join similar or dissimilar lightweight alloys are becoming more critical in manufacturing structures in engineering applications (Subramani et al., 2019) (Ulutaş, 2022) (Subramani et al., 2019). To consistently join these lightweight alloys, suitable joining methods are required. Joining methods include mechanical fastening, adhesive bonding, welding, brazing, and soldering (Thorny et al., 2007). Mechanical fastening and adhesive bonding result in no microstructural evaluation, as no heat is involved. Whereas fusion welding processes such as shielded metal arc welding, gas metal arc welding, and gas tungsten arc welding lead to metallurgical mismatching in the weld of dissimilar materials in metal-to-metal joints.

Emerging joining processes, such as friction stir welding, laser welding, and ultrasonic welding, are considered more feasible to join lightweight alloys (J. Yang et al., 2022) (Boopathi et al., 2022). Friction stir welding (FSW) is a solid-state joining method with low energy input. FSW enhances the fabrication and processing of lightweight alloy joints, which are challenging to do with other welding processes due to no melting involved (Sudhagar et al., 2019). In this method, coalescence is initiated by applying localized heat generated by the stirring action of the non-consumable rotating tool with a modest clamping force (Fig .1).

Tool rotation speed, tool traverse speed, and tool plunge fore are the significant process parameters of FSW (Sasikala et al., 2022)(S. M et al., 2021). Friction spot joining is similar to linear friction stir welding. However, the tool does not travel linearly, including plunging, stirring, and retracting (Suresh, Venkatesan, Natarajan, et al., 2019) (Suresh, Elango, et al., 2020). Recently, many researchers developed a new lightweight metal that welded with friction stir welded with the addition of ceramic nanoparticles that could pave the way for super-strong yet high-performance lightweight metals (Casati & Vedani, 2014)(V. Sharma et al., 2015) (Suresh S et al., 2022). They have concentrated on the use of nanoparticles in the FSW process, the production of composite matrixes, and the implications of these processes on mechanical properties. However, little study has been done on nanoparticle-reinforced dissimilar lightweight joints.

Figure 1. Schematics of friction stir spot welding process

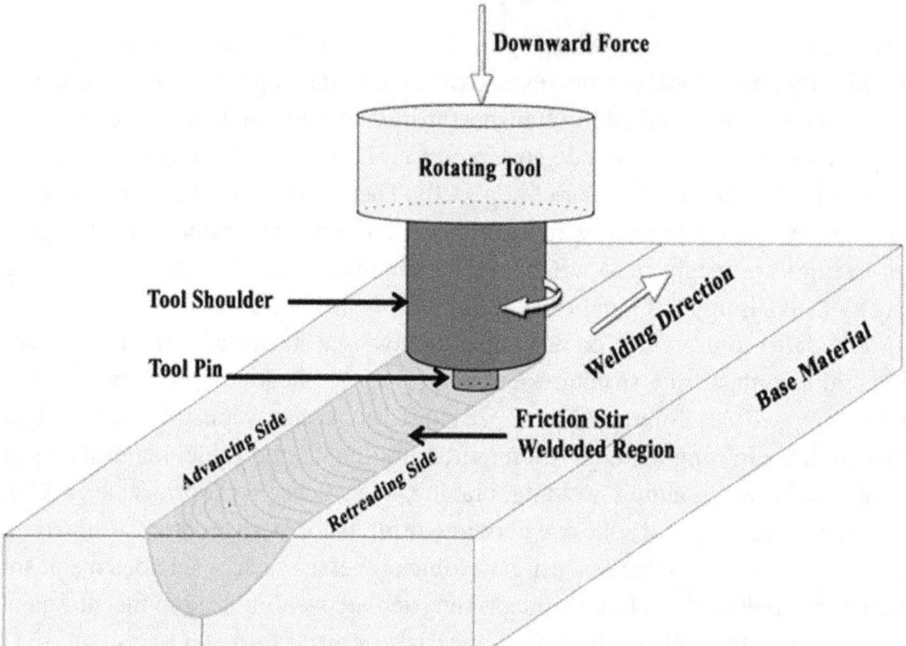

In the traditional process of making MMCs, ceramic particles are added to the matrices through casting or powder metallurgy (PM). Before being added to the matrix metal, the reinforcing particles used in MMCs are created. In this instance, the beginning particle size, often much bigger than one μm, limits the reinforcing phase scale. Due to the large particle size, the particles cannot use the Orowan mechanism since they are intergranular positioned in the metal matrix of traditionally treated MMCs. Because they prefer to cluster, it is challenging to disperse small ceramic particles into a metal matrix. However, the mechanical characteristics of the material might be significantly improved by dispersing small ceramic particles produced in situ more evenly throughout the matrices of MMCs (Dayanand et al., 2021)(Dayanand et al., 2021). Friction stir processing (FSP), which combines the hot-working aspect and the exothermic reaction between aluminium and a transition metal or metal oxide, has recently shown promise in creating in situ aluminium matrix composites with improved mechanical properties. In FSP, the sample is moved by a rotating tool with a pin and shoulder put into a single piece of material (Dayanand et al., 2021)(R. Yang et al., 2015). The friction between the revolving tool and the sample results in localised heating, which raises the material's local temperature to a level where it is easily plastically deformable. The material is agitated and forged behind the trailing face of the pin as the tool moves along the sample, where it condenses and cools (Singh Sidhu et al., 2020)(Mehdi & Mishra, 2021).

In addition, some research indicates that the conventional welding method could be more reliable for combining Aluminium composite materials. Laser welding (Guo et al., 2012), electron beam welding (Guo et al., 2012), diffusion welding (Guo et al., 2012), gas tungsten arc welding (Guo et al., 2012), tungsten inert gas welding (Guo et al., 2012), ultrasonic brazing (Guo et al., 2012), and metal inert gas welding (Guo et al., 2012) are some of the conventional welding methods for welding AMCs. The required joint characteristics are challenging to achieve using these conventional welding procedures. Conventional joining methods for MMCs encourage an unfavorable interaction between the matrix and reinforcement phase, eventually creating brittle joints. The melting of these composites during the joining process must be prevented to prevent the development of brittle joints. Thus, friction stir welding (FSW) can fuse such materials. Because FSW executes workpiece joining without melting, it is also known as the solid-state joining process. FSW is a technique for joining various grades of alloys, including copper alloys, lead, titanium alloys, magnesium alloys, zinc, aluminium alloys, MMCs, dissimilar alloys, metallic glass, lithium alloys, and many more. Interestingly, during FSW of different alloys, all the earlier research described above developed in-situ MMCs in the butt joint arrangement. However, more research must be conducted on MMC fabrication in situ during friction stir lap welding. More recently, lap welding of aluminium alloys through FSW with zinc interlayer was reported to have enhanced actual sheet thickness, greater lap width, increased SZ width, and higher tensile shear load (Niu et al., 2019)(Suresh S et al., 2022). To replace riveted lap joints, lap joints are frequently utilized during the assembly several parts and products in the automotive and aviation industries. Substantial efforts must be made to improve the welding efficiency of the joints in this particular design. To replace riveted lap joints, lap joints are frequently utilized during the assembly of several parts and products in the automotive and aviation industries. Substantial efforts must be made to improve the welding efficiency of the joints in this particular design.

The paper focuses on how welding parameters such as tool traverse speed, tool rotating speed, and the number of passes affect the welding process of friction-stir welded aluminium alloys. Moreover, the effects of nanoparticle addition on weld properties and microstructural formation have also been highlighted as well as recent trends in friction stir welding of aluminium alloys with nanoparticle incorporation.

2. TYPES OF REINFORCING NANOMATERIALS AND DEPOSITION METHODS

Many different kinds of particles for reinforcement have been used in FSW to produce metal matrix composite joints. When an incoherent reinforcement/matrix interface is achieved coupled with an excellent distribution of reinforcing particles, the initial size of the reinforcing particles is a critical parameter to the characteristics of metal matrix composite (MMC), providing an extra benefit to the strengthening process (Bafakeeh et al., 2021) (Faraji & Asadi, 2011).

Due to its superior capabilities at basic structures, nano-ceramic reinforcement is widely employed in many ways to create composites. Improved mechanical characteristics, high thermal stability, and superior chemical and tribological properties are benefits of such nano-reinforced composites (Bafakeeh et al., 2021). Because of these characteristics, ceramic nanoparticles are utilized in FSW as reinforcement for bettering common characteristics (Bafakeeh et al., 2021). Silicon Carbide (SiC) is the ceramic with the most significant reinforcing nanoparticles because of its vast capabilities to generate stronger joints (Bafakeeh et al., 2021). These characteristics of SiC include remarkable wear resistance, excellent chemical resistance, high hardness, and excellent oxidation resistance. Because of this property, there is a lower chance of cracking due to the great coherency between the nanoparticles and matrix. The physical characteristics of ceramic particles are summarised in Table 1.

The carbon nanotubes (CNTs) and graphene, commonly known as wonder materials, revolutionized research in the field of carbon. Researchers rapidly became aware of the global strengthening properties of graphene and CNTs. Due to their inherent qualities (planar geometry, high aspect ratio) and their remarkable physical, mechanical, electrical, and thermal capabilities, CNTs and graphene have become the ideal reinforcement for composite constructions (Dorri Moghadam et al., 2015). Owing to these excellent properties study was conducted in the field of FSP to make composites. Carbonaceous nanomaterial's large surface area accommodates its adequate interfacial bonding with the matrix of the composite. A tremendous decline in the wear rate and an increase in thermal conductivity has been observed during the processing of Al-Graphene composite through FSP (Dorri Moghadam et al., 2015). Table 2 shows the significant properties of CNTs and graphene.

Over the last decade, graphene has emerged as a potential reinforcement for various composites made of aluminium. It is the perfect reinforcement for many low-density alloys due to its vast aspect ratio, low density, and excellent mechanical and thermal properties. Graphene sheets' enormous surface area encourages a robust interfacial interaction with the metal component of composites (Latief et al., 2011). Although graphene can significantly boost an alloy's or composite's strength, its

Table 1. Properties of Ceramic particles

Particles	Density ($g*cm^{-3}$)	Elastic Modulus (GPa)	Melting Point (°C)	Coefficient of Thermal Expansion ($x10^{-6}$ K^{-1})
SiC	3.19	430	2697	4.63
Al_2O_3	3.95	380	2072	8.4
B_4C	2.51	445	2447	4.78
TiB	4.5	425-480	2200	8.60
TiC	4.99	440	3160	6.42-7.15
TiN	3.97	420	3290	8.30

Table 2. Properties of CNT and Graphene

Particles	Tensile Strength (GPa)	Thermal Conductivity (W/mK)	Electrical Conductivity (s/m)
CNTs (Tube form)	60-150	3500	3000-4000
Graphene (Platelet form)	130 ± 10	4.84 x 10^3 to 5.3 x 10^3	7200

application is constrained by the low corrosion resistance of aluminium alloys. Due to galvanic corrosion, the graphene layers cause Al to corrode (Latief et al., 2011).

The introduction of nanoparticles into the FSW process, their dispersion in the weld, and the selection of nanoparticles are minor concerns with nanoparticle reinforcement. For instance, groove preparation was necessary to reinforce the nanoparticles in the FSW process. Since the open groove preparation made it easier to incorporate nanoparticles as reinforcement, squeezing the FSW tool caused a spatter of nanoparticles to be distributed unevenly over the weld zone (Singh et al., 2019). Research has been done to create nanoparticle paste by combining the nanoparticles with an ethanol solution to prevent uneven nanoparticle distribution and limit nanoparticle spatter from the groove [33,39]. According to some research, a pinless tool was used to avoid nanoparticle expulsion by passing along the groove with nanoparticle reinforcement before welding (Abdollahzadeh et al., 2019). It is necessary to conduct more research that employs this technique of evenly dispersing nanoparticles throughout the weld. Fig. 2 represents the nanoparticles reinforcing method using cylindrical drilled holes in faying surfaces. Nano-particle deposition system (NPDS) is developed by feeding the graphite power through a nozzle to produce an FSSWelded Aluminium alloy 6061-T4 alloy sheet (Hong et al., 2017) (Suresh,

Figure 2. FSW process representing the location and method of addition of nanoparticles into the base matrix
(Singh, 2021b)

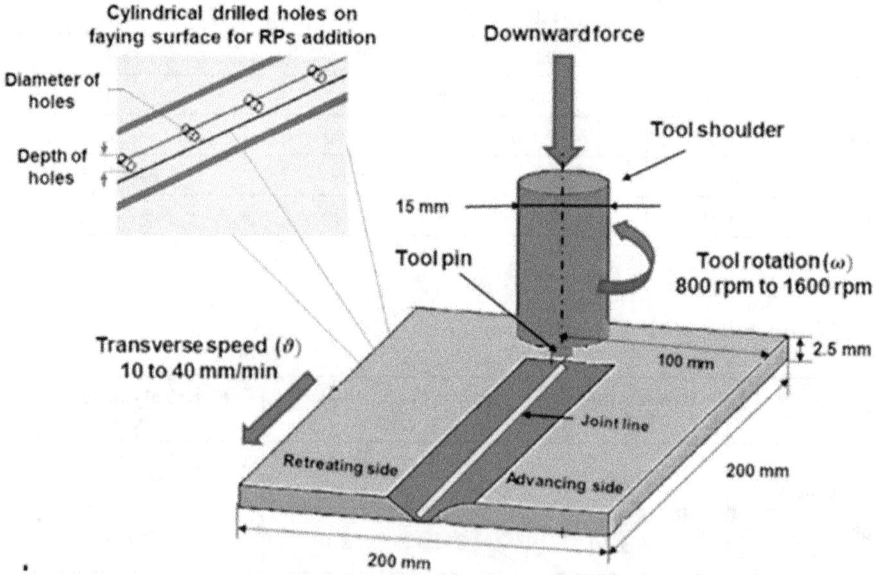

Venkatesan, et al., 2020a). The deposition method of carbon-based reinforcement is in the form of an interlayer, a slurry, or a powder.

One such investigation of graphene-reinforced FSW prepares a graphene interlayer on the top surface of the bottom plate. First, 20 mL of distilled water are ultrasonically sonicated with 0.1 gm of GNP for 60 minutes. Separately, 0.1 gm of PVA is dissolved in 20 mL of distilled water using a magnetic stirrer at 80 C for 60 minutes (A. Sharma et al., 2020).

3. FSW OF ALUMINIUM ALLOYS USING NANOPARTICLES

Aluminium alloys result in poor weldability and can be subject to defects when employed by the fusion process in the welding of similar and dissimilar alloys. Moreover, coarse-grain microstructural have formed due to high heat input during the process which results in low mechanical strength (Kah et al., 2015). For example, the study of GTAW on the Al alloy results in lower tensile strength due to the formation of porosity. However, FSW-processed Al alloys show better hardness and tensile properties than joints made by GTAW (Vasu et al., 2019). FSW processing below the melting temperature of the Al alloys is the significant reason for the properties enrichments (Suresh, Venkatesan, Natarajan, et al., 2019) (Grong et al., 2019) (Sudhagar & Gopal, 2021).

Recent studies focus on introducing nanoparticles in the Stir Zone (SZ) of the Al welds to improve the grain refinements and joint properties. The most comely used nanoparticles as reinforcements to fabricate aluminium friction stir welded composite joints are Alumina (Al_2O_3) (Suresh, Venkatesan, & Rajesh, 2019), Silicon Carbide (SiC) (Suresh et al., 2022), Titanium Dioxide (TiO_2) (Singh, 2021a) and Titanium Carbide (TiC) (Karakizis et al., 2018a). Some studies show that the reinforced nanoparticles are agglomerated in the SZ after the FSW process. The agglomerated particles cause a reduction in the tensile strength of the joints (Suresh et al., 2021). FSW is carried out in Al alloys by changing the features of the tool and changing the number of passes along the welding direction in order to reduce the agglomeration of reinforced nanoparticles in the SZ (Raja et al., 2020a; Singh, 2021a). Bahrami et al. 2015 (Bahrami et al., 2015) reported that the size of the agglomerated particles could be significantly reduced by changing the tool configuration and two pass-along grooves.

3.1 Effect of Nanoparticles on Mechanical Properties

Many factors, such as metallurgical bonding between nanoparticles and base material under various process parameters, grain size, and dislocation density, influence the mechanical properties of nanoparticle reinforcements FSWed joints. To enhance mechanical properties, including tensile strength, hardness, toughness, and fatigue, nanoparticles are added during FSW. One of the factors used to determine whether or not the weld quality is satisfactory is the mechanical characterization.

Initially, Bahrami et al. (Bahrami, Helmi, et al., 2014) experiments with SiC nanoparticle reinforcement during a similar FSW of AA7075-O alloy were effective. The authors noted a better UTS in specimens with SiC reinforcement, which the HallePetch equation can account for. Additionally, the fracture of the SiC-reinforced specimen occurred from the base metal as opposed to the TMAZ for the SiC-free specimen. The fact that TMAZ only experiences deformation rather than recrystallization may be the cause of this. Fractography samples reveal fine dimples in SiC-reinforced specimens, indicating a ductile mode of fracture, while SiC-free specimens show no dimples or coarse ones, indicating a brittle mode of fracture.

The macro and micro Silicon Carbide particles were incorporated to fabricate AA6061 and AA2024 dissimilar FSW joints (Moradi et al., 2018). It is found that the nugget grain size of nano-size SiC reinforced joints has decreased more than that of micro SiC reinforced weld nugget zone's grain size during the FSW process. Further, the mechanical properties such as tensile strength and hardness improved in nano SiC reinforced joints as compared to that micro SiC. It is noticed that the agglomerations of micro-scale particles are visible more in the nugget than the nano SiC. By using three different sizes of reinforcement particles, a similar finding was achieved from an Al_2O_3-reinforced FSW of aluminium composite (Nosko et al., 2019). This analysis confirms the Zener pinning effect mentioned in Eq (1). Recent research has revealed that incorporating reinforcing particles in FSW can enhance weld characteristics, enable the development of microstructures, and decrease the formation of joint defects (Anand & Sridhar, 2021; Moradi et al., 2017). The amount of addition of nanoparticles in the joint area has significantly influenced the tensile strength of the joint as represented in Fig. 3 (Suresh et al., 2018) (Suresh, Venkatesan, et al., 2020b), and also noted that the hardness of the joints also increases with increasing the volume fraction of nanoparticles. SiC reinforced welded samples reveal significantly improved fatigue life compared to the unreinforced weld samples due to the grain refinements in the SZ. The SiC nanoparticle-reinforced specimen fractured away from the stir zone during the fatigue test, whereas the unreinforced specimen fractured in the stir zone. In addition, it has been noted that SiC reinforcements have an excellent impact on impact energy absorption, which increases toughness (Bahrami, Helmi, et al., 2014).

The FSSW procedure revealed the feasibility of nanomaterial reinforcement. SiC nanoparticles were employed to reinforce the magnesium alloy during the FSSW process. The tensile shear load for reinforced joints was determined to be 3117 N, more significant than the 2562 N for unreinforced FSSW joints. The authors claimed that the fine grain size and homogeneous distribution of the SiC particles were primarily responsible for the increase in the strength of the joints. Compared to HAZ, the microhardness values in the nugget zone are higher. These improved characteristics of reinforced joints are likely a result of the grain boundary strengthening brought on by refined grains and the strengthening process (Wu et al., 2017). In another study, SiC particles were incorprated using FSW while reinforced with AA6351 alloy (Karthikeyan & Mahadevan, 2015). They claimed that SiC pinning action prevented grain boundary expansion, which led to improved tensile characteristics . Recently, in addition to employing SiC particles, researchers have also looked at metal powder, B_4C, and Al_2O_3 nanoparticles as reinforcement in the FSW process.

Residual stress is frequently viewed as an unfavorable characteristic of the welded specimen. The proper operation of the welded structure depends significantly on controlling residual stress. It is attempt to use SiO_2 nanoparticles as reinforcement to lower residual stress of dissimilar FSWed AA7075-T6 and AA6061-T6 joints (Jafari et al., 2019). They used the contour method to examine the residual stress distribution in the Z-direct unreinforced and nanoparticle-reinforced FSWed joints. Compared to the unreinforced specimen, a reduction in residual stress has been seen in the reinforced with SiO_2 nanoparticles. The temperature mismatch between the

Figure 3. Tensile shear load with varying amounts of SiC nanoparticles
(Suresh et al., 2018)

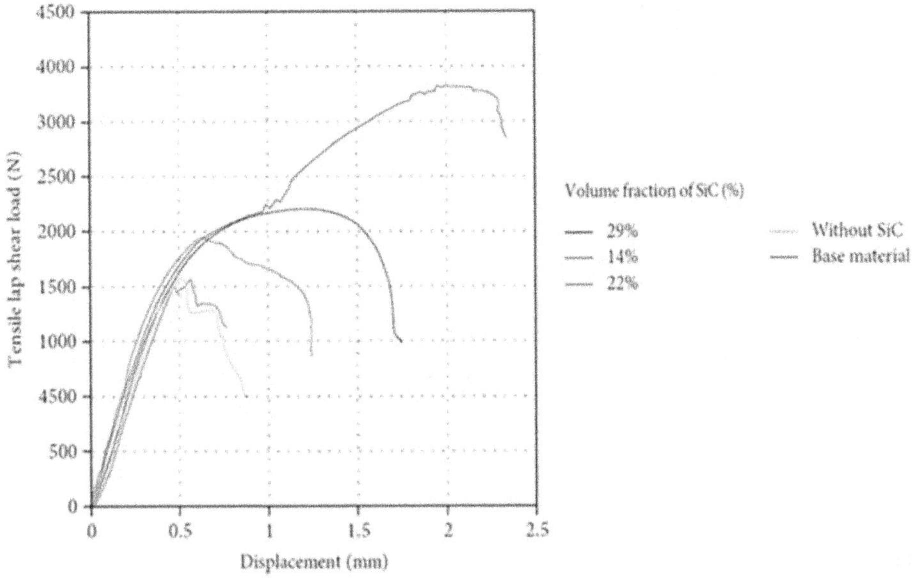

base materials and the SiO_2 nanoparticles has caused residual stress in the reinforced FSWed specimen.

Therefore, the studies also show that the effect of the amount of nanoparticles addition and welding parameters such as rotational tool speed and tool traverse speed play a critical roll in the distribution of nanoparticles in the weld area to improve the mechanical properties of the joints. And adding nanoparticles through FSW significantly improved mechanical properties like hardness, tensile strength, and toughness.

3. 2. Effect of Nanoparticles on Microstructural Formation

Fabricating a joint in an FSW reinforced by nanoparticles is a difficult process comprising thermal-mechanical interaction between work material, tool, and reinforcement material (Raja et al., 2020b).

The reinforcing particle size and volume addition play a crucial role in the refinement of the grains by reinforced particles. In accordance with the Zener pinning effect, grain boundary displacement occurs in the presence of reinforcing particles as a result of grain growth and recrystallization. The grain size using the Zener limit is expressed in Eq. (1) (Rohrer, 2010).

Table 3. Grain Size of nanoparticles reinforced aluminium alloy through FSW

Sl. No.	Base Material	Reinforcement Nanoparticles	Avg. Grain Size		% of Reduction in Grain Size	Ref.
			With Reinforcement	Without Reinforcement		
1	SiC	AA6082-T6	5.5	4.2	23	(Karakizis et al., 2018a)
2	TiO_2	AA 5083	18.5	8.5	54	(Mirjavadi et al., 2017)
3	Al_2O_3	AA 2021-T3	2.2	2.1	4.5	(Enami et al., 2019)
4	SiC	AA 5083 H111	6.6	5.3	19.6	(Karakizis et al., 2018b)
5	TiC	AA 5083 H111	6.6	4.7	28.78	(Karakizis et al., 2018b)
6	SiC	AA 7075	4.8	3.8	20	(Bahrami, Dehghani, et al., 2014)

$$R(Lim)=4r/3f \tag{1}$$

Where r and f represent the mean radius (size of the nanoparticles) and volume fraction (Volume addition of nanoparticles) of pinning particles, respectively.

FSW on AA7075-T6 and AA 7475-T7 dissimilar joints were effectively fabricated by incorporating SiC and Al_2O_3 with varying volume fractions (Anand & Sridhar, 2021). It is reported that the addition of SiC and Al_2O_3 nanoparticles enhances the mechanical properties of the dissimilar Al joints due to the uniform and homogenous distribution of reinforced particles in the stir zone. However, SiC-reinforced joints show improved joint performance than Al_2O_3 joints because of the SiC nanoparticles increase in stir zone grain refinements. Table 3 further summarises the influence of nanoparticle strengthening on weld nugget grain size during FSW compared to weld nugget grain size with no reinforcement.

SiO_2 nanoparticle distribution in aluminium matrix composites processed using FSP processes at different tool rotational speeds is shown in Fig. 4. Most of the nanoparticles were evenly dispersed throughout the weld that was produced at a lower speed, whereas the sample that was processed at a higher tool rotational speed exhibits nanoparticle aggregation, as seen in Fig. 4. (b).

The reinforcement effect of Al_2O_3 and TiO_2 nanoparticles on the friction stir welded 6061-T6 aluminium alloy was analysed respectively to microstructural changes in the stir zone by sing et al. (Singh et al., 2020). It is reported that, in the case of Al_2O_3, the stir zone shows fine recrystallized grains through friction heat and plastic deformation between the matrix and reinforced particles. Whereas, TiO_2

Figure 4. TEM images of Al-SiO_2 friction stir processed samples by the tool rotation speed of (a) 500 rpm and (b) 1400 rpm
(You et al., 2013)

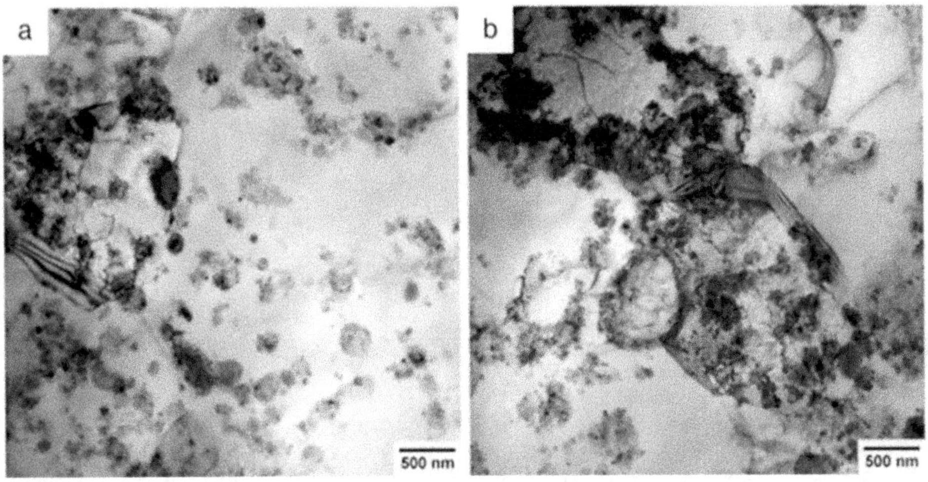

reinforced samples show agglomerated in the Heat affects Zone, which leads to a heat generation between deformed material and the rotating tool.

Sharifi et al. (Sharifi Asl et al., 2019) also reported the effect of friction heat on grain morphology of nanoparticles reinforced dissimilar joint on AA6063-T4 and AZ31-B alloy. It is observed that the average grain size (12-16 μm) is reduced sufficiently as compared to the base material grain size, due to the pinning effect of reinforced nanoparticles in the SZ. The effect of heat input on the microstructure can be seen at a constant tool traverse speed and increase in the tool rotational speed leads to high heat input resulting in better grain refinement in the composite joints. Graphene nanoplatelets are effectively used with the aluminium alloy 6061 as an interlayer to fabricate FSW joints (A. Sharma et al., 2020). Lap shear strength, percentage of elongation, and grain size of the joint significantly improved compared to the weld without interlayer. The graphene interlayer improves strengthening mechanisms such as grain refinement, load transfer, and thermal mismatch. It is noticed that the prevention of the formation of the Al_2O_3 layer at the sheets interface plays a significant role in welding strengthening.

The number of passes during friction stir welding/processing plays a critical role in the distribution of nanoparticles in the stir zone. Patel et al. (Patel et al., 2022) used single, two, four, and six passes during FSP on AA6063-T4 by adding nanoparticles. The study concluded that the sample made with six passes produced better fatigue strength than others due to the uniform distribution of nanoparticles as shown in Fig. 5.

Figure 5. SEM images showing the distribution of the reinforcement particles in the AA6063-T4 with (a) single pass, (b) two passes, (c) four passes, and (d) six passes (Patel et al., 2022)

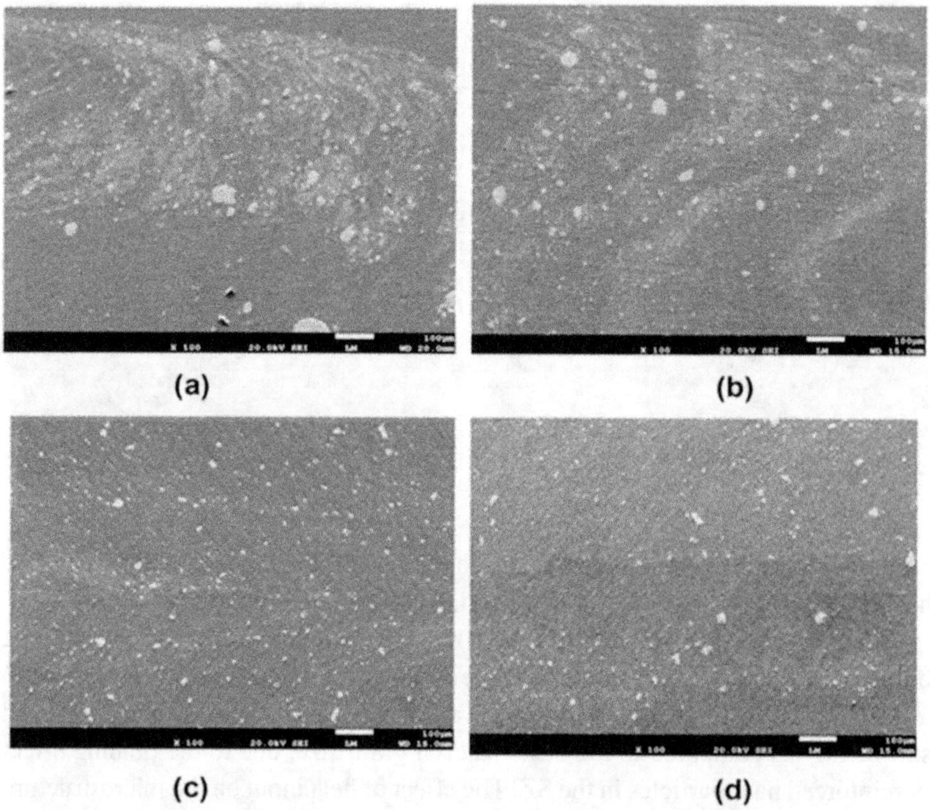

3.3 Effect of Nanoparticles on Tribological Properties

The ceramic nanoparticles, widely used as reinforcements in FSW joints, can improve the tribological behavior where wear is considered in application. The wear characteristics of a SiC-reinforced FSWed AA2024-T3 alloy joint were studied (Paidar & Sarab, 2016). The authors found that the wear rate dropped as the sliding distance increased when a joint strengthened with SiC was incorporated through a pin-on-disk wear test. Compared to the unreinforced joint's wear rate of 0.96 g/m, the joint with SiC's wear rate is significantly lower at 0.37 g/m. This outcome might be the result of reinforced SiC nanoparticles increasing hardness. As was previously stated, the existence of SiC particles causes fine grain size due to the pinning effect and further increases hardness in line with the Halle-Petch relation. Another investigation investigated the tribological characteristics of the AA5182 alloy employing tungsten carbide (WC) nanoparticles as reinforcement (Paidar & Sarab, 2016). The authors evaluated the wear rate variation for the various tool traverse speeds for WC-reinforced and unreinforced joints. It has been found that joints without WC nanoparticles wear out much more quickly when compared to those.

The FSW joints reinforced with TiO_2 and Al_2O_3 nanoparticles show significant increase in the wear rate agenized resistance compared to without reinforced joints (Mirjavadi et al., 2017). The increased wear resistance for reinforced joints is attributable to the high microhardness value, which comes from the process's refining of the grain. Additionally, it has been found that reinforced and unreinforced joints have improved coefficients of friction (COF) as compared to base aluminium (Suresh et al., 2022). A low COF for an Al-reinforced joint exposed it to the abrasive wear mechanism. In contrast, the joint reinforced with TiO_2 nanoparticles experienced lower wear resistance and a higher COF than those reinforced with Al_2O_3 nanoparticles. This results from poor adhesion between the aluminium matrix and TiO_2 nanoparticles caused by nanoparticle aggregation in the HAZ on the advancing side. Therefore, a detachment of the surface would occur instead of erosion (Singh et al., 2020).

Tribological characteristics of FSWelded AA6061–AA5083 dissimilar aluminum alloys by the reinforcing nanoparticles such as La_2O_3 and CeO_2. The La_2O_3 reinforced joints reportedly have higher mechanical and tribological properties. Due to the reinforcing effect of $La2O_3$ particles and the tiny grains created by the FSW process, the FSW dissimilar $La2O_3$ reinforced joints showed a lower wear rate of $7.37x10^3$ mm^3/m and low friction coefficient of 0.31 compared to all other materials (A. M. et al., 2021). According to several research, the direct load between the specimen surface and the disc is reduced when ceramic particles are present in the weld SZ. The aluminium matrix's strong reinforcing particle's capacity to carry load reduces direct load, reducing the wear rate at which strengthened welds wear out (Sudhagar et al., 2019). Furthermore, slowing down the tool traverse speed causes a noticeable variation in wear rate. Due to a reduction in tool traverse speed, this occurrence can be attributed to the correct dispersion of nanoparticles in the weld (A. M. et al., 2021) (Suresh et al., 2023).

3.4 Effect of Electrical and Electro-Chemical Properies

Other than excellent mechanical, thermal and tribological properties, ceramic particles, graphene and CNTs also possess excellent electrical properties. Researchers are trying to improve the electrical properties of welded joints without compromising any other Properties. CNTs with grapheme can meet the mechanical, thermal anr electrical requirements (Montazerian et al., 2019) (N. Sharma et al., 2018).

The FSP of pure aluminium reinforcing with CNTs was studied by Liu et al (Liu et al., 2014). They observed a slight decrease in electrical conductivity in the Al/

CNT composite by FSP. The two most important contributing elements to this drop are the uneven distribution of CNTs inside the structure and the propagation of the electron at the CNT/Al interface. The CNT's uneven dispersion prevents electrons from moving through its axis direction efficiently. Additionally, the situation is made worse by the electron's scattering at the CNT/Al intersections As a result, base electrical conductivity was somewhat lowered compared to composite CNT/Al electrical conductivity. Compared to base metal Al6061, the thermal conductivity of the CNT/AA6061 composite increased somewhat from 27.3 to 28.9 MS/m.

Recently, graphene reinforcement was used in FSW of copper and aluminium by Montazerian et al (Liu et al., 2014). The electric resistance of the joint was reduced by 40% by adding graphene nanoparticle reinforcement to the Al/Cu dissimilar weld at a tool rotational speed of 1200 rpm. As a result of the composite's development of a graphene conductive network and the high electron mobility of graphene, it is thought that the electrical resistance has decreased. The researcher noticed that adding graphene to a similar Al/Al weld reduced the joint's electrical resistance. In contrast, equivalent Cu/Cu welds with and without graphene reinforcement showed no apparent distinction in resistance.

The FSWed joints sensitivity to corrosion has additionally been taken into consideration and ceramic and carbonaceous nanomaterial could be a potential inhibitor for mitigate corrosion susceptibility. The thermal cycle that occurs during welding and the coarsening of precipitates are two factors that make FSWed joints more susceptible to corrosion. Kumar et al. (Vijaya Kumar et al., 2015) carried out an electrochemical study on the FSW of the AA7075 alloy reinforced with boron carbide nanoparticles.They investigated two different types of post-weld heat-treated materials with and without the addition of B_4C nanoparticles, as well as the impact of pitting corrosion on the FSW of the metal as received. The resistance to pitting corrosion was improved in the stir zone of the B_4C nanoparticles reinforced specimen. This is most likely caused by the hard ceramic particles insulating properties, which prevent a galvanic couple between the precipitates and matrix. Additionally, the possibility of segregation at grain boundaries increases due to the increased density of finer B_4C nanoparticle particles, enhancing the resistance to pitting and corrosion. Additionally, it has been found that when compared to all other specimen conditions, retrogression and re-aging (RRA) heat-treated specimens with B_4C nanoparticle reinforcement show the most remarkable improvement in resistance to pitting corrosion. Precipitate Free Zones (PFZs) have been eliminated, grain boundary precipitates have been relocated, and precipitate coherence in the matrix has increased.

An electrochemical examined dissimilar FSW specimens, AA6082-T6 and AA5083-H111, was looked at by Kartsonakis et al. (Kartsonakis et al., 2016). This investigation used ethanol to slurry up the TiC nanoparticles and MWCNT before they were applied to the grooves. Before the FSW, hybrid organic-inorganic coatings were sprayed using a roller applicator technique to the face of the aluminium plates, loading the CeMo containers with MBT as reinforcing nano additives. EIS The FS-CeMo sample had a greater total impedance modulus on bode plots of joint samples after exposure to a corrosive environment for 24 hours than samples containing TiC particles, MWCNT, or samples without nano additives. Additionally, it has been found that using TiC and MWCNT as reinforcement particles do not improve the reinforced product's corrosion resistance compared to the control sample without nano additions. This result shows that galvanic corrosion occurs due to the formation of galvanic couples between the TiC nanoparticles or the MWCNT, which both serve as cathodic electrodes for the aluminium at the weld nugget area. Carbon nanotubes (CNTs) and boron carbide (B4C) were combined to produce friction stir-welded Al-5083 plates to improve the corrosion characteristics. After FSW and the addition of reinforcements, there was an increase in corrosion resistance.

4. SUMMARY AND CONCLUSION

This review provides an in-depth review of state-of-the-art friction stir welding used to join aluminium alloys reinforced with ceramic and carbonaceous nanomaterial. It has been reported that welded joints have been evaluated critically based on their macrostructure, microstructure, mechanical characteristics, residual stresses, and tool wear. The followings are the summary and the key findings from the existing studies:

- The nanoparticles-reinforced FSWed aluminium joints have exhibited high strength, high hardness, and wear resistance.
- The intention of using nanomaterials as reinforcement is to facilitate the ability to reduce the grain size of joints to improve mechanical characteristics.
- Nanoparticle reinforcement, especially SiC, and Al_2O_3 have been revealed their constructive effect on mechanical properties enrichments.
- Despite the recent decade's concern in FSW technology, the fundamental physics understanding of this process is still lacking. It is necessary to comprehend several significant problems, including the influence of tool geometry features, the influence of nanoparticles on the mechanical characteristics of the FSW process, the stability of the microstructure, welding, resistance tools, and the welding of aluminium alloys.
- Thus, it emerged that a reduction in the electrical resistance of the stir zone with graphene reinforcements mainly caused the electrical resistance reduction of the aluminium joints. Additionally, researchers concluded that

an unbalanced distribution of the graphene reinforcement could raise the joint's electrical resistance.

- The tendency of nanomaterial to aggregate is caused by its large surface area. In the literature, there are two ways to deagglomerate the application of vibrational energy during welding and the increase in the number of passes. More research has to be done on this. In addition to these approaches, a new strategy must be created.
- Lack of research in stabilizing and optimizing process parameters of ultrasonic-assisted, laser-assisted, and electrical current-assisted FSW.

REFERENCES

Abdollahzadeh, A., Shokuhfar, A., Cabrera, J. M., Zhilyaev, A. P., & Omidvar, H. (2019). In-situ nanocomposite in friction stir welding of 6061-T6 aluminum alloy to AZ31 magnesium alloy. *Journal of Materials Processing Technology*, *263*, 296–307. doi:10.1016/j.jmatprotec.2018.08.025

Anand, R., & Sridhar, V. G. (2021). Effects of SiC and Al2O3 Reinforcement of Varied Volume Fractions on Mechanical and Micro Structure Properties of Interlock FSW Dissimilar Joints AA7075-T6-AA7475-T7. *Silicon*, *13*(9), 3017–3029. doi:10.100712633-020-00630-y

Bafakeeh, O. T., Shewakh, W. M., Abu-Oqail, A., Abd-Elaziem, W., Abdel Ghafaar, M., & Abu-Okail, M. (2021). Synthesis and Characterization of Hybrid Fiber-Reinforced Polymer by Adding Ceramic Nanoparticles for Aeronautical Structural Applications. *Polymers*, *13*(23), 4116. doi:10.3390/polym13234116 PMID:34883619

Bahrami, M., Dehghani, K., & Besharati Givi, M. K. (2014). A novel approach to develop aluminum matrix nano-composite employing friction stir welding technique. *Materials & Design*, *53*, 217–225. doi:10.1016/j.matdes.2013.07.006

Bahrami, M., Farahmand Nikoo, M., & Besharati Givi, M. K. (2015). Microstructural and mechanical behaviors of nano-SiC-reinforced AA7075-O FSW joints prepared through two passes. *Materials Science and Engineering A*, *626*, 220–228. doi:10.1016/j.msea.2014.12.009

Bahrami, M., Helmi, N., Dehghani, K., & Givi, M. K. B. (2014). Exploring the effects of SiC reinforcement incorporation on mechanical properties of friction stir welded 7075 aluminum alloy: Fatigue life, impact energy, tensile strength. *Materials Science and Engineering A*, *595*, 173–178. doi:10.1016/j.msea.2013.11.068

Boopathi, S., Jeyakumar, M., Singh, G. R., King, F. L., Pandian, M., Subbiah, R., & Haribalaji, V. (2022). An experimental study on friction stir processing of aluminium alloy (AA-2024) and boron nitride (BNp) surface composite. *Materials Today: Proceedings*, *59*, 1094–1099. doi:10.1016/j.matpr.2022.02.435

Casati, R., & Vedani, M. (2014). Metal matrix composites reinforced by nano-particles—A review. *Metals*, *4*(1), 65–83. doi:10.3390/met4010065

Daehn, G. S. (2014). 13 - Sustainable design and manufacture of lightweight vehicle structures. In R. Folkson (Ed.), *Alternative Fuels and Advanced Vehicle Technologies for Improved Environmental Performance* (pp. 433–461). Woodhead Publishing. doi:10.1533/9780857097422.2.433

Danish, M., Ginta, T., Yasir, M., & Abdul-Rani, A.-M. (2018). Light alloys and their machinability. *RE:view*, 1–34. doi:10.1201/b22153-1

Dayanand, S., Boppana, S. B., Auradi, V., Nagaral, M., Udaya Ravi, M., & Bharath. (2021). Evaluation of Wear Properties of Heat-Treated Al-AlB2 in-situ Metal Matrix Composites. *Journal of Bio- and Tribo-Corrosion*, *7*(2), 40. doi:10.100740735-021-00476-w

Dorri Moghadam, A., Omrani, E., Menezes, P. L., & Rohatgi, P. K. (2015). Mechanical and tribological properties of self-lubricating metal matrix nanocomposites reinforced by carbon nanotubes (CNTs) and graphene – A review. *Composites. Part B, Engineering*, *77*, 402–420. doi:10.1016/j.compositesb.2015.03.014

Enami, M., Farahani, M., & Farhang, M. (2019). Novel study on keyhole less friction stir spot welding of Al 2024 reinforced with alumina nanopowder. *International Journal of Advanced Manufacturing Technology*, *101*(9), 3093–3106. doi:10.100700170-018-3142-z

Faraji, G., & Asadi, P. (2011). Characterization of AZ91/alumina nanocomposite produced by FSP. *Materials Science and Engineering A*, *528*(6), 2431–2440. doi:10.1016/j.msea.2010.11.065

Grong, Ø., Sandnes, L., & Berto, F. (2019). Progress in Solid State Joining of Metals and Alloys. *Procedia Structural Integrity*, *17*, 788–798. doi:10.1016/j.prostr.2019.08.105

Guo, J., Gougeon, P., Nadeau, F., & Chen, X.-G. (2012). Joining of AA1100–16 vol.-%B 4 C metal matrix composite using laser welding and friction stir welding. *Canadian Metallurgical Quarterly*, *51*(3), 277–283. doi:10.1179/187913951 2Y.0000000003

Hong, S.-T., Das, H., Oh, H.-S., Al Nasim, M. N. E. A., & Chun, D.-M. (2017). Combination of nano-particle deposition system and friction stir spot welding for fabrication of carbon/aluminum metal matrix composite joints of dissimilar aluminum alloys. *CIRP Annals*, *66*(1), 261–264. doi:10.1016/j.cirp.2017.04.115

Jafari, H., Mansouri, H., & Honarpisheh, M. (2019). Investigation of residual stress distribution of dissimilar Al-7075-T6 and Al-6061-T6 in the friction stir welding process strengthened with SiO2 nanoparticles. *Journal of Manufacturing Processes*, *43*, 145–153. doi:10.1016/j.jmapro.2019.05.023

Kah, P., Rajan, R., Martikainen, J., & Suoranta, R. (2015). Investigation of weld defects in friction-stir welding and fusion welding of aluminium alloys. *International Journal of Mechanical and Materials Engineering*, *10*(1), 26. doi:10.118640712-015-0053-8

Karakizis, P. N., Pantelis, D. I., Fourlaris, G., & Tsakiridis, P. (2018a). Effect of SiC and TiC nanoparticle reinforcement on the microstructure, microhardness, and tensile performance of AA6082-T6 friction stir welds. *International Journal of Advanced Manufacturing Technology*, *95*(9), 3823–3837. doi:10.100700170-017-1446-z

Karakizis, P. N., Pantelis, D. I., Fourlaris, G., & Tsakiridis, P. (2018b). The role of SiC and TiC nanoparticle reinforcement on AA5083-H111 friction stir welds studied by electron microscopy and mechanical testing. *International Journal of Advanced Manufacturing Technology*, *94*(9), 4159–4176. doi:10.100700170-017-1147-7

Karthikeyan, P., & Mahadevan, K. (2015). Investigation on the effects of SiC particle addition in the weld zone during friction stir welding of Al 6351 alloy. *International Journal of Advanced Manufacturing Technology*, *80*(9–12), 1919–1926. doi:10.100700170-015-7160-9

Kartsonakis, I. A., Dragatogiannis, D. A., Koumoulos, E. P., Karantonis, A., & Charitidis, C. A. (2016). Corrosion behaviour of dissimilar friction stir welded aluminium alloys reinforced with nanoadditives. *Materials & Design*, *102*, 56–67. doi:10.1016/j.matdes.2016.04.027

Latief, F. H., Sherif, E.-S. M., Almajid, A. A., & Junaedi, H. (2011). Fabrication of exfoliated graphite nanoplatelets-reinforced aluminum composites and evaluating their mechanical properties and corrosion behavior. *Journal of Analytical and Applied Pyrolysis*, *92*(2), 485–492. doi:10.1016/j.jaap.2011.09.003

Liu, Z. Y., Xiao, B. L., Wang, W. G., & Ma, Z. Y. (2014). Tensile Strength and Electrical Conductivity of Carbon Nanotube Reinforced Aluminum Matrix Composites Fabricated by Powder Metallurgy Combined with Friction Stir Processing. *Journal of Materials Science and Technology*, *30*(7), 649–655. doi:10.1016/j.jmst.2014.04.016

M., A., M., M., & S., B. (2021). Tribological characterization of friction stir welded dissimilar aluminum alloy AA6061–AA5083 reinforced with CeO and LaO nanoparticles. *Industrial Lubrication and Tribology, 73*(5), 783–788. doi:10.1108/ILT-01-2021-0009

M, S., S, N., & Davidson, B. S. (2021). Friction stir welding parameters and their influence on mechanical properties of welded AA6061 and AA5052 aluminium plates. *Materials Research Express, 8*(10), 106525. doi:10.1088/2053-1591/ac2daf

Madesh, S., Arul, T. G., & Perumal, V. (2022). *Advances in Welding Techniques for Similar and Dissimilar Materials.*, doi:10.4018/978-1-7998-9574-9.ch011

Mayyas, A. T., Mayyas, A. R., & Omar, M. (2016). 11 - Sustainable lightweight vehicle design: A case study in eco-material selection for body-in-white. In J. Njuguna (Ed.), *Lightweight Composite Structures in Transport* (pp. 267–302). Woodhead Publishing. doi:10.1016/B978-1-78242-325-6.00011-6

Mehdi, H., & Mishra, R. S. (2021). Effect of multi-pass friction stir processing and SiC nanoparticles on microstructure and mechanical properties of AA6082-T6. *Advances in Industrial and Manufacturing Engineering, 3*, 100062. doi:10.1016/j.aime.2021.100062

Mirjavadi, S. S., Alipour, M., Emamian, S., Kord, S., Hamouda, A. M. S., Koppad, P. G., & Keshavamurthy, R. (2017). Influence of TiO2 nanoparticles incorporation to friction stir welded 5083 aluminum alloy on the microstructure, mechanical properties and wear resistance. *Journal of Alloys and Compounds, 712*, 795–803. doi:10.1016/j.jallcom.2017.04.114

Montazerian, M. H., Movahedi, M., & Jondi, M. R. (2019). Effect of graphene and process parameters on mechanical performance and electrical resistance of aluminum to copper friction stir joint. *Materials Research Express, 6*(4), 046561. doi:10.1088/2053-1591/aafe1a

Moradi, M. M., Aval, H. J., & Jamaati, R. (2018). Microstructure and mechanical properties in nano and microscale SiC-included dissimilar friction stir welding of AA6061-AA2024. *Materials Science and Technology, 34*(4), 388–401. doi:10.1080/02670836.2017.1393976

Moradi, M. M., Jamshidi Aval, H., & Jamaati, R. (2017). Effect of pre and post welding heat treatment in SiC-fortified dissimilar AA6061-AA2024 FSW butt joint. *Journal of Manufacturing Processes, 30*, 97–105. doi:10.1016/j.jmapro.2017.08.014

Niu, S., Ji, S., Yan, D., Meng, X., & Xiong, X. (2019). AZ31B/7075-T6 alloys friction stir lap welding with a zinc interlayer. *Journal of Materials Processing Technology, 263*, 82–90. doi:10.1016/j.jmatprotec.2018.08.009

Nosko, M., Štepánek, M., Zifčák, P., Orovčík, L., Nagy, Š., Dvorák, T., Oslanec, P., Khodabakhshi, F., & Gerlich, A. P. (2019). Solid-state joining of powder metallurgy Al-Al2O3 nanocomposites via friction-stir welding: Effects of powder particle size on the weldability, microstructure, and mechanical property. *Materials Science and Engineering A, 754*, 190–204. doi:10.1016/j.msea.2019.03.074

Paidar, M., & Sarab, M. L. (2016). Friction stir spot welding of 2024-T3 aluminum alloy with SiC nanoparticles. *Journal of Mechanical Science and Technology, 30*(1), 365–370. doi:10.100712206-015-1241-4

Patel, M., Chaudhary, B., Murugesan, J., Jain, N. K., & Patel, V. (2022). Enhancement of tensile and fatigue properties of hybrid aluminium matrix composite via multipass friction stir processing. *Journal of Materials Research and Technology, 21*, 4811–4823. doi:10.1016/j.jmrt.2022.11.073

Raja, S., Muhamad, M. R., Jamaludin, M. F., & Yusof, F. (2020a). A review on nanomaterials reinforcement in friction stir welding. *Journal of Materials Research and Technology, 9*(6), 16459–16487. doi:10.1016/j.jmrt.2020.11.072

Ravikumar, M., Sivashankar, A., & Suresh, S. (2020). Materials Today : Proceedings Effects of rice bran oil blends and engine load on emission characteristics of a variable compression ratio diesel engine. *Materials Today: Proceedings*. Science Direct. doi:10.1016/j.matpr.2020.03.680

Rohrer, G. S. (2010). "Introduction to Grains, Phases, and Interfaces—An Interpretation of Microstructure," Trans. AIME, 1948, vol. 175, pp. 15–51, by C.S. Smith. *Metallurgical and Materials Transactions. A, Physical Metallurgy and Materials Science, 41*(5), 1063–1100. doi:10.100711661-010-0215-5

Sasikala, G., Jothiprakash, V. M., Pant, B., Subalakshmi, R., Thirumal Azhagan, M., Arul, K., Alonazi, W. B., Karnan, M., & Praveen Kumar, S. (2022). Optimization of Process Parameters for Friction Stir Welding of Different Aluminum Alloys AA2618 to AA5086 by Taguchi Method. *Advances in Materials Science and Engineering, 2022*, 1–9. doi:10.1155/2022/3808605

Sharifi Asl, N., Mirsalehi, S. E., & Dehghani, K. (2019). Effect of TiO2 nanoparticles addition on microstructure and mechanical properties of dissimilar friction stir welded AA6063-T4 aluminum alloy and AZ31B-O magnesium alloy. *Journal of Manufacturing Processes, 38*, 338–354. doi:10.1016/j.jmapro.2019.01.023

Sharma, A., Sharma, V. M., Gugaliya, A., Rai, P., Pal, S. K., & Paul, J. (2020). Friction stir lap welding of AA6061 aluminium alloy with a graphene interlayer. *Materials and Manufacturing Processes*, *35*(3), 258–269. doi:10.1080/10426914.2020.1718694

Sharma, N., Khan, Z. A., Siddiquee, A. N., Shihab, S. K., & Wahid, M. A. (2018). Effect of process parameters on microstructure and electrical conductivity during FSW of Al-6101 and Pure Copper. *Materials Research Express*, *5*(4), 046519. doi:10.1088/2053-1591/aabbe0

Sharma, V., Prakash, U., & Kumar, B. V. M. (2015). Surface composites by friction stir processing: A review. *Journal of Materials Processing Technology*, *224*, 117–134. doi:10.1016/j.jmatprotec.2015.04.019

Singh, T. (2021a). Processing of friction stir welded AA6061-T6 joints reinforced with nanoparticles. *Results in Materials*, *12*, 100210. doi.org/10.1016/j.rinma.2021.100210

Singh, T. (2021b). Processing of friction stir welded AA6061-T6 joints reinforced with nanoparticles. *Results in Materials*, *12*, 100210. doi:10.1016/j.rinma.2021.100210

Singh, T., Tiwari, S. K., & Shukla, D. K. (2019). Friction-stir welding of AA6061-T6: The effects of Al2O3 nano-particles addition. *Results in Materials*, *1*, 100005. doi:10.1016/j.rinma.2019.100005

Singh, T., Tiwari, S. K., & Shukla, D. K. (2020). Mechanical and microstructural characterization of friction stir welded AA6061-T6 joints reinforced with nano-sized particles. *Materials Characterization*, *159*, 110047. doi:10.1016/j.matchar.2019.110047

Singh Sidhu, H., Singh, B., & Kumar, P. (2020). To study the corrosion behavior of friction stir processed magnesium alloy AZ91. *Materials Today: Proceedings*, *44*, 4633–4639. doi:10.1016/j.matpr.2020.10.920

Subramani, V., Jayavel, B., Sengottuvelu, R., & Lazar, P. (2019). Assessment of Microstructure and Mechanical Properties of Stir Zone Seam of Friction Stir Welded Magnesium AZ31B through Nano-SiC. *Materials (Basel)*, *12*(7), 1044. doi:10.3390/ma12071044 PMID:30934939

Sudhagar, S., & Gopal, P. M. (2021). Investigation on Mechanical and Tribological Characteristics Cu/Si3N4 Surface Composite Developed Through Friction Stir Processing. *Silicon*, 1–10.

Sudhagar, S., Sakthivel, M., & Ganeshkumar, P. (2019). Monitoring of friction stir welding based on vision system coupled with Machine learning algorithm. *Measurement*, *144*, 135–143. doi:10.1016/j.measurement.2019.05.018

Sukkasamy, S., & Gopal, P. M. (2022). Monitoring of the Friction Stir Welding Process. In Advanced Manufacturing Techniques for Engineering and Engineered Materials (pp. 173–182). doi:10.4018/978-1-7998-9574-9.ch010

Suresh, S., Elango, N., Venkatesan, K., Lim, W. H., Palanikumar, K., & Rajesh, S. (2020). Sustainable friction stir spot welding of 6061-T6 aluminium alloy using improved non-dominated sorting teaching learning algorithm. *Journal of Materials Research and Technology*, *9*(5), 11650–11674. doi:10.1016/j.jmrt.2020.08.043

Suresh, S., Natarajan, E., Franz, G., & Rajesh, S. (2022). Differentiation in the SiC Filler Size Effect in the Mechanical and Tribological Properties of Friction-Spot-Welded AA5083-H116 Alloy. *Fibers (Basel, Switzerland)*, *10*(12), 109. doi:10.3390/fib10120109

Suresh, S., Natarajan, E., & Shanmugam, R. (2022). Strategized friction stir welded AA6061-T6/SiC composite lap joint suitable for sheet metal applications. *Journal of Materials Research and Technology*, *21*, 30–39. doi:10.1016/j.jmrt.2022.09.022

Suresh, S., Natarajan, E., Vinayagamurthi, P., Venkatesan, K., Viswanathan, R., & Rajesh, S. (2023). Optimum Tool Traverse Speed Resulting Equiaxed Recrystallized Grains and High Mechanical Strength at Swept Friction Stir Spot Welded AA7075-T6 Lap Joints. In E. Natarajan, S. Vinodh, & V. Rajkumar (Eds.), *Materials, Design and Manufacturing for Sustainable Environment* (pp. 547–555). Springer Nature Singapore. doi:10.1007/978-981-19-3053-9_41

Suresh, S., Venkatesan, K., & Natarajan, E. (2018). Influence of SiC Nanoparticle Reinforcement on FSS Welded 6061-T6 Aluminum Alloy. *Journal of Nanomaterials*, *2018*, 1–11. doi:10.1155/2018/7031867

Suresh, S., Venkatesan, K., Natarajan, E., & Rajesh, S. (2020a). Influence of tool rotational speed on the properties of friction stir spot welded AA7075-T6/Al2O3 composite joint. *Materials Today: Proceedings*, *27*, 62–67. doi:10.1016/j.matpr.2019.08.220

Suresh, S., Venkatesan, K., Natarajan, E., & Rajesh, S. (2021). Performance Analysis of Nano Silicon Carbide Reinforced Swept Friction Stir Spot Weld Joint in AA6061-T6 Alloy. *Silicon*, *13*(10), 3399–3412. doi:10.100712633-020-00751-4

Suresh, S., Venkatesan, K., Natarajan, E., Rajesh, S., & Lim, W. H. (2019). Evaluating weld properties of conventional and swept friction stir spot welded 6061-t6 aluminium alloy. *Materials Express*, *9*(8), 851–860. doi:10.1166/mex.2019.1584

Suresh, S., Venkatesan, K., & Rajesh, S. (2019). Optimization of process parameters for friction stir spot welding of AA6061/Al2O3 by Taguchi method. *AIP Conference Proceedings*, *2128*(July), 1–10. doi:10.1063/1.5117961

Thorny, C., Wirth, A., Kreimeyer, M., Wagner, F., & Vollertsen, F. (2007). Joining of dissimilar materials - New perspectives for lightweight design in the transportation industries. *Welding in the World*, *51*, 311–326.

Ulutaş, A. (2022). *Joining Techniques Like Welding in Lightweight Material Structures*., doi:10.4018/978-1-7998-7864-3.ch006

Vasu, K., Chelladurai, H., Ramaswamy, A., Malarvizhi, S., & Balasubramanian, V. (2019). Effect of fusion welding processes on tensile properties of armor grade, high thickness, non-heat treatable aluminium alloy joints. *Defence Technology, 15*(3), 353–362. https://doi.org/10.1016/j.dt.2018.11.004

Vijaya Kumar, P., Madhusudhan Reddy, G., & Srinivasa Rao, K. (2015). Microstructure and pitting corrosion of armor grade AA7075 aluminum alloy friction stir weld nugget zone – Effect of post weld heat treatment and addition of boron carbide. *Defence Technology*, *11*(2), 166–173. doi:10.1016/j.dt.2015.01.002

Wu, D., Shen, J., Lv, L., Wen, L., & Xie, X. (2017). Effects of nano-SiC particles on the FSSW welded AZ31 magnesium alloy joints. *Materials Science and Technology*, *33*(8), 998–1003. doi:10.1080/02670836.2016.1254891

Yang, J., Oliveira, J. P., Li, Y., Tan, C., Gao, C., Zhao, Y., & Yu, Z. (2022). Laser techniques for dissimilar joining of aluminum alloys to steels: A critical review. *Journal of Materials Processing Technology*, *301*, 117443. doi:10.1016/j.jmatprotec.2021.117443

Yang, R., Zhang, Z., Zhao, Y., Chen, G., Guo, Y., Liu, M., & Zhang, J. (2015). Effect of multi-pass friction stir processing on microstructure and mechanical properties of Al3Ti/A356 composites. *Materials Characterization*, *106*, 62–69. doi:10.1016/j.matchar.2015.05.019

You, G. L., Ho, N. J., & Kao, P. W. (2013). In-situ formation of Al2O3 nanoparticles during friction stir processing of AlSiO2 composite. *Materials Characterization*, *80*, 1–8. doi:10.1016/j.matchar.2013.03.004

Chapter 3
Impacts of Nanomaterials on Enhancing Concrete Quality: A Comprehensive Analysis

M. Harikaran
https://orcid.org/0000-0003-2229-7127
Muthayammal Engineering College, India

P. Kulanthaivel
Kongu Engineering College, India

P. Keerthana Devi
Muthayammal Engineering College, India

M. Sanchaya
Muthayammal Engineering College, India

ABSTRACT

Our perceptions, expectations, and ability to control over the world of building materials have transformed as a result of nanotechnology. The field of construction materials will be significantly impacted by advances in nanoscience. This study examines the steps used to stop or lessen concrete's deterioration, which contrasts with concrete with greater compressive strength. The mechanism of reaction clearly lays forth rules for how the likelihood of this disastrous reaction should be completely minimized. The microstructure of concrete became more homogeneous and less permeable as a result of the Nano filler effect and the pozzolanic reaction, which reduced permeability. The corrosion of the steel reinforcement, which is accelerated by the development of micro cracks, is one of the main inherent characteristics that contribute to the deterioration of structural elements. Impact of nano-silica particles on a blended cement paste's hydration, rheology, and strength development. An isothermal calorimetry reveals that nano silica particles cause cement to hydrate more quickly.

DOI: 10.4018/978-1-6684-9135-5.ch003

INTRODUCTION

Concrete is the most widely used building material, and cement is used most frequently in concrete design. The usage of cement contributes to both increasing CO2 emissions and greenhouse gas emissions. Using silica fines is one way to reduce the amount of cement in concrete mixtures. Nano-silica is a silica fine that has a great deal of potential as a concrete additive and cement substitute (nS).

The manufacturing process for commercial nS is very complex, resulting in high purity and labor-intensive processes that render them unusable for the construction industry (Montgomery et al., 2016). Additionally, nS's use and impact in concrete are yet not well understood. In a recent research project, a nano-silica will be used to increase the strength of concrete. The nS that is offered for sale will be used and tested. Using nano-silica enhances the characteristics of hardened concrete and lowers the CO_2 footprint of made concrete products. Consequently, concrete with improved performance, lower prices, and sustainability of the environment is designed. Concretes can be used anywhere, in both structures and infrastructure. Our perceptions of, and capacities for, controlling the material world have transformed as a result of nanotechnology. The field of construction materials will be significantly impacted by advances in Nanoscience. A new generation of concrete that is stronger and more resilient with the necessary stress-strain behavior will undoubtedly be produced as a result of improved engineering knowledge of the intricate nanoscale structure of cement-based materials. The methods used to avoid or limit concrete failure, which affects concrete's strength, are the main topic of this study. Although its strength, concrete's endurance is influenced by its surroundings and capacity to withstand damage. Both conventional methods and cutting-edge ones, such as nanotechnology, can be used to make concrete resistant to acid. The micro-level does not currently offer sufficient insights into construction materials. As a result, a growing amount of research funding is being directed at the nanoscale, which is seen to have enormous future promise. One of the most active research fields is nanotechnology, which has numerous applications in practically every field. Concrete's performance on a nanoscale affects the fundamental processes that control its physical characteristics. Since concrete is the most versatile building material, quality improvement is necessary.

A combination of finely ground cement, different-sized aggregates, and water with inherent physical, chemical, and mechanical qualities results in the very versatile substance known as concrete. Many mineral admixtures, like fly ash, silica fume, etc., that share some of the same qualities as cement, can substitute it in part. Concrete composites with enhanced characteristics can be created by using nanomaterials. Concrete is subjected to nanotechnology by the use of nano-materials like nano-silica, which enhances the microstructure and lowers the water permeability of concrete, increasing its density and durability (Jaya., 2016). Concrete particles are more cohesive and bleeding and segregation are reduced when HPC with Nano-silica is used. Nano-silica can be added to solve several issues, such as longer setting times and weaker compressive strength at larger percentages. Moreover, the workability, variation in setting times, and durability of the concrete will all be enhanced by the inclusion of supplemental cementitious elements. The mechanical qualities of high strength concrete with different replacements of Nano-silica, 1% have been taken into consideration (Sarade Rutuja et al., 2017). These parameters include stress-strain characteristics, splitting tensile strength, compressive strength, and flexural strength. A study found that silica nanoparticles work as fillers to fill holes or empty regions. The evenly distributed nS serve as the nucleation or crystallization centres of the hydrated products, boosting the rate of hydration. In other words, the nS contributed to the creation of homogenous clusters of C-S-H composition and smaller-sized CH crystals. In addition, they discovered that nS enhanced the structure of the zone between aggregates and paste. The M40 grade concrete with 1% of the cement replaced by nano-silica was the primary parameter examined in this investigation. A thorough experimental research on compressive strength, flexure strength, split tensile strength, impact strength, and oxygen permeability strength at ages 7 and 28 days is presented in this work. According to test results, adding nano silica to concrete has enhanced both the strength and longevity of the material. The microstructure analysis of the concrete supports the durability claim. Moreover, nowhere in the preceding research work has the mix design of the concrete containing nS been determined or mentioned. Without regard to strength, durability, or mixed design, the current research work has addressed the research gap identified by the numerous research articles.

The Impact of nano-silica in concrete review. More water or plasticizers are required to keep cement mixes workable when high surface mineral particles are added the dosage, type (powder or colloidal), amount, and water-to-cement ratio all affect how much nS affects concrete (Hasan Biricika & NihalSarierb., 2017). The mechanical and durability qualities could be enhanced by nS dosages of up to 2 to 3%. This could be brought on by poszolanic activity, the filling effect, or finely tailored pore structure (Mahender & Ashok, B., 2017). As nS concentration rises, hydration is accelerated and compression strength rises as a result. If the nS dosage is greater than 3%, however, the strength can diminish (Vasanthi et al., 2017). The engineering qualities may change with a higher dosage, such as 3% or more, due to the nS particles' tendency to aggregate and cause high porosity, micro cracking, decreased compressive strength, etc. In comparison to silica fume, nS exhibits a much higher level of pozzolanic activity. Concrete with nS has greater compressive strength than concrete without nS at all ages. The addition of nS-particle also improved the flexural and split tensile strengths of the concrete. Concrete developed long-term strength and endurance due to the filler effect, pozzolanic action, and reactivity of nS with CH crystals to reduce their size and amount. This was accomplished by creating a denser ITZ microstructure between aggregates and cement paste. If the nS particles were evenly dispersed throughout the concrete, the microstructure would benefit more.

Concrete's compressive strength as a result of nano silica. The results for compressive strength show that adding a specific minimal amount of Nano SiO_2 to concrete causes its compressive strength to increase. The strength increases most when nS is 1% bwc and least when nS is 0.3% bwc. The early-age strength of concrete has significantly increased in comparison to the 28-day increase in strength, in addition to Nano SiO_2 (Rajkumar et al., 2016). The results of the UPV test show that, despite a slight reduction in quality due to the inclusion of Nano SiO2, the overall quality of concrete is maintained. The C-S-H gel's interconnected pores were filled with Nano SiO_2, which increased the microstructure's density and uniformity.

Impact of adding nano-silica to high-performance concrete on mechanical and microstructure studies. The impact of additional cementitious materials and other cementitious elements in concrete along with cement. There have been significant advancements in permeability, pore filling effects, microstructure analysis, and strength. The review as a whole demonstrated the impending doom of adopting nanotechnology in general and Nano silica in particular (Harikaran &Balasundaram., 2022). Yet, there is a gap or space for additional research in order to successfully apply Nano silica for building with various nanostructure characterization tools, which will be able to explain a number of mysteries about concrete. The amount of Nano-silica had a growing tendency in relation to the compressive strength of concrete, however the increase was halted once it reached 3%. Concrete containing Nano silica at 3% increases compressive strength and split tensile strength by 14% more than control concrete. When 3% of control concrete is replaced with Nano silica, the increase in flexural strength is merely 2%. When the nano-silica content in concrete is 4%, the strength has been dramatically reduced by 50%. Also, it has been shown that when 4% Nano-silica is added, the strength decreases since the dosage may have beyond the ideal range. Hence, 3% of Nano-Silica is the ideal percentage to replace some cement. Because of the smaller pores caused by nano-silica in high-performance concrete, the structures are more dense and long-lasting. Analyzed experimentally the effects of nano-silica on the characteristics of concrete (Dushyanth et al., 2022). The primary goal of the current work was to ascertain how nano-silica affected the flexural and mechanical properties of concrete. The results of the current inquiry were used to study the performance of nano-silica concrete both with and without fly ash and to compare it to the performance of the control mix. Using the use of the workability test, compacting factor, vee-bee, and flow test, the fresh qualities of the concrete were ascertained. The mechanical qualities of a material are influenced by its compressive strength, split tensile strength, elastic modulus, flexural strength, and impact resistance. The durability is evaluated using bulk diffusion, acid attack, and sulphate attack tests. Also, a two-point loading flexure test was conducted to examine how a nano-silica beam behaved when bent (Sasivarman et al., 2022).

Behavior of high performance concrete incorporating Nano silica in terms of mechanics and flexure. Concrete's compressive and tensile characteristics can be improved by the inclusion of silicon dioxide nanoparticles. The inclusion of Nano Silica content affects the rise in characteristics. A concrete mix with 3% Nano silica was determined to have the best mechanical characteristics. Results from the experiment and the computation showed a strong connection. For RC beams with 3% additional Nano Silica content, the first fracture load and ultimate load were determined to be maximum. The Energy absorption to ductility ratio for the RC beam with higher-order replacement of Nano Silica was also found to be high. The inclusion of Nano Silica improves the microstructure of the concrete and reduces crack width by acting as a packing material (Sanchaya et al., 2022). From the total discussion, it can be determined that the optimal proportion for adding Nano silica to concrete is 3%, which results in an improvement in the material's compressive strength, tensile strength, and flexural behavior.

Effect of nano-SiO_2 and nano-Fe_2o_3combined on cement mortars' electrical resistivity, porosity, compressive strength, and flexural strength. If the curing conditions and the nano addition dispersing technique are ineffective, the expected positive effect on compressive and flexural strength, as well as durability, obtained when nanoSiO_2 or nanoFe_2o_3 are added to Portland cement mortars, can be reversed. The rise in nanoSiO_2 results in a decrease in compressive strength for all ages when the nanoFe_2o_3 content is kept constant. A rise in holes smaller than 50 nm will have also been noticed in this scenario.

Impact of nano silica on the compressive strength of cured concrete paste at various hydration stages. The findings indicate that both methods result in comparable aging-related increases. Throughout all testing ages, the compressive strength of micro silica dramatically increased in all percentages of cement substitutes. The optimal percentage of cement replacement, according to the findings of comparing the amounts of nano-silica replacement, is 1%, closely followed by the addition of 3% nano-silica. Results from the 5% nano-silica substitution showed less strength than those from the 1% and 3% additions. With better overall strength values than vacuum curing, water curing demonstrated to be the most advantageous curing process (Rahmanzadeh et al., 2018). Nonetheless, when compared to water-cured samples, the compressive strength of vacuum-cured specimens revealed a slight drop (often within 5%).

Impact of nano-silica particles on a blended cement paste's hydration, rheology, and strength development. An isothermal calorimetry reveals that nano silica particles cause cement to hydrate more quickly. Its impact is explained quantitatively in a hydration model and is related to their unique surface. Furthermore, it is shown that 5% nS cement paste contains a reduced but significant amount of Portlandite throughout the course of 90 days. Even if nS particles make up only 5% of the total weight of the combination, the pore solution is still likely to be strongly alkaline even though nS particles are a more potent pozzolanic reactant than silica fumes. As a result, using such cement pastes to build or repair steel-reinforced concrete members may be an option.

Concrete durability research using nano-silica. The unique concrete had a lower water absorption value when tested than regular Portland cement concrete, according to the results of the test (Sridher et al., 2019). The unique concrete exhibits less capillary rise than regular Portland cement concrete in a sorptivity test. The unique concrete exhibits a lower percentage of water loss during an acid attack test than regular Portland cement concrete.

According to the aforementioned research papers, it is evident that there has been previous research on both concrete with micro silica and concrete with nS, but the specific Mix Design of concrete with nS has been carried out in this research work .Using it, various tests on strength and Microstructure analysis parameters have been carried out for the research work to support its scope and goal.

IMPORTANCE OF ALTERNATIVE MATERIALS FOR CEMENT

- Cement replacements are substances that can be used in place of cement to enhance properties like strength and durability.
- The removal of cement substitutes from the waste stream, the reduction in energy needed to repurpose them (when compared to the manufacture of cement), and the preservation of raw materials like silica, alumina, and iron oxide are all reasons why the use of cement substitutes is typically encouraged.
- Portland cement production is not only expensive and energy-intensive, but it also produces a significant amount of CO_2. Around one tonne of CO_2 is released into the environment for every ton of Portland cement produced.
- Cement production has increased significantly in recent years, making it the third-largest source of human CO_2 emissions after fossil fuels and land use.
- The global cement industry is responsible for around 6% of all human-caused CO2 emissions, which contributes to about 4% of all global warming.
- Future CO_2 emissions trading will undoubtedly have a significant effect on the business. Alternate cementitious materials are finely divided materials

that can be used instead of or in addition to Portland cement. As a result, the cement industry will concentrate on them. Its use lowers the price of concrete and/or enhances one or more of its technical characteristics.

- Fly ash, powdered granulated blast furnace slag, condensed silica fume, limestone dust, cement kiln dust, and natural or synthetic pozzolans are a few of these materials.

These cementitious materials have additional advantages besides boosting the capacity of cement plants, decreasing fuel consumption, lowering greenhouse gas emissions, managing alkali-silica reactivity, and enhancing durability. Depending on the substitute cementitious material, these advantages vary.

Characteristic of Nano-Silica

The interface transition zone might become denser as a result of the nanoparticles' potential to reduce cement porosity. Moreover, the use of nanoparticles in reinforced cement can be used to create durable, high-strength concrete buildings, reducing the need for maintenance or early replacement.

Micro-silica and silica fume can be replaced with nano-silica, a nanomaterial used in civil engineering applications. During the cement hydration process, nano-silica and lime react, creating a C-S-H gel that could increase the concrete's longevity and mechanical strength. Cement-based materials with effective dispersion of nano-silica can hydrate cement paste more quickly, resulting in a denser microstructure. However, an excessive amount of nanoparticles may aggregate because of their high surface energy, resulting in a non-uniform dispersion.

The white, fluffy, and soft powder known as nano-silica contains amorphous SiO2. Due to its nanoscale particle size, there is a significant amount of surface absorption. Due to its significant dispersion, increased chemical purity, and vast surface area, amorphous nano-silica is a crucial pozzolanic activator. It is divided into hydrophilic and hydrophobic nano-silica based on particle sizes. Cement concrete uses hydrophilic nano silica because of its superior water dispersion.

MATERIALS REQUIRED FOR INCREASING CONCRETE QUALITY WITH NANOMATERIALS

Cement: Cement is a necessary prerequisite for every building activity and often provides the distinct materials with a binding medium. In the current analysis, concrete was prepared using regular Portland cement grade 53 in accordance with Indian Standards: 12269-1987;

Fine aggregate: The actual sand of the river that passes through IS sieve 4.75 mm is used for concreting. According to IS: 383-1970 natural river sand has been graded under grading zone

Coarse aggregate: It went through a 20 mm sieve and held between a 20 to 10 mm sieve and a 10 to 4.75 mm sieve for IS: 383-1970 was used for concreting.

Properties of Nano Silica

OPC, or regular Portland cement, was utilized exactly as received. Tables 1 to 4 display the cement's chemical and physical characteristics. The employed OPC's particle size distribution pattern has been shown. SiO2 nanoparticles have a 15 nm average particle size. The N-SCC series was produced using a range of SiO2 nanoparticle concentrations, with an average particle size of 15 nm. Cement was replaced in the creation of the mixes with SiO2 nanoparticles ranging from 1.0 to 2.0 weight percent and a 1.0 weight percent polycarboxylate addition. For three minutes, the nano-SiO_2 was introduced and vigorously mixed.

Table 1. Chemical properties nano silica

Type	Composites
Chemical Symbol	SiO_2
Group	Silicon 14
	Oxygen 16
Electronic Configuration	Silicon [Ne] $3s^2 .3p^2$
	Oxygen [He] $2s^2 .2p^4$

Table 2. Chemical compositions nano silica

Element	Content
Silicon	46.77
Oxygen	53.30

Table 3. Mechanical properties nano silica

Properties	Metric
Density	2.3 gm/cm^3
Molar Mass	55.90 gm/mol

Table 4. Thermal Properties Nano Silica

Properties	Metric
Melting Point	1670° C
Boiling Point	2250° C

Table 5. Chemical properties of opc cement

Material	SiO_2	Al2O3	Fe2O3	CaO	MgO	SO3	Na2O	K2O	Loss of Ignitious
Cement	21.89	5.3	3.34	53.27	6.45	3.67	0.18	0.98	3.21

Fresh Concrete Property: Workability

Theoretical calculations based on the aforementioned factors will not result in the perfect conditions for maximal strength. Concrete must be completely compacted in order for it to contribute to the greatest strength. Absence of compaction will cause air gaps, which can be just as harmful to strength and durability as the presence of capillary cavities.

In order for the concrete to be properly compacted with the provided efforts, normal conditions may require a greater water/cement ratio than that indicated by theoretical considerations. Water serves the additional purpose of lubricating the concrete, enabling it to be compacted at the jobsite with the required force. For concrete to be handled without segregating, placed without losing homogeneity, compacted with the anticipated labor, and finished with sufficient ease, a certain amount of water must be present. Workable concrete is defined as concrete that satisfies the aforementioned criteria. In contrast to the other phrase "consistency," which is frequently used haphazardly for workability, "workability" or "workable concrete" denotes a much larger and deeper meaning. In a general sense, the word "consistency" refers to how fluid or dynamic something is. Concrete with a higher consistency and more mobility doesn't always need to have the right workability for a particular application. Every task has a certain level of workability requirements. Concrete that is deemed workable while the vibrator is used is not workable when the concrete is to be compacted by hand, even in the construction of roofs. For the purpose of building a roof, concrete that is deemed workable for the mass concrete base is not suitable. Similar to this, thin parts of concrete that are thought to be workable when used in large sections are not. Figure 1 depicts the workability of concrete. The term "workability" now encompasses the whole significance of the type

Figure 1. Workability of concrete

of work, section thickness, level of reinforcing, and compaction method. To develop a mix, a concrete technologist needs to have in-depth understanding of workability. A mixed designer must thoroughly understand the type of work, transportation distance, loss of slump, placement method, and many other elements involved in the mixed design process before indicating a workability parameter. If the required degree of workability, comprehension, and experience is expected, the concreting operation will be both affordable and long-lasting. The term "workability" has been defined by numerous researchers. However, because it refers to a broader range of concrete's characteristics and features and does not convey a single meaning, it defies any attempts at precise definition. The U.K.'s Road Research Laboratory defined workability as "the feature of concrete which determines the amount of meaningful internal work necessary to accomplish full compaction" in their comprehensive research on compaction and workability. It is sometimes defined as "the ease with which concrete can be compacted to a thickness of 100 percent, taking into account the manner of compaction and place of deposition," which encompasses a larger meaning.

Harden Concrete Property: Compression Tests

The cubes are poured into a 150x150x150 mm mold using a concrete mix design ratio and allowed to cure for 7, 14, and 28 days. On concrete, 45 different numbers of cubes are cast, both with and without nS. After 7, 14, and 28 days, the specimen is taken out of the healing tank and dehydrated. Cubes are examined using a

Figure 2. Compression test

compressive testing equipment with a 1000 kN capacity, and test results are produced in accordance with IS 516. The load is slowly added to the specimen, until it fails. The maximum load at which the specimen is subjected and the corresponding loads are noted. Concrete crushing intensity may be determined by assigning the ultimate load to cross- sectional space. Testing of the specimen shown in Figure 2.

Flexural Strength Test

The goal of the flexural strength test is to determine the load at which a part made of concrete containing nano-silica may crack under stress. It provides a general understanding of the concrete with nano-silica tensile strength. For many applications, including buildings, oil wells, nuclear power plants, seashores, ports, etc., the flexural strength of unreinforced cement-based materials and their nano-material composites is crucial.

Typically, beam samples are tested under either third- or central-point stress to determine the flexural strength. Third-point loading is used here for testing normal concrete beams since it has less variability in findings is shown in Figure 3. To fully understand the function of sand and water in the material characterization of typical concrete under flexural stress, a large number of samples are prepared, and their results are thoroughly evaluated in the following sections.

It is chosen to prepare specimens that are 100 mm thick in order to measure flexural strength. Iron moulds with dimensions of 100 mm in width and 500 mm in length are prepared for this use. First, it was decided to cast the aforementioned samples of concrete using nano-silica. Before testing, all samples are cured for 28 days, and then they are coated with a whitewash of lime to make it easier to see crack formation. The length of the specimens is 500 mm. Nonetheless, 450 mm of effective length is used for testing, and 25 mm of space is left between both edges. Once more, 450mm is divided into three sections, and specimens are marked before testing. During the flexural test all the details such as first crack load and

Figure 3. Four point loading arrangement for flexure test

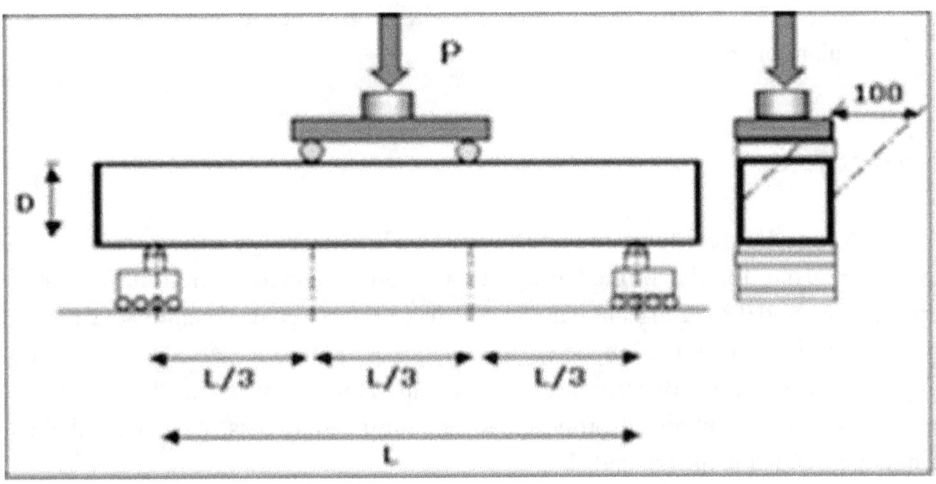

displacement, crack patterns, crack locations, number of cracks, ultimate load, displacement, etc. are noted clearly.

Split Tensile Test

Originally created in Brazil, an indirect test to measure the tensile strength of concrete has been standardized by ASTM. Because of its ease of use, it is widely used. In this test, specimens are put through their paces while being compressed along their length and diameter. The direction of tension development is at a right angle to the applied load's line of action.

It has been determined to utilize a cylindrical mould that is 75 mm in diameter and 150 mm long to compare the indirect tensile strength of conventional concrete with concrete that contains nano-silica. Initially, specimens are cast in regular concrete, and then specimens are cast in nano-silica concrete. Samples are evaluated using a Universal Testing Machine (UTM) with a 50T capacity as per IS: 516 - 1959 underload control after 28 days of curing. To obtain a uniformly distributed strip load, two strips are put, one at the top and the other at the bottom of the cylinder. Mould and Test setup are shown respectively in Figure 4. Here, the indirect to direct tensile strength ratio is assessed so that it can be used to distinguish between direct and indirect tensile strength.

- As anticipated, using concrete with nano-silica boosts the indirect tensile strength. In contrast to concrete failure, the failure pattern of the splitting tensile test demonstrates that specimens do not separate after first cracking (Shah et al., 2019). A splitting plane is surrounded by widely spaced micro cracks that cause a significant damage zone. The cause of this improved ductile failure pattern is fibre bridging activity and the stress transfer mechanism that results from it. Concrete reinforced with nano-silica is more durable than regular concrete.

Impact Test

The material's low resistance to impact loads and lack of energy absorption capacity are the reasons for the limited usage of reinforced concrete in some potential applications and the significant amount of infrastructure repair. The impact strength of a material is determined by an abruptly large quantity of external energy being applied to a structure or structural component (Harikaran et al., 2022). Impact resistance is the mechanical property of the composite that is most strengthened when nano-silica is introduced.

Figure 4. Split tensile test setup

In order to assess the impact strength of cement-based composites, relatively small samples are often prepared, and the following equipment or methods are frequently used:

Machines for 103 MPa/s or higher hydraulic pressure testing.

- Impact velocity of up to 30 m/s using a rotating hammer mechanism. Hopkinson Bar technique with a 106 MPa/s loading rate.
- Charpy Hammer with Specific Loading Equipment As much as 102 MPa/s
- The falling weight (drop test) method, which is described in ASTM publications for loading rates up to 104 MPa.

The number of hits required to inflict the test specimen with the required level of suffering is determined by the "repeated impact," drop weight test. This test method can be used to examine the relative effectiveness of several matrices and is the easiest and most realistic test method available.

It should be underlined that impact strength is determined by how many strikes must be administered to a test specimen until it stops rebounding, or when it splits into separate pieces rather than remaining a single, solid body. Because there are no widely approved testing protocols or standardized specimens, all reported results have a relative and conventional value.

Cylindrical moulds with dimensions of 150 mm in diameter and 300 mm in length are used to create samples for the impact test. In order to evaluate the impact resistance of matrices, three specimens of each type of concrete normal concrete and concrete with nano-silica—are constructed. The cure period for cylindrical specimens is 28 days. After 28 days of drying, specimens are chopped with a concrete cutter machine while maintaining a 64 mm height. In order to show how a fracture develops at the top under impact stress, samples are then painted white.

Samples are maintained in the casing provided at the bottom of the impact machine after being prepared for the test. The specimen is subjected to impact loading by being struck by a 4.54 kg hammer that has been released from a height of 457 mm in accordance with ASTM D 698 or ASTM D 1557. Blows are delivered until the specimen's top begins to crack.

- The number of strikes that were necessary to create the first break to appear on the surface of the specimens. Compared to concrete containing nano-silica, normal concrete performed poorly [35].
- Due to the impact load that is reflected, a significant depression is noticed at the specimen's centre, where the ball was held.
- I used regular concrete as a point of comparison with concrete with nS.
- Concrete's impact resistance at nS is 297 blows.
- While Normal Concrete listed 59 blows of strength.

With concrete containing nano-silica, we attempt to hit the specimen from the first crack to the specimens' failure, thus there were obviously multiple cracks, and we only count the strength of the first crack.

Oxygen Permeability Test

Permeation happens when a liquid permeates through the pores of concrete that are permeable to pressure from the outside and saturated with the particular fluid. To evaluate the permeability of concrete, numerous tests have been created. These tests can be broadly divided into two categories: flow testing and permeability tests, which aim to determine the permeability's Darcy coefficient through flow. In order to do this, a sustained pressure head is used to track the pressure gradient (flow rate) through concrete. It can take a lot of time and effort to pass these assessments. The depth of fluid penetration following a time of applied pressure is measured by inflow permeability tests. A falling head permeameter applies pressure to a concrete sample at first, then as permeation progresses, the pressure is permitted to diminish.

Because pressure can be tested regularly throughout time, this approach maintains a high level of precision. While inflow permeability tests are simpler to carry out than through flow permeability testing, the falling head gas permeameter test, created by Ballim at the University of the Witwatersrand, is utilized in this instance.

By monitoring the pressure depreciation over time while keeping the initial pressure value at 100 kPa, the permeability is ascertained. Data loggers that read the pressure every minute were used to measure the pressure decrease.

The ability of concrete to transport fluids or gases through permeation is consequently measured by permeability. Concrete's permeability is influenced by its microstructure, moisture content, and the properties of the penetrating agent (Abhilash et al., 2021). The Permeation characteristics are frequently determined while assessing the qualities of concrete by measuring the gaseous flow through a specimen under an externally imposed pressure gradient (Harikaran et al., 2023). In order to forecast the entry of carbon dioxide into concrete members, in terms of durability specifications, gas permeability characteristics are utilized.

The samples determined oxygen permeability index (OPI). Most of the samples had permeability indexes between 9 and 9.5, which are considered to be insufficient for concrete permeability. The overall sample's average OPI value was in this range. The oxygen permeability test analyzes the entire micro- and macrostructure of the cast concrete's outside surface, significant variation in test results is anticipated (Sankar et al., 2023). Fissures and macro-voids are particularly sensitive to it because they allow gas to enter.

OPI Observation of Normal Concrete

The permeability index (OPI) was found to range between 9.0 to 9.25 for silica contents of 1.0, 1.5, 2.0, and 2.5% in hot-cured concrete and 1.0, 1.5, 2.0, and 2.5% in normal-curried concrete. Hence, 3.0% for hot and cold curried Regular Concrete was also classed as concrete with high pore permeability. As a result, it had low permeability properties (Sankar et al., 2023). Also, it can be seen that 1.0% nS and its best application in normal concrete were increased microstructure formation or densification considerably more quickly in hot-cured concrete than in cold-cured concrete.

Water Sorptivity Test

Water Sorptivity Test measures the speed at which water penetrates concrete while being pulled up by capillaries. Since it is especially sensitive to the microstructural properties of the concrete's near-surface zone, it reflects the type and effectiveness of curing. Lower water sorptivity indices increase the concrete's potential durability.

Sorptivity ratings for well cured Grade 30–50 concretes range from 5–10 mm/h to 15-20 mm/h for improperly cured Grade 20 concretes (Harikaran et al., 2022). Linear correlations are detected when the mass of water absorbed is plotted against the square root of time, and the slope reflects the sorptivity (S). Four test specimens must be used for each test. A concrete disc that is 68 mm in diameter, 25 mm thick, and cut in line with the concrete Durability Index Testing Monograph must be used as the test sample. The oxygen permeability test may also be performed on samples that have already been used.

Condition of Specimens

a) Immediately after being cut into pieces, the specimens must be roasted for 7 days and 4 hours at 50°C. Alternately, after the oxygen permeability test is complete, remove the samples from the oxygen permeability cells. Mark the samples with the same reference as soon as you take them from the desiccators or oxygen permeability cells (Harikaran et al., 2023). Hot curry water sample on the inner face (historically non-exposed or non-test face): 1.0, 1.5, 2.0, 2.5, and 3.0; regular curry: 1.0%, 1.5%, 2.0, 2.5%, and 3.0%.

b) Place the samples on a steel tray in the desiccator and chill them to 232 c for no less than 2 hours or no longer than 4 hours. Alternately, the specimens may be cooled on a steel tray in a room with controlled ambient temperatures of 23.2-degrees Celsius and 60% relative humidity for no less than two hours or no more than four hours. You can gauge the temperature of the specimens by inserting a thermometer between the test face.

c) Mark the samples from the reference immediately after removing them from the desiccators or oxygen permeability cells. 1.0 H%, 1.5 H%, 2.0 H%, 2.5 H%, and 3.0 H% for hot curried water sample on the inner face and 1.0%, 1.5%, 2.0%, 2.5%, and 3.0% for regular curried (historically non-exposed or non-test face).

d) Apply a waterproof seal to the specimens' vertically curved sides using one of the specified products in accordance with the product's instructions. If epoxy paint is used, care must be taken to avoid getting it on the flat test surface; otherwise, pinholes will be present. Return the specimens to the desiccators or room after spraying with epoxy paint so that the ambient (Gokulakannan et al., 2023). The epoxy can dry for 18 to 24 hours when the temperature and relative humidity are kept at 23.2 °C and 60%, respectively. If packaging tape is to be used, attach it now to the curved vertical edges. In this instance, no further drying period is required.

The majority of the well-stratified samples were present, and all of the samples water sorptivity and porosity were examined. The normal concrete's porosity was higher than that of the normal concrete with Nano silica specimens (Harikaran et al., 2023). With the Normal Concrete Samples, we found that the sorptivity and porosity varied depending on the composition. In relation to the degree of stratification and the materials being employed, sorptivity and porosity were also discovered. The porosity of normal concrete, which is significantly lower than that of concrete with Nano-silica, can be calculated from the test results as (%). So, in the future, we will try this experiment with a higher Nano-silica percentage.

Findings From This Study

- The Mix design was created for concrete with nS, using concrete with an M40 grade and a w-c ratio of 0.40 as per IS:10262-2009.
- According to the Microstructure Analysis, nS improves homogeneity by reducing voids and making the concrete sound, which raises the various hardened strengths of the concrete.
- Usage of nano-silica in concrete can boost the material's compressive strength.
- According to the results of the microstructure investigation, nS provides higher homogeneity, which decreases voids in the concrete and improves sound quality.
- Normal concrete has a compressive strength that is comparable to medium strength concrete.
- Material strength and structural strength are not directly connected, however structural strength seems to rise with ductility. Hence, balanced material strength and ductility are necessary for good structural performance, which was made achievable from concrete with nS and proven through the current experimental examination.
- It is more accurate to measure the beam's flexural strength in a two-point bending configuration. The reproducibility of the result is greater compared to any other arrangement because of the pure bending in the middle third portion. However, normal concrete produces unsatisfactory results when compared to specimens of the same size made of concrete with nano silica, and this specimen was unable to exhibit the strain hardening property.
- Due to its strain-softening behaviour, the split tensile test in the case of concrete with nS does not provide an accurate estimation of direct tensile strength.
- The "Repeated Impact" drop weight test is straightforward to carry out and produces accurate findings for comparison. However, unlike any other

mechanical qualities in Normal Concrete and concrete with nS, a significant variance in the findings of impact strength is seen between three specimens. According to the study, Normal Concrete is not recommended for use in impact-resistant structures because it lacks the strain hardening capabilities of Concrete with nS content.

- The water sorptivity test and the oxygen permeability test, which were introduced for the first time for the Concrete with Nano Silica practise and Concrete with nS, show good results like smaller pore space compared to regular concrete.
- As a result, the strength parameters and concrete with nS in the mix design that were derived in this work as well as the research gap, Microstructure analysis leads to a detailed understanding of Nano silica behaviour that was not justified in earlier research papers that were performed and explained in detail.

Scope for Further Investigation

A variety of additives can be used to improve concrete's performance under long-lasting conditions, together with concrete that contains more than 1.5% nS.

REFERENCES

Abhilash, P.P, Dheeresh, N., Bhaskar, S., Rajesh, K., & Veerendra, K. (. (2021). Effect of nano-silica in concrete: A review. *Construction & Building Materials*, *278*, 122347. doi:10.1016/j.conbuildmat.2021.122347

Ashok, B. & Mahender, B. (2017). Effect of Nano Silica on the Compressive Strength of Concrete. *International Journal of Professional Engineering Studies.*, *8*(2), 222–226.

Dushyanth, V., Babu, R., Mukesh, P., Menka, P., Harikaran, M., Balaji, G., & Monisha, K.M. (2022). Experimental Study of Concrete By Replacement of Cement With Polymer Resin. *IOP Conf. Series: Earth and Environmental Science*. IOP. . doi:10.1088/1755-1315/1125/1/012014

Gokulakannan, S., Harikaran, M., Mounesh, K., Sankaranarayanan, S., Sathishkumar, R., & Sivakumar, R. (2023). Experimental studies on durability and deflection behavior of concrete in metakaolin cement concrete. *Materials Today: Proceedings*. doi:10.1016/j.matpr.2023.05.236

Harikaran, M. & Balasundaram, N. (2019). Evaluation of Concrete Using Reshaped Waste Tyre Rubber as Partial Replacement of Coarse Aggregate. *International Journal of Recent Technology and Engineering (IJRTE), 7*(6s5), 2277-3878.

Harikaran, M., Gokulakannan, S., Loganathan, A., Chandiran, R. D., Ajith, M., & Dhanasekar, V. (2023). Metakaolin cement concrete evaluation using industrial by-products as fine aggregate. *Materials Today: Proceedings*. doi:10.1016/j.matpr.2023.04.586

Harikaran, M., Keerthanadevi, P., Inbarasan, S., Tamilan, N., Manimaran, R., & Jayavikram, S. (2023). Experimental examination of concrete using hazardous plasti-fiber and rub – fiber. *Materials Today: Proceedings*. doi:10.1016/j.matpr.2023.04.598

Harikaran, M., Kulanthaivel, P., Krishnaraja, A. R., & Murugan, P. C. (2023). Performance enhancement of concrete by using ceramic waste as a partial replacement for coarse aggregate. *Journal of Ceramic Processing Research, 24*(2), 266–273.

Harikaran, M., Mukesh, P., Suganth, R. S., Nithishkumar, R., & Vijay, V. (2022). Study of structural bonding strength with tyre rubber instead of coarse aggregate and water hyacinth ash instead of cement. *Materials Today: Proceedings, 68*, 1330–1334. doi:10.1016/j.matpr.2022.06.359

Harikaran, M., & Sampath Boopathi, S. (2023). *Gokulakannan & M. Poonguzhali*. Study on the Source of E-Waste Management and Disposal Methods. Sustainable Approaches and Strategies for E-Waste Management and Utilization. doi:10.4018/978-1-6684-7573-7.ch003

Hasan, B., & Nihal, S. (2014). Comparative Study of the Characteristics of Nano Silica, Silica Fumeand Fly Ash Incorporated Cement Mortars. *IJESE, 17*(3), 570–582. doi:10.1590/S1516-14392014005000054

Jaya, R. P. (2016). A Review of Microstructure Properties of Porous Concrete Pavement Incorporating Nano Silica. *Journal of Engineering and Applied Sciences (Asian Research Publishing Network), 11*(20), 1819–6608.

Montgomery, J., Abu-Lebdeh, T., Hamoush, S. A., & Picornell, M. (2016). Effect of nano-silicaon the compressive strength of hardening cement paste at different stages of hydration. *American Journal of Engineering and Applied Sciences, 9*(1), 166–177. doi:10.3844/ajeassp.2016.166.177

Rahmanzadeh, B., Rahmani, K., & Piroti, S. (2018). Experimental study of the effect of water-cement ratio on compressive strength, abrasion resistance, porosity and permeability of Nanosilica concrete. *Fratturaed Integrit Struttural., 12*(44), 16–24. doi:10.3221/IGF-ESIS.44.02

Rajkumar, R., Akkineni, S., & Ramya, S. (2016). Experimental Study on the Strength and Durability of Nano Concrete. *International Journal of Applied Engineering Research, 11*(4), 2854-2858.

Rutuja, S., Rajguru, N., Madhuri, P., Suraj, S., Rohan, W., Kishor, Z. & Ghogare, R. (2017). Review Paper on Multi-functional use of Nano Silica in Concrete. *International Journal of Engineering Science and Computing.* 10780-10782.

Sanchaya, M., Harikaran, M., Cheran, K., Varoon, M., Jayaprakash, P., & Navin, G. (2022). Experimental investigation on rice husk ash and coconut shell used in building materials. *Materials Today: Proceedings*, *68*(1), 1697–1702. doi:10.1016/j.matpr.2022.08.264

Sankar, P., Suresh, P., Aravind, T., Buvanesvari, V. K., Harikaran, M., & Raja, R. (2023). Investigation on Supply Chain Distribution Network Problem using RFID Technique [IJIRSET]. *International Journal of Innovative Research in Science, Engineering and Technology*, *12*(1). doi:10.15680/IJIRSET.2023.1201094

Sankar. P, Suresh, T., Aravind, V., Buvanesvari, K., Harikaran, M., & R. Raja. (2023). Analysis and Optimization of Supply Chain Travelling Salesman Problem Using K-Means Cluster based Genetic Algorithm. *IJIRSET, 12*(1), doi:. doi:10.15680/IJIRSET.2023.1201095

Sasivarman, B., Nagalakshmi, R., & Harikaran, M. (2022). Treatment of wastewater by slow sand process with polystyrene. *Materials Today: Proceedings*, *68*(1), 1648–1653. doi:10.1016/j.matpr.2022.08.082

Shah, M., & Shah, D. Kaushal, V., & Niragi, D. (2019). Effect of GGBS and Nano Silica on the Strength Properties of Ternary Concrete. *4th world congress on civil, structural and environmental engineering, 522*. IEEE. 10.1088/1757-899X/522/1/012011

Sridher, J., Vivek, D., & Jegatheeswaran, D. (2019). Mechanical and Flexural Behavior of High-performance Concrete Containing Nano Silica. [IJEAT]. *International Journal of Engineering and Advanced Technology*, *9*(2), 2249–8958. doi:10.35940/ijeat.B3164.129219

Vasanthi, P, Senthil selvan S, Benita Hexton D, Mohamed Salman Farooq I & Suresh P. (2017). Durability Study on Nano Concrete. *International Journal of Chemtech Research*, *10*(3), 693–698.

Chapter 4
Nanoparticle Characterization and Bioremediation:
Prospects for Ecological Advantages

Jyoti Ahlawat
School of Agricultural Sciences, G. D. Goenka University, Gurgaon, India

Digvijay Pandey
https://orcid.org/0000-0003-0353-174X
IET Lucknow, India

Ravish Chaudhary
Division of Seed Science and Technology, ICAR, Indian Agricultural Research Institute, New Delhi, India

Nidhi Verma
Government P.G. College for Women, Rohtak, India

Binay Kumar Pandey
https://orcid.org/0000-0002-4041-1213
G.B. Pant University of Agriculture and Technology, India

Shailesh Somnath Parkhe
https://orcid.org/0000-0003-0785-1142
Department of Mechanical Engineering, SNJB's Shri H.H.J.B. Polytechnic, India

Pankaj Dadheech
https://orcid.org/0000-0001-5783-1989
Swami Keshvanand Institute of Technology, Management, and Gramothan, India

ABSTRACT

Nanomaterials reduce biodegradable pollutants before promoting standard levels. Thus, nanomaterials could efficiently and sustainably treat environmental

DOI: 10.4018/978-1-6684-9135-5.ch004

contaminants. However, additional research is needed to determine the destiny of environment remediation nanomaterials. This review covers biological and plant-based bioremediation nanotechnologies. Nanomaterials reduce waste and harmful material degradation costs. Nanomaterials/nanoparticles immediately catalyse waste and toxic material breakdown, which is hazardous to microorganisms, and enable microorganisms degrade waste and toxic materials more efficiently and sustainably.

INTRODUCTION

Nanoparticles are increasingly used in various ðelds, including medical, food, health care, consumer, and industrial purposes, due to their unique physical and chemical properties. These include optical, electrical, and thermal, high electrical conductivity, and biological properties. Due to their peculiar properties, they have been used for several applications, including as antibacterial agents, in industrial, household, and healthcare-related products, in consumer products, medical device coatings, optical sensors, and cosmetics, in the pharmaceutical industry, the food industry and in diagnostics. This enables them to be an object of intensive studies due to their academic interest and the prospective technological applications in various fields (Faivre et al., 2016). Such nanostructures may be synthesized by a wide number of methods, which involve mechanical, chemical and other pathways. So here in this chapter we describe extensively the use of different methods for the characterization of NPs. Research sometimes favours the synthesis of inorganic nanoparticle (NPs) by biological systems makes more biocompatible and environmentally benign. Microorganisms, plant extracts could be an alternative to chemical and physical synthesis of NPs. In physical synthesised silver nanoparticles by laser ablation method the properties of particles synthesised depends upon the wavelength and duration of the laser, the ablation time duration and the effective liquid medium which may or may not contain the surfactant. The chemical synthesis involves the use of hazardous chemicals such as reductants, stabilizers and organic solvents and special requirements for the techniques like high energy radiation, microwave irradiation and inert gas condensation are important (Mittal *et al.,* 2014). The physical and chemical processes often involve high temperatures/pressure for the reaction and the use of hazardous chemicals due to which the biological methods of synthesising metal nanoparticles is gaining importance. Now a day, green nanoparticles synthesis from plant extracts can be used to reduce metal ions to nanoparticles in a single step. This green approach is non-toxic and environment friendly as compared to physical and chemical methods. Silver nanoparticles have several effective application in spectrally selective coating for solar energy absorption, interchelation material for electrical batteries, optical receptors, catalysis for biolabelling, antimicrobial agents etc.Nanoparticles (1-100 nm) exhibits new or improved properties which is based on specific characteristics like size, distribution and morphology (Malabadi *et al.,* 2012).Considering the drawbacks of physio-chemical methods,cost-effective and energy efficient new alternative for AgNP synthesis using microorganisms (Sharma *et al.,* 2009) plant extracts (Song *et al.,* 2009) and natural polymers (Huang *et al.,* 2004) as reducing and capping agents are emerging very fast. The association of nanotechnology and green chemistry unfold the biologically and cytologically compatible metallic nanoparticles which is a strong benevolent from lab to land approach. The biological reduction of silver ions into silver nanoparticles have been reported in *Capparis decidua*(Ahlawat et al., 2015, 2017). The use of plant and plant extract in nanoparticle synthesis is considered advantageous over microbial based system because it reduces the elaborate process of maintaining cell cultures.The particle size, growth can also be controlled by altering synthesis conditions like pH, reductant concentration, temperature, mixing ratio of the reactants etc. The plant-based synthesis can be carried out either extracellularly or intracellularly. Intracellular synthesis takes place inside the plant whereas the extracellular synthesis occurs *in vitro*. The findings reported that extracellular synthesis using plant extracts has been considered better as compared to intracellular because it abolishes the extraction and purification methods. Green synthesis of nanoparticles have been reported in Corynebacterium species (Gowramma et al. 2015). The photochemical which are responsible for reduction may be terpenoids, flavonoids, ketones, aldehydes, amides, and carboxylic acids. The water-soluble metabolites such as flavones, organic acids, and quinones are solely responsible for the bioreduction ions. The biological activity of AgNPs depends on factors including surface chemistry, size, size distribution, shape, particle morphology, particle composition, coating/capping, agglomeration, and dissolution rate, particle reactivity in solution, efðciency of ion release, and cell type, and the type of reducing agents used for the synthesis of AgNPs are a crucial factor for the determination of cytotoxicity (Murdock et al., 2008). To evaluate the physically, chemically and biosynthesized nanomaterials, many analytical techniques have been used. In this chapter we elaborate the physicochemical and structural characterization of nanoparticles.

PROCESS OF NANOPARTICLES CHARACTERIZATION

Physicochemical Characterization

Typically, engineered materials with dimensions in the nanometre scale are intermediates between isolated small molecules and bulk materials (Duan et al., 2013). Nanomaterials, which are similar to biological moieties in scale, can be used as diagnostic and therapeutic nanomedicines (Hachanj et al., 2013; Kim et al., 2010). Compared to their bulk material counterparts, the distinct physicochemical properties of the nanomaterials, such as size, surface properties, shape, composition, molecular weight, identity, purity, stability and solubility, are critically relevant to particular physiological interactions. The different techniques used for characterization of nanomaterials, based on their different features, are described in the following sections.

Optical Spectroscopy

ptical spectroscopy uses the interaction of light with matter as a function of wavelength or energy in order to obtain information about the material. Absorption or emission experiments with visible and UV light tend to reveal the electronic structure. Vibrational properties of the lattice (i.e., phonons) are usually in the IR and are studied either using IR absorption or Raman spectroscopy. Optical spectroscopy is attractive for materials characterization because it is fast, non-destructive and of high resolution. It is further classified into the following:

Optical Absorption Spectroscopy (OAS)

Optical absorption spectra of small single metal nanoparticles are measured using a far-field technique combining a spatial modulation microscope with a broadband light source. Quantitative determination of the spectral and polarization dependencies of the absorption cross section of individual gold nanoparticles permits precise determination of their geometrical properties in excellent agreement with transmission electron microscopy measurements. This technique works on the principle of the UV-Visible spectroscopy and this technique involves the absorption of near-UV or visible light Absorption spectroscopy is employed to determine the presence of a particular substance in a sample and, in many cases, to quantify the amount of the substance present. Preliminary characterization of the AgNPs was carried out using UV–Visible spectroscopy. The reduction of silver ions to the nanoparticle form was monitored by measuring the UV–Visible spectra of solutions after diluting the sample with Millipore water 20 times. The spectra of AgNPs solution was monitored by a UV-Vis spectrophotometer and a UV-Vis spectrograph of the colloidal solution of silver nanoparticles was recorded as a function of time by using quartz cuvette with water as reference and scanning the spectra between 200-800 nm at the resolution of 0.1nm (Ahlawat et al., 2017).

Fourier Transform Infrared Spectroscopy (FTIR)

FTIR is able to provide accuracy, reproducibility, and also a favorable signal-to-noise ratio. By using FTIR spectroscopy, it becomes possible to detect small absorbance changes on the order of 10^{-3}, which helps to perform difference spectroscopy, where one could distinguish the small absorption bands of functionally active residues from the large background absorption of the entire protein. FTIR spectroscopy is frequently used to ðnd out whether biomolecules are involved in the synthesis of nanoparticles, which is more pronounced in academic and industrial research. Furthermore, FTIR has also been extended to the study of nano-scaled materials, such as conðrmation of functional molecules covalently grafted onto silver, carbon nanotubes, graphene and gold nanoparticles, or interactions occurring between enzyme and substrate during the catalytic process. Finally, the advantages of FTIR spectrometers over dispersive ones are rapid data collection, strong signal, large signal-to-noise ratio, and less sample heat-up (Lin et al., 2014). Recently, further advancement has been made in an FTIR method called attenuated total reñection (ATR)-FTIR spectroscopy. Therefore, FTIR is a suitable, valuable, non-invasive, cost effective, and simple technique to identify the role of biological molecules in the reduction of silver nitrate to silver.

ATR (Attenuated Total Reflection)

Attenuated total reflectance Fourier transform infrared (ATR-FTIR) spectroscopy is a technique capable of *in situ* characterization of the liquid–solid interface to probe surface adsorption on nanoparticle surfaces in environmentally and biologically relevant media. Furthermore, given that the interfacial region in these media is dynamic, ATR-FTIR spectroscopy facilitates monitoring these dynamics by interrogating a layer of immobilized nanoparticles coated on the ATR element while changing the overlying aqueous phase. The molecular information acquired from this technique allows for the determination of the adsorption mode, including conformational and structural changes of the coordinating ligand, and can directly measure ligand displacement reactions (Hind et al., 2001). Rather, in some cases, ATR-FTIR spectroscopy can be used as a quantitative surface analytical tool. ATR-FTIR spectroscopy to probe nanoparticle surfaces in general with respect to: (i) the adsorption of different environmentally and biologically relevant coordinating ligands; (ii) competitive ligand adsorption and; (iii) the determination of kinetic and thermodynamic parameters. Using ATR-FTIR, we can determine the chemical properties on the polymer surface, and sample preparation is easy compared to conventional FTIR. Electron Spectroscopy: The illustration given in Table 1.

Table 1. Comparative data of the various Electron spectroscopy techniques used to characterise the nanoparticles

Energy Dispersive X-ray Spectroscopy (EDS)	X-ray Photoelectron Spectroscopy (XPS)	Auger Electron Spectroscopy (AES)
It is widely used analytical methods for the analysis of elemental composition of solid matter, energy-dispersive X-ray spectroscopy (EDS) has recently gained significant importance regarding its application to the chemical analysis of nanoparticles, especially in conjunction with the use of a scanning electron microscope (SEM) and the use of the transmission operation mode of SEM (STEM-in-SEM). This development was mainly driven by the technological progress with highly sensitive EDS detectors, such that individual nanoparticles can be quickly inspected with EDS at a SEM. Qualitative information on elemental composition with about 10-nm spatial resolution can be achieved complementary to the high-resolution information of the sample surface morphology within the same scanned area as provided by the electron microscope.	The method has been extensively used to identify elements, their chemical state, and their concentration at surfaces for many years. It is the most accurate method for measuring the composition and thickness of thin (< 10 nm) surface coatings. The information depth of the method arises from the small distance that electrons can travel through materials without losing energy and these dimensions are similar to those of importance in nanoparticle analysis. This chapter provides an overview of the physical principles of XPS, the instrumentation, and the methods to interpret the data. That interpretation depends upon prior knowledge, or inference, of the physical structure of the sample. Therefore, for nanomaterials, XPS is best used in combination with other methods that are capable of measuring particle size and particle shape. Due to the importance of surfaces in the performance, toxicity, and behaviour of nanoparticles, XPS is an important technique to ensure that particles are completely characterized.	Auger electron spectroscopy is applicable to surface chemical analysis of nanoparticles & it is a mature method in the field of surface chemical analysis. the principal design of recent instruments together with modes of operation and options for the presentation of spectra, and different approaches for qualitative (including identification of chemical species) and quantitative surface analysis of elements. An application paragraph on surface chemical analysis of nanoparticles by AES or SAM introduces the different measurement approaches and sample preparation strategies applied by analysts. The analysis of nanoparticle ensembles, the so-called selected point analysis where a narrow primary electron beam is centred on an individual nanoparticle, and chemical mapping of individual nanoparticles (or a line scan across) is addressed.

Rutherford Backscattering Spectrometry (RBS): This method is used in many situations when it is desirable to estimate not only a diameter of the nanoparticles but also their composition and the composition of the interface between nanoparticles and surrounding matrix. Ion beam analysis methods, such as medium energy ion scattering spectroscopy and low energy ion scattering, offer distinctive advantages.

Secondary Ion Mass Spectrometry(SIMS): Secondary ion mass spectrometry is as one of the leading surface chemical analysis and imaging technique with molecular specificity in the field of material sciences. The nanoparticle analysis by SIMS specifically focuses the application in terms of surface spectrometry, imaging, analysis in organic and complex media, and depth profiling.

Structural Characterization

X-ray diffraction technique: XRD can be used to look at various characteristics of the single crystal or polycrystalline materials using Bragg's Law (equation (1)).

$$n\lambda = 2d \sin\theta \quad (1)$$

The use of XRD is often compared to the microscopy techniques. XRD avoids issues of representative samples and determining crystals as opposed to particles as shown in Figure 1. XRD is time consuming and requires a large volume of sample. This works on the fundamental principle which relates the wavelength of electromagnetic radiation to the diffraction angle and the lattice spacing in crystalline sample. These diffracted X-rays are then detected, processed and then counted. By scanning the sample through a range of 2θ angle, all possible diffractions direction of the lattice should be attained due to the random orientation of the powered material. A key component of all diffraction is the angle between the incident and the diffracted rays.

Figure 1. X-Ray diffraction technique

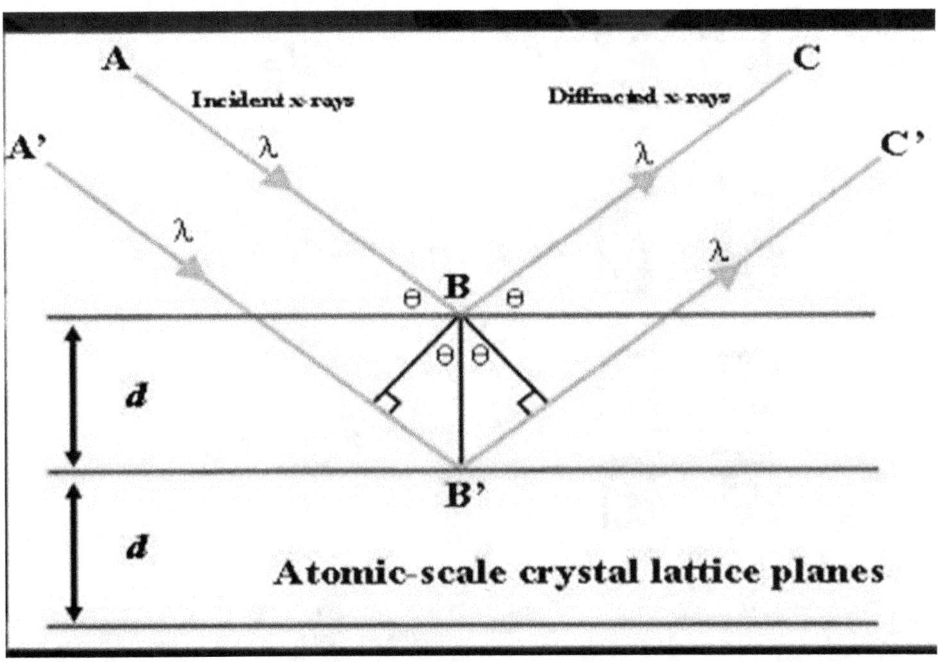

Electron Microscopy

Transmission Electron Microscopy (TEM)

The crystalline sample interacts with the electron beam mostly by diffraction rather than by absorption. The intensity of the diffraction depends on the orientation of the planes of atoms in a crystal relative to the electron beam. A high contrast image can be formed by blocking deflected electrons which produces a variation in the electron intensity that reveals information on the crystal structure. This can generate both „bright or light field" & „dark field "images.

TEM enables Direct 2-D imaging of particle size, shape & surface characteristics.

Figure 2. TEM (transmission electron microscopy)

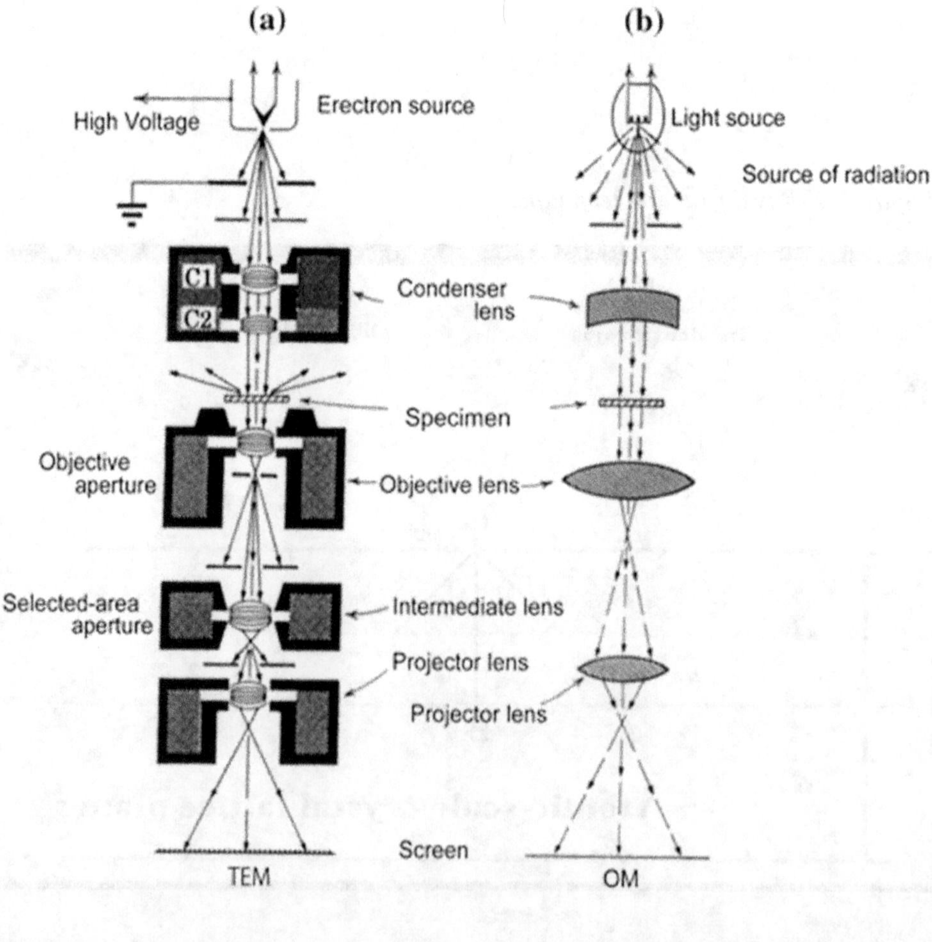

- Changes in nanoparticle structure as a result of interactions with gas, liquid & solid-phase substrates can also be monitored.
- Sample must be able to withstand the electron beam & also the high vacuum chamber.
- Time consuming.
- It needs an analysis by image treatment & must be performed on a statistically significant large no. of samples (Figure 2)

Scanning Electron Microscopy (SEM)

When the beam of electrons strikes the surface of the specimen & interacts with the atoms of the sample, signals in the form of secondary electrons, back scattered electrons & characteristic X-rays are generated that contain information about the samples" surface topography, composition etc. There are 3 modes –

Primary: High resolution (1-5 nm); secondary electron imaging

Secondary: Characteristic X-rays; identification of elemental composition of sample by EDX technique

Tertiary: Back-scattered electronic images; clues to the elemental composition of sample.

In a typical SEM, the beam passes through pairs of scanning coils or pairs of deflector plates in the electron column to the final lens, which deflects the beam horizontally & vertically. Electronic devices are used to detect & amplify the signals & display them as an image on a cathode ray tube in which the raster scanning is synchronized with that of the microscope. The image displayed is therefore a distribution map of the intensity of the signal being emitted from the scanned area of the specimen (Figure 3). Atomic Force Microscopy (AFM) versus Scanning Tunnelling Microscopy (STM) are shown in Table 2.

The sample must be electrically conductive at the surface.

- Time consuming & expensive.
- Sometimes it is not possible to clearly differentiate nanoparticle from the substrate.
- SEM can't resolve the internal structure of these domains.
- SEM can yield valuable information regarding the purity as well as degree of aggregation.

Figure 3. SEM (scanning electron microscopy)

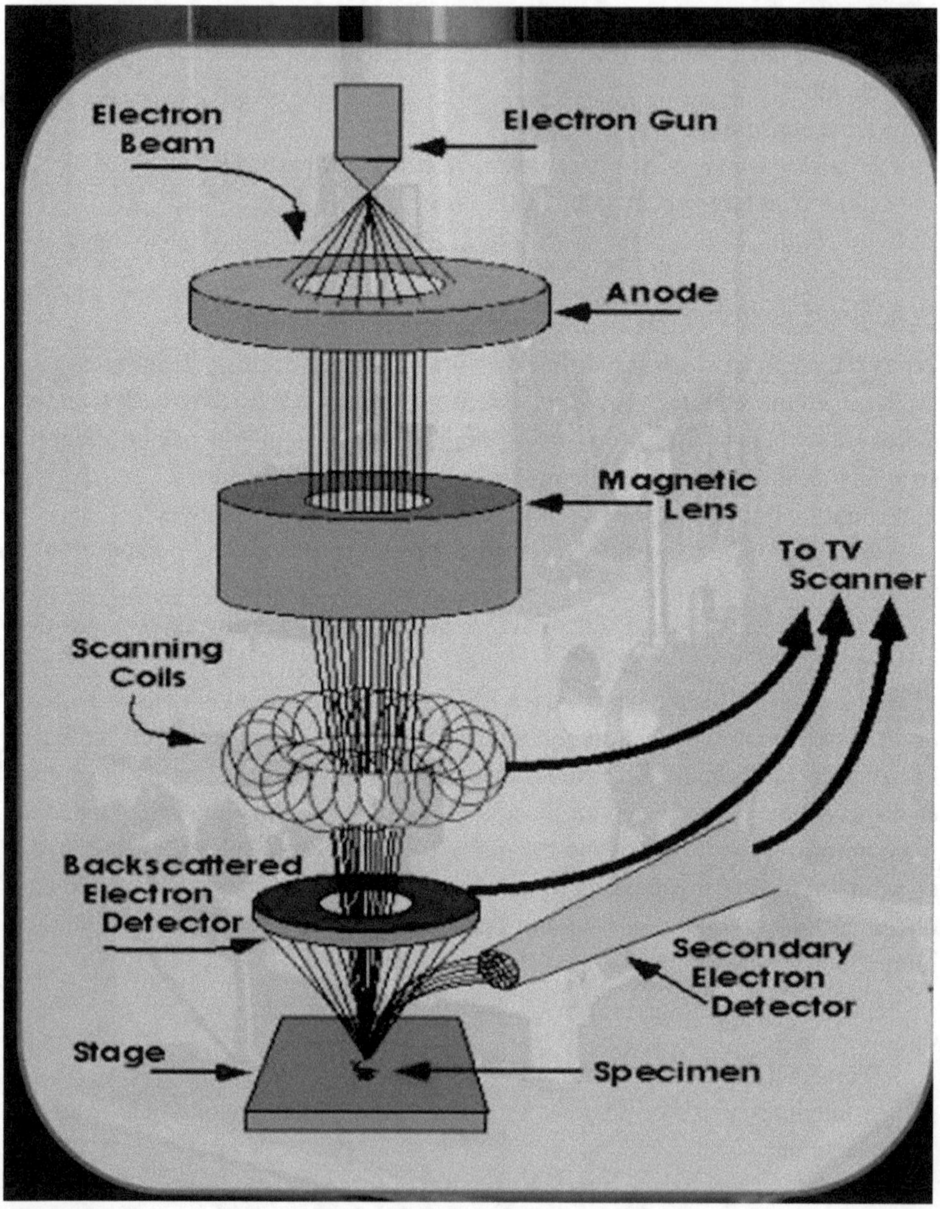

Table 2. Atomic force microscopy (AFM) versus scanning tunnelling microscopy (STM)

Atomic Force Microscopy (AFM)	Scanning Tunnelling Microscopy (STM)
In AFM, a probe consisting of a sharp tip(~ 10 nm) located near the end of a cantilever beam is raster scanned across the surface of a specimen using piezoelectric scanners. Changes in the tip specimen interaction are often monitored using an optical lever detection system, in which a laser is reflected off the cantilever & onto a position-sensitive photodiode.	It is based on the concept of quantum tunneling. When a conducting tip is brought very near to a metallic or semi-conducting surface, a bias between the two can allow electrons to tunnel through the vacuum between them. Variations in tunnelling current as the probe passes over the surface are translated into an image.
AFM modes: i. Contact mode: Tip scans in close contact with surface (repelled), Constant force, Highest resolution, May damage surface. ii. Non-contact mode: Tip hovers above the surface (attracted), Variable force measured, Lowest resolution, Non-destructive iii. Tapping mode: Intermittent tip contact, Variable force measured, Improved resolution, Non-destructive	They normally generate image by holding the current between the tip of the electrode & the specimen at some constant value by using a piezoelectric crystal to adjust the distance between the tip & the specimen surface.
The AFM offers visualization in three dimensions. Resolution in the vertical, or Z, axis is limited by the vibration environment of the instrument, whereas resolution in the horizontal, or X-Y, axis is limited by the diameter of tip utilized for scanning.	Lateral resolution ~ 0.1 nm Depth resolution ~ 0.01 nm
Typically, AFM instruments have vertical resolutions of less than 0.1 nm and X-Y resolutions of around 1 nm.	STM can be used not only in ultra high vacuum but also in air & various other liquid or gas, at ambient & wide range of temperature.
AFM can be performed in liquid or gas mediums. This capability can be very advantageous for nanoparticle characterization. For example, a major component of the combustion-generated nanoparticles are volatile components that are only present in ambient conditions.	STM can be a challenging technique, as it requires extremely clean surfaces & sharp tips.
In many industries, the ability to scan from the nanometre range into the micron range is important. With AFM, particles anywhere from 1nm to 5μm in height can be measured in a single scan. AFM can characterize nanoparticles in multiple mediums including ambient air, controlled environments & even liquid dispersions.	As for characterization of biomolecules using STM or EM techniques, the samples are usually embedded into a matrix to preserve their original conformations, followed by coating the samples with a thin metallic layer, such as gold, before acquiring images STM is a preferred tool for investigating conductive atomic structures of, for example, carbon nanotubes, fullerenes and grapheme(Wang & chu et al., 2013)

iii) Scanning Probe Microscopy (SPM): It can be discussed further in two subheadings:

Dynamic Light Scattering

Several physicochemical characteristics of nanomaterials including hydrodynamic size, shape, structure, aggregation state, and biomolecular conformation can be explored using radiation scattering techniques. DLS, one of the most popular light scattering modalities, can probe the size distribution of small particles, molecules or polymers at the scale from submicron down to one nanometer in solution or suspension using a monochromatic light source, e.g. a laser. The principle of DLS is to monitor the temporal fluctuation of the elastic scattering intensity of light, i.e., Rayleigh scattering, induced from the Brownian motion of the particles/molecules of a size much smaller than the incident light wavelength, at a fixed scattering angle. The intensity fluctuation trace comprises a mixture of the constructive and destructive interferences of the scattered light, through which the particle size can be derived from analysis of the motion-dependent autocorrelation function using the Stokes–Einstein equation. physicochemical characterization of nanomaterials, the main strengths of DLS include its noninvasive manner, short experiment duration (in minutes), accuracy in determining the hydrodynamic size of monodisperse samples, and capabilities of measuring diluted samples,

However, the functions of DLS are impacted by several disadvantages, such as difficulty in correlating size fractions with a particular composition when certain amounts of aggregates are present, dust particles interfering in the scattering intensity, and a relatively small range of particle or molecule size (1 nm–3 μm), although the scale limitation is not really a pitfall for characterization of nanomaterials. In addition, DLS has limited utility for analysis of samples with heterogeneous size distributions, and resolving the dimensions of a mixed sample population varying in size less than a factor of three; moreover, DLS is unsuited to accurately measuring the sizes of non-spherical nanomaterials because spherical nature of particles is already assumed in the analysis (LIN et al., 2014)

Small Angle X-Ray Scattering (SAXS):

In contrast to XRD, whose applications are limited to crystalline materials, SAXS provides information of several characteristics by examining either crystalline or amorphous materials from polymers, proteins to nanomaterials. In SAXS, a portion of an incident X-ray beam elastically scattered from the sample forms a scattering pattern on a two-dimensional flat X-ray detector perpendicular to the direction of the incident X-ray beam. By analyzing the intensity of the scattered X-ray collected within the scattering angle, ranging from 0.1 to 3°, SAXS can evaluate the size/size distribution, shape, orientation, and structure of a variety of polymers and nanomaterial-bioconjugate systems in solution (Sapsford et al., 2011).

Remediation and Bioremediation

Remediation is the field of study that focuses on investigating the clean up or removal of contaminants from the environment is "environmental remediation". Environmental remediation techniques use various methods to remove and/or break-down (degrade) environmental contaminants in polluted soil, surface water, ground water as well as in sediments.

Figure 4. Application of nanomaterials in BIOREMEDIATION

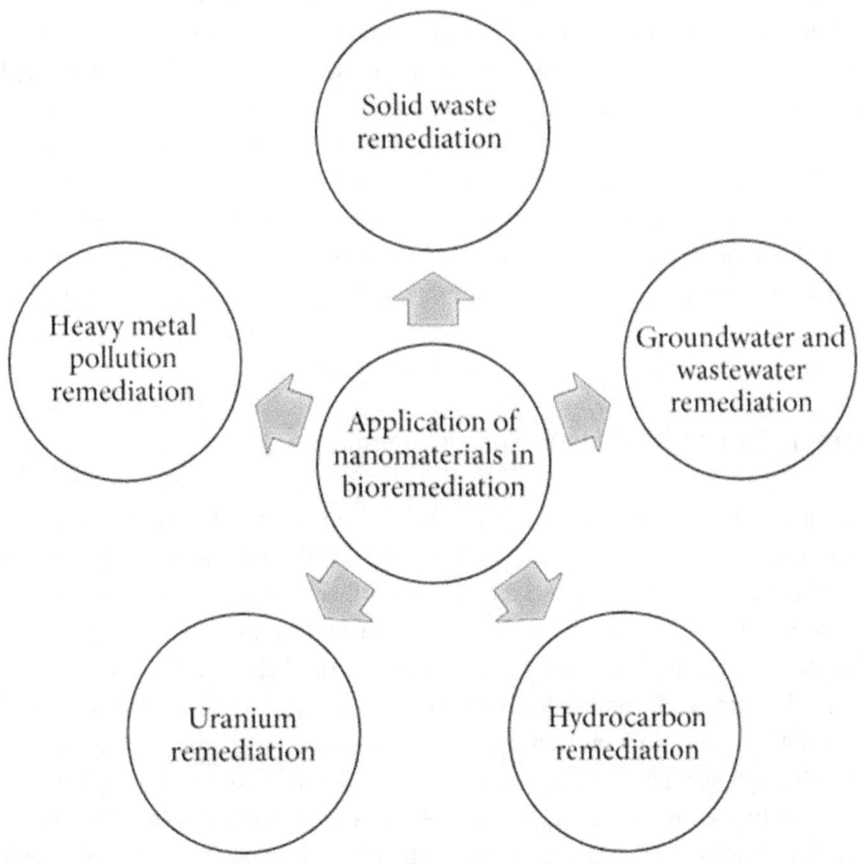

Ex situ techniques: Removing the contaminated soil, sediment, or water from the polluted sites and then treating the pollution above ground. In situ techniques: Clean up the contamination while it is still in the ground without the need of off-site treatment.Whereas, **Bioremediation**is the use of biological system for the reduction of pollution from air, aquatic or terrestrial systems.The process involves aerobic or anaerobic microorganisms that results in degraded products. Bioremediation is used to reduce or remove hazardous waste which has already polluted the environment(Figure 4).

Nanomaterials exhibit unique physical and chemical properties and, hence, they have received much attention from scientists and researchers in different areas of environmental sciences, specifically in bioremediation. Bioremediation provides a good clean-up strategy for some types of waste, but as it is expected, it will not be useful for all. For example, bioremediation may not provide a feasible strategy at sites with high concentrations of chemicals that are toxic to most microorganisms. These include heavy metals and salt. Further, the advancement in science and technology has increased standard of living which directly or indirectly contributes to the increase in waste and toxic material. Therefore, the remediation of contaminants by use of existing technology is not effective and efficient in cleaning up the environment. Hence, nanomaterials may be applied for bioremediation, which will not only have less toxic effect on microorganisms, but will also improve the microbial activity of the specific waste and toxic material which will reduce the overall time consumption as well as reduce the overall cost. In this paper we have briefly summarized the major types of nanomaterials that have been used so far in bioremediation of waste and toxic materials. Understanding the interaction between the contaminant, the microorganism, and the nanomaterials (NMs) is of crucial importance since positive and negative effects may be produced. For example, some nanomaterials are stimulants for microorganisms, while others are toxic.

Working Principle of Nano Bioremediation

Million tons of toxic chemical compounds are released in the environment by the industry (Avio et al., 2017). After release, these compounds may further react to form chemicals, for instance, polychlorinated dibenzo-p-dioxins or polychlorinated dibenzofurans, which are by-products of certain chemical processes involving chlorine. There is high variability in the physical and chemical properties of these chemical compounds, and their cytotoxicity and multiple interactions with biotic and abiotic environmental factors, i.e., microorganisms, plants, animals, water, minerals, organic matter, wind, etc., have complicated the successful implementation of remediation technologies. The combined use of Nanomaterials and Nanoparticles with biotechnologies could offer a step-change in remediation capabilities, avoiding process intermediates, and increasing the speed of degradation. Besides physical and chemical technologies to remediate polluted sites, biological treatments have become relevant due to their low cost and wide range of applications (Kuppusamy et al.,2017). Bioremediation includes biosorption, bioaccumulation, biotransformation, and biological stabilization, among others. These technologies use plants and some microorganisms including bacteria and fungus, as well as combinations of them. During recent years, NMs have been integrated with biological processes to accelerate and promote the removal of toxic compounds from the environment. Nano bioremediation term was first used by Cecchin et al. for processes where NPs and microorganisms or plants are used to remove contaminants (Koul et al., 2018). Based on the nature of the organism utilized for the remediation of contaminants, the practices are named speciðcically as phyto-nanoremediation, microbial nanoremediation, and zoo-nanoremediation (El-Ramady et al.,2017). In any case, since bioremediation uses living organisms to remediate contaminated environments, a proper interaction between nanoparticles (NPs) and living organisms is essential. In this context, some aspects are of paramount importance. For example, it is known that nanotoxicity, NPs size, and nanonutrition may affect the living organisms and this in turn may affect the whole bioremediation process.

Nano Bioremediation Tools: Nanomaterials and Nanoparticles Used in Bioremediation

Different NMs have been tested to determine their potential in reducing contaminants with the aid of living organisms. Further criteria may be applied for more speciðcity such as (a) nano iron and its derivatives; (b) dendrimers; (c) carbon-based NMs; (d) single-enzyme NPs; (e) engineered polymeric; (f) biogenic uraninite; and (g) metals other than iron. The selection of the type of NM would depend upon the nature of the contaminant. For example, magnetite, a nano iron material is used to separate heavy metals in soils or water through its magnetic properties. Besides, carbon-based NMs are used to trap organic pollutants or heavy metals from water, soil or air. As earlier mentioned, the type of organismis also important. Plants display certain advantages over microbial organisms. For example, they produce several molecules that are involved in the transformation of pollutants. These include gl utathione,ñavonoids,reactiveoxygenspecies and bioactive molecules that act as a response understress (Nunez et al., 2020). Besides, plants are easier to cultivate and handle than other organisms that need continuous nutrient supply as well as more controlled conditions. It is important to take into consideration that NPs not only aid in the remediation process. They also serve for the detection of contaminants and, in consequence, for pollution prevention.

Applications of Nano Bioremediation

1. Waste water and Industrial effluents
2. Drinking and process water
3. Soil and land treatment
4. Solid waste
5. Sensing of pollutants

Advances in Nano scale science and engineering suggest that many of the current problems involving water quality could be resolved or greatly uplifted using Nano sorbents, Nano catalysts, bioactive nanoparticles, nanostructured catalytic membranes and nanoparticle enhanced filtration among other products and processes resulting from the development of nanotechnology. According to above all application of Nanobioremediation it can be definitely concluded that, Nanoparticles, Nano technological instrument play efficient role in the process of Nanobioremediation.

Figure 5. Sensing of pollutants by NMs

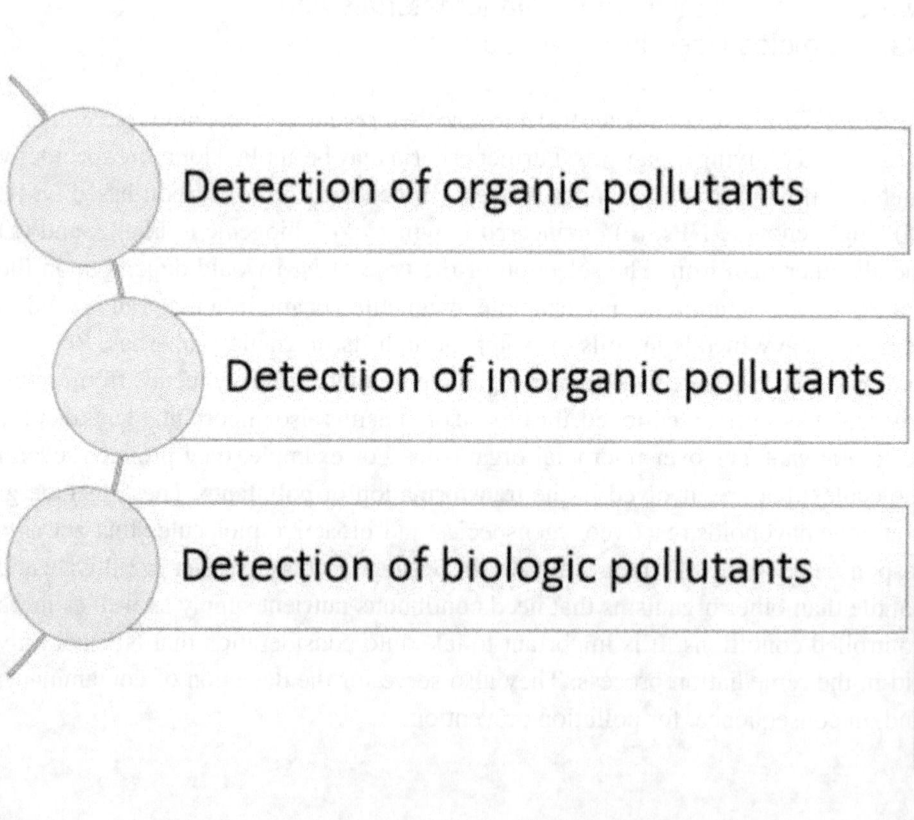

By applying the Nanobioremediation to environment hazards, it can clean them Faster and Safer than other methods and technology.

Sensing of organic, inorganic and biological pollutants

Nanotechnology has great potential for making a sensor for monitoring pollution. The ability of the nanoparticles to be coated with wide range of chemical and biological ligands helps in giving the specificity to the sensor (Figure 5). The surface to volume ratio of the nanoparticles can be easily controlled by varying the size and shape of the nanoparticles thus giving control over the interaction quality with the analyte molecule.

Porous silicon exhibits the phenomena of photoluminescence and this luminescence is quenched in the presence of organic molecule. By using this technology pesticide concentration as low as 1 ppm could be detected. The detection of various heavy metals, like Pb, Hg, Cd, using nanoparticles is either fluorescence based or calorimetric based. Binding of heavy metal ions to these metal chelators results in aggregation of the nanoparticles yielding a shift in wavelength absorption and ultimately resulting in colour change from red to blue. Biological contaminants include living organisms (generally bacteria) or their by-products that can be toxic to human health. The major bacterial pollutant belongs to Enterobacteriaceae family. Coated gold nanoparticles with IgG proved successful for sensing *S. aureus* and *S. saprophyticus*. Quantum dots (QD) are also used as a fluorescent system for detection of microorganisms like Giardia.

Prevention or Removal of Pollutants Mediated by Nanoparticles and Bio-Based Treatments

Pollution prevention is defined as the reduction of pollutants at the source. Metal oxide Nano catalysts (chiefly gold Nano catalyst) show promising results for preventing or reducing the pollution at the source.

Product innovation: Development of new products that have less impact on environment their predecessors for e.g. genetically improved plant varieties that resistant against insects/ diseases, thereby reducing the use of pesticides.

(GENETIC ENGINEERING: Involves the use of microorganisms genetically modified by recombinant DNA technology. GE is a powerful tool in creating environment friendly alternatives for products and processes.)

Different nanomaterials have been used in coordination with various biological systems for the pollutants degradation and varied results of removal efficiency were observed depending upon the nature of the NMs used (Table 3).

Table 3. Prevention/Removal of pollutants by utilising NMs

Nanomaterial	Biological Systems	Pollutants degraded	Removal efficiency	External system used
nanocrystals	Enzymatic degradation by bacterially overexpressed organophosphorus hydrolase	P-nitrophenol and acid orange 7	>80%	Water
nanopowders	Soil microorganisms	Azo dye direct red 23	98%	Water
nanomembranes	A biological extract of Cynomorium coccineum L.	Cyanide compounds	~20%	Industrial waste water
nanoparticles	Sphingomonas sp.	Decarbominated diphenyl ether; chlorinated hydrocarbons, pathogens	67% for diphenyl ether; >76.8% for chlorinated hydrocarbons	Water
nanocomposites	Arthrobacter globiformis D47	Herbicide(diuron)	>90%	Water
nanotubes	enzyme organophosphate hydrolase–MWNT paper	Organophosphates and heavy metal	~22%	Water & Soil
Nano sponge	Two organo-clays(Dellite67G and Dellite 43 B)	Triclopyr(3,5,6-Trichloro-2-pyridinyloxyaceticacid	92%	water

(Kos et al., 2014; Liu et al., 2018).

NMs (Nanomaterials)/NPs as an antimicrobial tool in Plant tissue culture

Use of biologically synthesized Silver nanoparticles from medicinal plants removes the contamination in tissue culture.

Silver Nanoparticle Synthesis Using Medicinal Plants

Nanotechnology is mainly concerned with synthesis of Nano particles of variable sizes within a range of 1-100 nm, shapes, chemical compositions and controlled dispersity and their potential use for human benefits (Zhang*et al.,* 2016). Nanotechnology is currently employed as a tool in medical sciences for different functions like imaging; sensing; targeted drug delivery; gene delivery systems and artificial implants. In present situation, silver nanoparticles are in great use in the medicinal, pharmaceutical, agricultural industry and in water purification. These nanoparticles can be synthesized either chemically or biologically. The chemical process involving hazardous chemicals, for synthesis of silver nanoparticles is more elaborate and possess various negative impacts leaving behind toxic effect, which may pose potential environmental and biological risks. Whereas, biological synthesis of silver nanoparticles is less time consuming, cheaper and eco-friendlier therefore in recent time scientist are looking forward to the possible biological methods for the synthesis of silver nanoparticles. The growing need to develop environmentally friendly and economically feasible technologies for material synthesis led to the search for biomimetic methods of synthesis. There are three major sources of synthesizing silver nanoparticles: bacteria, fungi, and plant extracts. The use of environmentally benign materials like plant leaf extract, bacteria and fungi for the synthesis of silver nanoparticles offers numerous benefits of eco-friendliness and compatibility for pharmaceutical and biomedical applications as they do not use toxic chemicals in the synthesis protocol. Use of biological organisms such as microorganisms, plant extract or plant biomass could be an alternative to chemical and physical methods for the production of nanoparticles in an eco-friendly manner. Biological methods based on microbes such as bacteria and fungi are able to absorb and accumulate

Table 4. Nanoparticles synthesis from medicinal plants and their antimicrobial potency

Botanical Name	Explant	Reference
Costus speciosus (Koen)	In vitro derived plants and callus culture	Malabadi et al., 2012
Trianthema decandra	Root extract	Geethalakshmi et al., 2012
Elettaria cardamomom	Seeds	GnanaJobitha et al., 2012
Clitoria ternatea	Invitro derived plants& callus culture	Malabadi et al.,2012
Boswellia ovalifolia	Stem bark extract	Savithramma et al., 2011
Catharanthus roseus	Leaf, Callus, Root extract of in vitro derived plants	Malabadi et al.,2012
Citrullus colocynthis	Stem derived callus	Satyavani et al., 2011
Azadirachta indica	Neem leaf extract	Gavhane et al., 2012
Syzygium cumini	Leaf extract	Behera et al.,2013
Cassia italica Lam.	Leaves	Sermakkani et al.,2012
Memecylon edule	Leaf extract	Elavazhagan et al., 2013
Parthenium hysterophorus	Leaf extract	Kumar et al.,2012
Jasminum grandiflorum	Leaf extract	Dwivedi et al.,2013
Cymbopogon citrullus	Leaf extract	Dwivedi et al.,2013
Ocimum basillicum	Leaf extract	Sivaranjani et al.,2013

metals and can be used in the reduction of metal ions and thus known to synthesis of nanoparticles. Biological approaches using microorganisms and plants or plant extracts for metal nanoparticle synthesis have been suggested as valuable alternatives to chemical methods. The use of plant and plant extract (Table 4) in nanoparticle synthesis is considered advantageous over microbial based system because it reduces the elaborate process of maintaining cell cultures, and is an advantage over resource availability, safety and faster reaction rate.

To remove bacterial contamination during *in vitro* propagation, different methods have been employed in the last few years(Pandey *et al.,* 2021).. The most common is Antibiotics which are used in the medium to removed unwanted contaminants. However, antibiotics are frequently phytotoxic otherwise may retard or inhibit plant tissue growth. Most antibiotics have been shown inhibitory effects in the plants. Antibiotics are expensive; their range of efficiency against types of bacteria is not broad, they are usually heat-labile, phytotoxic and only effective against bacteria and not fungi. Otherwise they are capable of altering the behavior of plant tissue cultured by retardation of plant growth (Soltanloo, 2010). Streptomycin and chloramphenicol are remarkable inhibitors of protein synthesis; nucleic acid synthesis is inhibited by rifampicin and cell-wall membrane synthesis disrupted by penicillin. Therefore, antibiotics used without application in the medium may decline mutation risks well as their inhibitory effect. Mercury chloride, HgCl2, 70% ethanol etc. as another antimicrobial chemical agents, has been used to remove infections in explants, but it is very toxic and should be used with necessary cautions. Nano silver has antimicrobial effects at low concentrations, It has antibacterial, antifungal and antiviral effects. (Tahmasbi *et al.,* 2011). The use of nano-sized silver particles as antimicrobial agents has become more prevalent. (Safavi *et al.,* 2011). This capability of Nano-silver is due to release of very small particles of silver ions, so it is capable of destroying not only bacteria, fungus but also the viruses. Application of Nano-silver in *Valeriana officinalis* can lessen the contamination rate of tissue culture and had no effects on leaf number, proliferation and number and length of explant roots. (Abdi *et al.,* 2008). The antimicrobial role of silver nano-particles in *Dianthus caryophyllus* flowers favours the usage of nano silver in tissue culture (Hashemabadi et al,. 2014). However, it is reported that the antimicrobial activities of silver nanoparticles are tested by supplementing the silver nanoparticles in tissue culture media and soaking the explants in nano silver solution to evaluate the best concentration and elimination of contamination in tissue culture for regeneration of explants of *Capparis decidua* (Ahlawat et al., 2020).

The resultant products may be further bio transformed by the biotic systems and reduce the pollutant concentration in the media. In addition, some enzymes produced by living organisms may degrade a variety of contaminants. Due to their size, NPs may even enter contaminated zones where other entities are not able to do so. Therefore, Nano bioremediation technologies may extend their application ðelds. This aspect represents an advantage over other remediation techniques. However, other considerations are necessary, for example, the standardization of protocols to evaluate the toxicity of nanoparticles and nanomaterials in soil and water, elucidation of their interactions with biotic and abiotic elements, and the applicable regulatory frame where these materials could be applied (Garcia et al., 2019).

CONCLUSION

In conclusion, the selection of the NPs and the living organism represents a challenge and is an area of opportunity for further research in terms of the medium and long-term effects of the synergistic use of nanomaterials and biotechnologies, the collateral effects of NMs and NPs on microorganisms, and the trophic transfer of NMs in the food chain and the effects on human health.

REFERENCES

Abdi, G., Salehi, H., & Khosh-Khui, M. (2008). Nano silver: A novel nanomaterial for removal of bacterial contamination in valerian (*Valeriana officinalis* L.) tissue culture. *Acta Physiologiae Plantarum*, *30*(5), 709–714. doi:10.100711738-008-0169-z

Ahlawat, J., & Sehrawat, A. R. (2017). Nano Dimensional (1-20nm) Silver Nanoparticles: Stem Extract of *Capparis decidua* (FORSK.) EDGEW Mediated Synthesis and its Characterization-A Lab to Land Approach'. *International Journal of Current Microbiology and Applied Sciences*, *6*(10), 1874–1883. doi:10.20546/ijcmas.2017.610.226

Ahlawat, J., Sehrawat, A. R., & Choudhary, R. (2020). Biologically synthesized silver nanoparticles eclipse fungal and bacterial contamination in micropropagation of Capparis decidua (FORSK.) Edgew: A substitute to toxic substances. *Indian Journal of Experimental Biology*, *50*, 336–343.

Ahlawat, J Sehrawat, A. R. (2015). Biological Synthesis of Silver Nanoparticles using Aqueous Leaf Extract of *Capparis decidua* (forsk.) edgew: A better alternative'. *Journal of Pharmacy Research*, *9*(4), 244–249.

Avio, C. G., Gorbi, S., & Regoli, F. (2017). Plastics and microplastics in the oceans: From emerging pollutants to emerged threat. *Marine Environmental Research*, *128*, 2–11. doi:10.1016/j.marenvres.2016.05.012 PMID:27233985

Awoke, N., Pandey, D., & Habtemariam, A. B. (2021). Synthesis of Tin(IV) Oxide Nanoparticles Using Plant Leaf Extracts of Vernonia amygdalina and Mentha spicata. *Regenerative Engineering and Translational Medicine*, 1–6.

Behera, S., & Nayak, P. L. (2013). *In vitro* Antibacterial Activity of Green Synthesized Silver Nanoparticles Using Jamun Extract Against Multiple Drug Resistant Bacteria. *World J. Nano Sci. Tech.*, *2*(1), 62–65.

Duan, X. P., & Li, Y. P. (2013). Physicochemical characteristics of nanoparticles affect circulation, biodistribution, cellular internalization, and trafðcking. *Small*, *9*(9-10), 1521–1532. doi:10.1002mll.201201390 PMID:23019091

Dwivedi, R. (2013). Silver nanoparticles ecofriendly green synthesis by using two medicinal plant extract. *International Journal of Bio-Technology and Research*, *3*(4), 61–68.

El-Ramady, H., Alshaal, T., Abowaly, M., Abdalla, N., Taha, H. S., Al-Saeedi, A. H., Shalaby, T., Amer, M., Fári, M., & Domokos-Szabolcsy, É. (2017). Nanoremediation for Sustainable Crop Production. In S. Ranjan, N. Dasgupta, & E. Lichtfouse (Eds.), *Nanoscience in Food and agriculture* (1st ed., Vol. 1, pp. 335–363). Springer. doi:10.1007/978-3-319-58496-6_12

Elavazhagan, T., & Arunachalam, K. D. (2011). Memecylon edule leaf extract mediated green synthesis of silver and gold nanoparticles. *International Journal of Nanomedicine*, *6*, 1265–1278. doi:10.2147/IJN.S18347 PMID:21753878

Faivre, D., & Bennet, M. (2016). Materials science: Magnetic nanoparticles line up. *Nature*, *535*(7611), 235–236. doi:10.1038/535235a PMID:27411627

Gavhane, A J., & Padmanabhan, P, Suresh P Kamble and Suresh N Jangle. (2012). Synthesis of silver nanoparticles using extract of Neem leaf and Triphala and evaluation of their antimicrobial activities. *International Journal of Pharma and Bio Sciences*, *3*(3), 88–100.

Geethalakshmi, R., & Sarada, D. V. L. (2012). Gold and Silver nanoparticles from *Trianthema decandra*: Characterization and antimicrobial properties. *International Journal of Nanomedicine*, *7*, 5375–5384. doi:10.2147/IJN.S36516 PMID:23091381

Gnana, D., Gnana, J., Gurusamy, A., & Chellapandian, K. (2012). Green synthesis of silver nanoparticle using *Elettaria Cardamomom* and assessment of its antimicrobial activity. *International Journal of Pharmaceutical Sciences and Research*, *3*, 323–330.

Gowramma, B., Keerthi, U., Rafi, M., & Muralidhara Rao, D. (2015). Biogenic silver nanoparticles production and characterization from native stain of *Corynebacterium* species and its antimicrobial activity. *Biotech.*, *5*, 195–201. PMID:28324578

Hachani, R., Lowdell, M., Birchall, M., & Thanh, N. T. (2013). Tracking stem cells in tissue-engineered organs using magnetic nanoparticles. *Nanoscale*, *5*(23), 11362–11372. doi:10.1039/c3nr03861k PMID:24108444

Hashemabadi, D. (2014). The role of silver nano-particles and silver thiosulfate on the longevity of cut carnation (*Dianthus caryophyllus*) flowers. *Journal of Environmental Biology*, *35*, 661–666. PMID:25004750

Hind, A. R., Bhargava, S. K., & McKinnon, A. (2001). At the solid/liquid interface: FTIR/ATR—The tool of choice. *Advances in Colloid and Interface Science*, *93*(1-3), 91–114. doi:10.1016/S0001-8686(00)00079-8 PMID:11591110

Huang, H., & Yang, X. (2004). synthesis of polysaccharide-stabilized gold and silver nanoparticles: A green method. *Caries Research*, *339*(15), 2627–2631. doi:10.1016/j.carres.2004.08.005 PMID:15476726

Kim, T. H., Lee, S., & Chen, X. (2013). Nanotheranostics for personalized medicine. *Expert Review of Molecular Diagnostics*, *13*(3), 257–269. doi:10.1586/erm.13.15 PMID:23570404

Kos, L., Sójka-Ledakowicz, J., Michalska, K., & Perkowski, J. (2014). Decomposition of azodye C.I. Direct Yellow 86by the fenton process in the presence of nanoparticles of iron oxides. *Fibres & Textiles in Eastern Europe*, *5*, 114–120.

Koul, B., & Taak, P. (2018). Chemical methods for soil remediation. In B. Koul & P. Taak (Eds.), *Biotechnological Strategies for Effective Remediation of Polluted Soils* (1st ed., Vol. 1, pp. 77–84). Springer. doi:10.1007/978-981-13-2420-8_4

Kumar, A. D. (2012). Rapid and green synthesis of Silver Nanoparticles using the leaf extracts of *Parthenium Hysterophorus*: A Novel Biological Approach. *Int. Res. J.Pharm.*, *3*(2), 169–173.

Kuppusamy, S., Thavamani, P., Venkateswarlu, K., Lee, Y. B., Naidu, R., & Megharaj, M. (2017). Remediation approaches for polycyclic aromatic hydrocarbons (PAHs) contaminated soils: Technological constraints, emerging trends and future directions. *Chemosphere*, *168*, 944–968. doi:10.1016/j.chemosphere.2016.10.115 PMID:27823779

Lin, P. C., Lin, S., Wang, P. C., & Sridhar, R. (2014). Techniques for physicochemical characterization of nanomaterials. *Biotechnology Advances*, *32*(4), 711–726. doi:10.1016/j.biotechadv.2013.11.006 PMID:24252561

Liu, J., Morales-Narváez, E., Vicent, T., Merkoçi, A., & Zhong, G. H. (2018). Microorganism decorated nanocellulose for efficient diuron removal. *Chemical Engineering Journal*, *354*, 1083–1091. doi:10.1016/j.cej.2018.08.035

Malabadi, R. B., Raju, K., Chalannavar, N., Meti, T., Gangadhar, S., Mulgund, K., Nataraja, S., & Vijaya, K. (2012). Synthesis of antimicrobial silver nanoparticles by callus cultures and *in vitro* derived plants of *Catharanthus roseus*. *Research in Pharmacy*, *2*(6), 18–31.

Mittal, J., Batra, A., Singh, A., & Sharma, M. M. (2014). Phytofabrication of nanoparticles through plant as nanofactories. *Adv. Nat. Sci: Nanosci. Nanotechnol.*, *5*(4), 043002. doi:10.1088/2043-6262/5/4/043002

Murdock, R. C., Braydich-Stolle, L., Schrand, A. M., Schlager, J. J., & Hussain, S. M. (2008). Characterization of nanomaterial dispersion in solution prior to in vitro exposure using dynamic light scattering technique. *Toxicological Sciences*, *101*(2), 239–253. doi:10.1093/toxsci/kfm240 PMID:17872897

Ramírez-García, R., Gohil, N., & Singh, V. (2019). Recent Advances, Challenges, and Opportunities in Bioremediation of Hazardous Materials. In P. V. Chandra & K. Bauddh (Eds.), *Phyto management of Polluted Sites* (1st ed., Vol. 1, pp. 517–568). Elsevier. doi:10.1016/B978-0-12-813912-7.00021-1

Vázquez-Núñez, E., Molina-Guerrero, C., Peña-Castro, J., Fernández-Luqueño, F., & Guadalupe de la Rosa-Álvarez, M. (2020). Use of Nanotechnology for the Bioremediation of Contaminants. *Processes* (Basel, Switzerland), 8, 826. doi:10.3390/pr8070826

Safavi, K., Esfahanizadeh, M., Mortazaeinezahad, F., & Dastjerd, H. (2011) The study of nano silver anti microbiaactivity and evaluation of using NS in tissue culture media. *International Conference on Life and Technology*. Research Gate.

Saifuddin, N., Wong, C. W., & Yasumira, A. A. N. (2009). Rapid biosynthesis of silver nanoparticles using culture supernatant of bacteria with microwave irradiation. *E-Journal of Chemistry*, *6*(1), 61–70. doi:10.1155/2009/734264

Sapsford, K. E., Tyner, K. M., Dair, B. J., Deschamps, J. R., & Medintz, I. L. (2011). Analyzing nanomaterial bioconjugates: A review of current and emerging purification and characterization techniques. *Analytical Chemistry*, *83*(12), 4453–4488. doi:10.1021/ac200853a PMID:21545140

Satyavani, K., Gurudeeban, S., Ramanathan, T., & Balasubramanian, T. (2011). Biomedical potential of silver nanoparticles from calli cells of *Citrullus colocynthis* (L.) Schrad. *Journal of Nanobiotechnology*, *43*, 1–8. PMID:21943321

Savithramma, N. (2011). M Linga Rao, K Rukmini and P Suvarnalatha devi (2011). Antimicrobial activity of silver nanoparticles synthesized by using medicinal plants. African. *Australian Journal of Basic and Applied Sciences*, *3*, 265.

Sermakkani, M. & Thangapandian, V. (2012). Biological synthesis of silver nanoparticles using medicinal plant(*cassia italica*) leaves . *Int. J. Curr. Res.* *4*(10), 053-058.

Sharma, V. K., Yngard, R. A., & Lin, Y. (2009). Silver Nanoparticles: Green Synthesis and Their Antimicrobial Activities. *Advances in Colloid and Interface Science*, *145*(1-2), 83–96. doi:10.1016/j.cis.2008.09.002 PMID:18945421

Sivaranjani, K., & Meenakshisundaram, M. (2013). Biological synthesis of silver nanoparticles by using *Ocimum basillicum* leaf extract and their antibacterial activity. *IJPR*, *4*(1), 225–229.

Soltanloo, S., Alimohammadi, M., Ramezanpour, S. S., & Najar, M. B. (2010). Nanosilver colloid: A novel antimicrobial candidate applicable in plant tissue culture medium. *Australian Journal of Basic and Applied Sciences*, *4*, 5338–5345.

Song, J. Y., & Kim, B. S. (2009). Rapid Biological Synthesis of Silver Nanoparticles Using Plant Leaf Extracts. *Bioprocess and Biosystems Engineering*, *32*(1), 79–84. doi:10.100700449-008-0224-6 PMID:18438688

Tahmassbi, D., Zharghami, R., Vatanpour Azghandi, A., & Chaeich, M. (2011). Effects of nanosilver and nitrogen biofertilizer on yield and yield components of potato microtubes. *International Journal of Agriculture and Biology*, *13*, 986–990.

Wang, H., & Chu, P. K. (2013). Chapter 4 — surface characterization of biomaterials. In: Amit B, Susmita B, (eds.) Characterization of biomaterials. Oxford: Academic Press.

Zhang, X.-F., Liu, Z.-G., Wei, S., & Sangiliyandi, G. (2016). Silver Nanoparticles: Synthesis, Characterization, Properties, Applications, and Therapeutic Approaches. *International Journal of Molecular Sciences*, *17*(9), 1534. doi:10.3390/ijms17091534 PMID:27649147

Chapter 5
Applications of Nanoparticles in the Cosmetic Field

M. Francis Luther King
https://orcid.org/0000-0001-6144-3137
Swarnandhra College of Engineering and Technology, India

S. Sherin Nithya
KGNC, Sri Balaji Vidyapeeth, India

ABSTRACT

Nanotechnology is a cutting-edge scientific discipline that focuses on manipulating matter at the nanoscale scale to create new and useful materials, gadgets, and systems (1–100 nm). One of the most rapidly expanding, and commercially viable niches in the cosmetics industry is the use of micellar nanoparticles in these products. The improved qualities of the nano-sized particles in terms of colour, transparency, solubility, etc., are largely responsible for the extensive impact of nanotechnology in the cosmetics sectors. Cosmetics use a wide variety of nano materials, including nanosomes, liposomes, fullerenes, solid lipid nanoparticles, etc. Recent worries about the potential dangers of nano cosmetics have prompted the cosmetics industry to restrict the use of nanotechnology in cosmetics and push for stricter regulations requiring a thorough safety review of the products before they hit store shelves. This chapter looks at the latest developments in nano-sized materials for use in cosmetics in an effort to address these problems despite the not-so-obvious hidden hazards.

DOI: 10.4018/978-1-6684-9135-5.ch005

INTRODUCTION

Nanotechnology is widely regarded as the century's most promising innovation and is under scrutiny for its novel approach in the beauty sector. The term "nanocosmetics" was coined to describe the use of nanotechnology in the cosmetics industry. The use of nanotechnology in the creation of cosmetics is often regarded as the most cutting-edge and promising new development in the industry. Nanoscale compounds are used in cosmetics for a variety of reasons, including improved UV protection, deeper skin penetration, prolonged effects, greater color, and higher-quality finishes. It's been described as a cutting-edge field of study where particles in the 10-1,000 nm range are managed via design, fabrication, characterizations, and practical applications (Logothethetidis, 1st ed., 2012; Kaul et al., 2018). The term "nano-cosmeceuticals" refers to a kind of cosmetic product in which nanotechnology is used as a delivery mechanism to improve the efficacy of bioactive ingredients (Hougeir et al. 2012; Kaul et al. 2018). Cosmetic compounds may be formed into smaller nanoparticles with this method, allowing for easier absorption by the skin, faster wound healing, and increased product efficacy (Singh et al. 2013).

The nanoscale is 80,000 times smaller than the width of a human hair, and nanotechnology is the study of manipulating atoms and molecules at this scale. The term "nanocosmetics" was used to describe the incorporation of nanotechnology into the cosmetics industry. This trend can be traced back to the widespread usage of nanotechnology in fields such as electronics and medicine. Because nanoparticles have improved color, transparency, solubility, etc., nanotechnology has had a significant impact on the cosmetics industry. Nanosomes, liposomes, fullerenes, solid lipid nanoparticles, etc. are only a few examples of the many nanomaterials used in the cosmetics industry. Recent worries about the safety of such nanocosmetics have prompted the cosmetics industry to restrict the use of nanotechnology in cosmetics and to implement legislation requiring a comprehensive safety study of nanocosmetics before they hit the market. Figure 1 shows the human skin anatomy.

Cosmetics are "materials intended to be placed onto human bodies or any portion thereof for washing, beautifying, boosting attractiveness, or modifying the look," as defined by the FDA (U. S. Food and Drug Administration, 2018). Products that improve the skin's look, enhance its natural attractiveness, and significantly boost its cleanliness are all considered cosmetics (Gautam et al. 2011). These categorizations are what were used to label the many categories of cosmetics: skin moisturizer, anti-aging products, face makeup, shampoo, toothpaste, deodorant, hair color, and so on (Thornfeldt, 2005).

Figure 1. Human skin anatomy
(Copyright free Image from istockphoto.com)

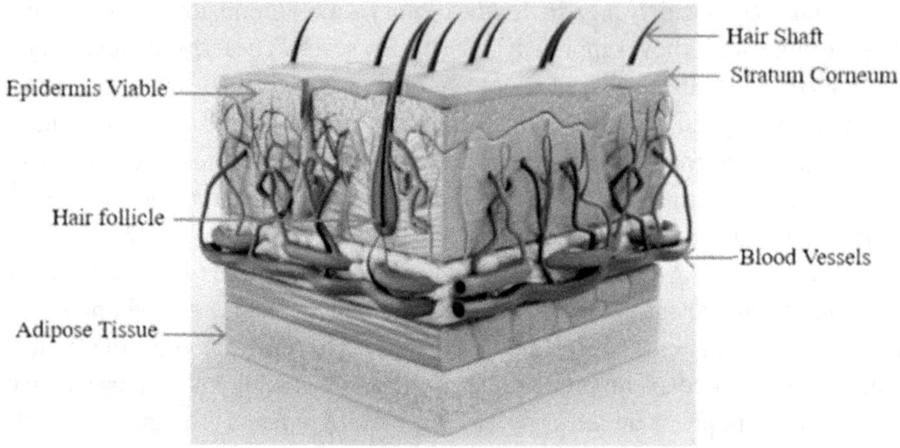

COSMETICS, COSMECEUTICALS, AND NANOCOSMETICS

Nearly nine cosmetics are used on a daily basis by the typical adult (Nigam, 2009). Makeup is, without question, one of the most popular consumer goods worldwide. Because they help men fulfill their deepest urge to feel and look attractive, they have universal appeal. The Roman playwright Plautus put it this way: "A lady without paint (cosmetics) is like bread without salt." What a wonderful expression of the significance of cosmetics! Cosmetics and associated products have been used for eons to do anything from boost self-esteem to enhance natural attractiveness (Sharma, P. 2008). Cosmetics are defined by the Food and Drug Administration (FDA) of the United States of America (USA) as "any material or substance intended for use in washing, beautifying, enhancing attractiveness, or modifying the appearance of the human body or any portion thereof, other than soap" (US Food and Drug Administration, Cosmeceuticals).

The term "cosmetics" is used to describe a broad variety of hygiene and beauty aids. Cosmetics may be broken down into subcategories like

1) Skin-care items (like lotions and cleansers) and
2) hair-care items (such as dyes, mousses, and shampoos) based on where they are used. three) cosmetics intended for use on the face, such as lip colors, eye liners, mascara, powders, and foundations.
3) Nail care items such as polish removers and nail varnishes
4) UV light screening preparations like sunscreen,

5) scent items like cologne, deodorant, aftershave, and perfume.

When seen through the lens of this concept, it becomes clear that cosmetics are not drugs. The word "cosmeceutical" (Sharma, P. 2008) is used to describe cosmetic goods that may include substances that have a biological impact, modest therapeutic activity, or drug-like benefit on the skin. The Food and Drug Administration (FDA) does not apply any legal classification to this term since it does not have any legal status in the United States. However, those in the cosmetics industry (such as skin scientists, physicians, product manufacturers, skin care professionals, and authors of academic or cosmetic journals) have used it in order to encourage consumer patronage of marketed cosmetic products and to differentiate cosmetics with a biological effect from those without (Lohani et al. 2014). Both cosmetics and cosmeceuticals are used interchangeably in this article. It's only appropriate to call cosmeceuticals containing nanosized components nanocosmeceuticals.

BRIEF HISTORY

The desire to look nice and feel good about one's appearance (the motivation for using cosmetics) has been around as long as humanity has, making it difficult to pinpoint when the practice of applying makeup first emerged. In the past, the fields of cosmetics, medicine, and religion were all intertwined and interwoven; nevertheless, thinkers like Henri de Mondeville and Hippocrates helped establish a clear separation between them (Sharma, P. 2008). As early as 10,000 BC, the Egyptians allegedly used oils, lotions, and ointments to attain cleanliness and preserve their skin from the harsh heat and dry air. Many ancient Egyptian and Mesopotamian tombs have colored art, frescoes, and mosaics that attest to the widespread utilization of cosmetics at the time. In reality, the discovery of Pharaoh Tutankhamen's tomb and the depiction of extensively made-up, almond-shaped eyes on Egyptian women's art both proved the widespread use of cosmetics in ancient times (Watchtower Bible and Tract Society of Pensylvania, A Report, 2012).

Since at least 3000 BC (Nicole, F. 2019), the Chinese upper class could be identified by their use of brilliantly colored nail polishes, while the lower class was forbidden from doing so. The elite of China during the T'ang dynasty were known for their excessive use of heang, a phrase referring to perfume, scent, or incense, both at home and in temple (Chaudhri et al. 2009). It is believed that makeup was introduced to Greece from the Far East by merchants and tourists. Face powders and false brows were popular beauty aids among ancient Greek women. As early as the 5th century C.E., Athenian women used lead to lighten their skin tone, while they colored their lips with rouge produced from seaweed or plant roots (Nigam, 2009).

Soot was used to draw attention to the eyebrows; kohl (like powdered antimony sulfide) was applied to the eyelids; and mascara was created from cow dung, egg whites, and gum. Many objects associated with women's grooming have been unearthed from ancient Greek palaces, graves, and towns. Face mirrors, combs, pins, razors, and tiny containers for cosmetics like scents, lotions, and pigments are all examples (Watchtower Bible and Tract Society of Pennsylvania, 1995). However, it is clear from the following that people have, throughout history, used variously harmful homemade cosmetics in their pursuit of beauty, putting their health and safety at risk. Women in several societies, especially in the 1800s, ingested poisons like lead, leeches, arsenic, and mercury in order to achieve the pasty white complexion that was considered attractive at the time (Stephanie, S.G. History of Makeup). These cosmetics shoppers often spend a lot of money. There is still a strong desire to seem attractive and youthful, but thankfully, people no longer resort to the use of potentially dangerous or even lethal concoctions in an attempt to accomplish this (Stephanie, S.G. History of Makeup). Nanotechnology is an example of a cutting-edge technology used in the creation of cosmetics.

NANOMATERIAL PROPERTIES

Researchers and manufacturers have hypothesized that improved discharge of cosmetic compositions into the skin is one reason for the recent uptick in interest and use of nanomaterial in the cosmetics industry. It's a reasonable assumption based on what we know about nanomaterial characteristics. However, the findings of the research don't seem to back up such an explanation. Possible causes include non-reproducible findings (Katz et al. 2015) due to differences in study conditions. Nanoemulsions, for instance, may allow cosmetic materials to be more easily absorbed into the skin due to the small size of lipid vesicles seen in these systems (Cevc et al., 1992). However, human skin biopsy studies by Fellipe et al. 2009. Titanium dioxide and zinc oxide nanoparticle delivery systems are used in commercial sunscreen products. found that these ingredients were unlikely to penetrate deeper than the stratum corneum, despite what the sunscreen's manufacturers claimed. There is no doubt, though, that nanoscale materials exhibit characteristics that set them apart from their macroscale counterparts.

These characteristics are present to a certain extent, like

- Structural integrity
- Improved solubility
- Optical activity and improved appearance
- Increased chemical reactivity

- increased surface area and better dispersibility

The characteristics of a substance change depending on the particle size. Color, structure, transparency, optical activity, solubility, and chemical reactivity are only some of the ways in which nanoparticles differ from their parent or larger-sized counterparts. Nanoscale materials, for instance, have the distinct size that allows for improved solubility, size matching of cosmetic ingredients with the biological structures of skin cells, and consequently easy interaction, allowing them to a greater extent than usual affect cellular activities on their normal scale (Sivanskar et al. 2010). Nanoparticles in cosmetics have these benefits: increased resistance to UV rays, a longer duration of action, a higher quality of application, and deeper skin penetration (Katz et al. 2015). Nanomaterials are also advantageous since they have a greater surface area to mass or volume ratio than larger particles. For any two particles of the same volume, the one with a higher surface-to-volume ratio is the one with a greater concentration of molecules or atoms on its surface. As particles shrink in size, their surface area grows in relation to their volume.

A mathematical derivation of the percentage increase in surface area of a spherical nanosized particle from a micrometer was attempted by Sivanskar et al. (2010). The dispersibility (the rate of sedimentation is lowered to practically zero) and dissolve rate of medicinal or cosmetic substances are both increased by an increase in surface area. Due to the increased number of contact sites enabled by the increased surface imperfections, nanoscale materials exhibit superior adhesion. If something sticks well to the skin, it will seem more natural. Nevertheless, surface area has an even greater impact on particle reactivity. Due to their distinct chemical composition, nanoparticles employed in cosmetics may exhibit novel bioactivity, and their potential biological response may be predicted with some accuracy (Castranova, 2011).

In suspension, nanoscale materials of any solid density form highly distributed systems that resist settlement due to Brownian motion. It follows that cosmetics containing nanoscale components will remain suspended for much longer than those containing microscale ingredients. Because of this quality, nanosized skin preparations may be stored for a longer period of time before showing signs of instability such as creaming, cracking, Ostwald ripening, or phase inversion. In this light, the need to acquire and use readily accessible nanoparticles in cosmetic manufacturing is understandable. The increased reactivity of nanoscale materials is a direct result of their larger surface area, which offers more active sites to facilitate chemical interaction. Nanomaterials provide aesthetic value to personal care items because of their optical activity and transparency. Some examples of this are lotions and sunscreens. The main ingredients of sunscreens are zinc oxide and titanium oxide. In contrast to a similar product containing nanosized particles, which do not leave any apparent white tint on the skin after application, the cosmetic value of the nanoproduct is enhanced (Hilton, L. Cosmetic Dermatology). Low manufacturing costs relative to larger-sized equivalents have also been cited as an additional incentive to employ nanoparticles in cosmetics. This is particularly common in cases where nanoparticles are accidentally created as a consequence of reducing bigger particles (Hilton, L. Cosmetic Dermatology).

TYPES

Our culture is fast shifting to recognise the importance of cosmetics as a vital part of maintaining physical and mental health. Water-based cosmetics including make-up remover, face cleanser, anti-aging lotion, and sun-screen all benefit greatly from oil in water nanoemulsion as an effective formulation. Cubosomes, nanodots, liposomes, dendrimers, and nano-emulsions are just few of the nano-sized materials that are showing up regularly in the cosmetics industry. Nanoparticle-based cosmetics or nano-cosmeceuticals have broadened the potential uses of cosmetics to treat ageing skin's wrinkles, dryness, and loss of elasticity, as well as scattered hyperpigmentation. Giant cosmetics companies make a lot of promises about the potential benefits of goods containing them, but there are still many unanswered concerns.

Inorganic Nanoparticles

In comparison to organic nanoparticles, inorganic ones have the advantages of being non-toxic, hydrophilic, biocompatible, and very stable. Beyond these differences,

Figure 2. Proportions of different inorganics nanoparticles in cosmetic formulation (Gupta et al. 2022)

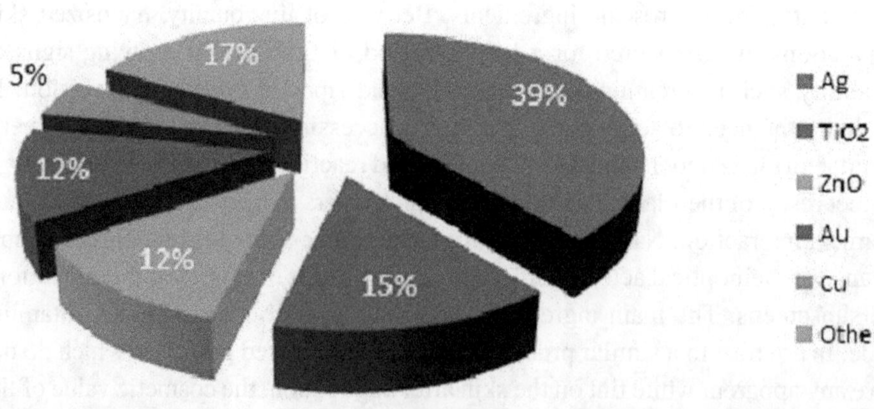

the primary distinction between inorganic and organic nanoparticles is that the former are synthesised from inorganic elements (Ag, Au, Ti, etc.), while the latter are produced using polymers.

Figure 2 shows the proportions of different inorganic nanoparticles in cosmetic formulations.

Titanium Dioxide and Zinc Oxide

TiO_2 is one of the most popular inorganic nanoparticles used in sunscreens because it is both more effective at blocking the sun's rays (thanks to its high sun protection factor, or SPF) and more cosmetically desirable (thanks to its transparency). The usage of nanoscale TiO_2 or ZnO is often advertised as "sheer" or "invisible" by manufacturers. TiO_2 and ZnO at the nanoscale are said to have significant benefits over many goods with dimensions greater than the nanoscale (Cornier et al. 2019). Sunscreens use the Ultrviolet A (UVA) and Ultrviolet B (UVB) absorption properties of micro-TiO_2 and ZnO, respectively. Both zinc oxide (ZnO) and titanium dioxide (TiO_2) nanoparticles (20 nm) are often employed as UV filters in sunscreens. They have a more even distribution and a more aesthetically pleasing aftereffect. It has been shown that combining X-ray powder diffraction (XRD) with Transmission Electron Microscopy (TEM) is a useful method for analysing ZnO and TiO_2 in cosmetic items (Lu et al. 2015). Concerning their safety, it has been noted that inhalation of large quantities of ZnO nanoparticles might cause health harm (Monsé et al. 2018). ZnO concentrations in standard sunscreen formulations are thought to be safe since there is no evidence for penetration into the viable epidermis or toxicity concerns (Mohammed et al., 2019). However, exposure through a different route (the dermal route) is not without risk.

Gold and Silver Nanoparticles

The antibacterial and antifungal qualities of gold and silver nanoparticles have made them popular in a variety of cosmetic products, including deodorants, anti-aging lotions, and face masks. In ancient Egypt, gold was used as part of a regimen to improve and preserve the skin's overall health and appearance. The ancient Egyptians thought gold made their skin more supple and healthy. These days, you can find gold in anything from salves and lotions to full-on facial treatments. Colloidal gold, or more accurately nanogold, if its particle size is between 5 nm and 400 nm, is the term most often used to describe the gold found in skincare products. Depending on size and overall surface area, its hue may be anywhere from red to purple (Irshad et al. 2020). Gold nanoparticles come in many forms, including spheres, rods, clusters, stars, shells, cubes, and triangles, and their optical properties and cellular uptake are dependent on their condition. They are better suited to use in skincare and cosmetics because of their stability and biocompatibility. In addition, they have well-established antifungal, antibacterial, and anti-aging effects, which are particularly important in the cosmeceutical and wound healing sectors (Akturk et al. 2016). Skin damage may be repaired, and the skin's surface, beauty, and pliability can all be enhanced with the help of gold nanoparticles. Gold's anti-inflammatory and anti-sunburn characteristics make it a valuable remedy for sunburn and other forms of skin irritation. This means it works well in face masks and other cosmetic applications.

Antimicrobial applications for silver nanoparticles have been shown. Different formulations of silver and silver-based combinations may be used to inhibit bacterial growth (Simonetti et al. 1992). Silver precipitates easily in silver-based combinations, which may be an issue for silver's use in cosmetics, but this can be avoided by using silver nanoparticles. Colloidal silver in nanostructures used in oral and cutaneous cosmetics is controversial in Europe due to questions about its safety (Scientific Committee on Consumer Safety). (Melatonin, 2019). In the United States, where the Food and Drug Administration does not regulate the cosmetics industry, there is a common misconception that these products do not have any effective antibacterial characteristics (Katz et al. 2015). According to studies, adding silver nanoparticles to cosmetics keeps the formula from settling for more than a year. Also, silver nanoparticles provided sufficient protection against bacteria and their development and did not penetrate human skin (Kokura et al. 2010).

Silica (SiO_2)

The cosmetics industry has taken an interest in crystalline silica nanoparticles because their hydrophilic surface promotes prolonged circulation and they are inexpensive to produce (Filipe et al. 2009). Cosmetics benefit from nano-sized silica in terms of their longevity, texture, and overall efficacy. Its absorbency and anti-caking properties are two of its many benefits. Evidence suggests that the addition of silica nanoparticles to lipsticks can help even out the color and keep it from migrating into the lip line (Poland et al., 2016). The encapsulation of lipophilic and hydrophilic substances within silica nanoparticles, which are stabilized nanodispersions in the size range of 5–100 nm (Santos et al. 2019). Hair, skin, lip, face, and nail products, both those that are left on and those that are rinsed off, already contain silica nanoparticles, and their presence is expected to grow (Mebert et al. 2017). There is conflicting evidence about the safety of silica-based nanoparticles, so it's important to consider variables like particle size and surface modifications when gauging their toxicity (Mebert et al. 2017). As a result, there is still some debate about whether or not silica nanoparticles should be used in cosmetics, and more long-term exposure studies are required.

The Colorless Pitch of Carbon Black (Nano)

Carbon black, also known as CI 77266, is a widely used cosmetic ingredient that provides a black color to a broad range of products, including those designed for the eyes, face, and lashes. The European Union has given its nanoform permission to be used as a colorant at 10% concentration or below. Carbon black nanoparticles have been shown to be more dangerous, to cause more inflammation, and to affect phagocytosis in human monocytes than nanoparticles in the micron size range (Sahud, 2014). In the European Union (EU), cosmetic products can contain it if there is no danger of inhalation (Chaudhry, 2016).

Carbon Nanotubes

Because of its high efficiency as a filter and its photostability, tris-biphenyl triazine is an excellent addition to sunscreens (Couteau et al. 2015). As a broad-spectrum UV filter, tris-biphenyl triazine (nano) is an excellent addition to sunscreens and anti-aging facial care products. In addition to having exceptional photostability, it is also a European Union-approved UV filter. The German chemical company BASF SE markets it as TINOSORB® A2B (Triazine B. TINOSORB® A2B). Methylene bis-benzotriazolyl tetramethylbutylphenol (nano), also known as MBBT, is an approved UV filter in the European Union market and may be used in dermally applied cosmetic products at concentrations of up to 10% w/w. The Scientific Committee on Consumer Safety (SCCS) has concluded (Doak, S. 2016) that MBBT poses no threat to people when applied to unharmed skin. But SCCS was worried about irritant effects and bioaccumulation in certain tissues.

Nano-Hydroxyapatite

In order to cure dental hypersensitivity and promote enamel remineralization, a number of companies have included nano-hydroxyapatite in their products (Coelho et al. 2019). With its remineralization and desensitization properties, nano-hydroxyapatite has previously been included in dental care products, including dentifrices and mouthwashes, as a potential alternative to fluoride toothpaste (Ramis et al. 2019).

Nanoliposomes

Nanoliposomes are liposomes of microscopic size. Phospholipid bilayers surround the water volume in these concentric bilayered vesicles (History of Cosmetics from Ancient Times). Because of their biodegradability and biocompatibility, nanoliposomes are a very flexible component class in the cosmetics industry (Nicole,

F. 2019). They are applied to the skin to increase permeability, hydrate, and shield the skin from potentially harmful active components (such as vitamins). Lipsticks, body deodorants, and antiperspirants may all use them to distribute scent. Their commercial uses in the cosmetics industry have been restricted by difficulties such as poor drug loading, low repeatability, and physicochemical instability [Xu et al. (2012); Joseph et al. (2018)].

Similar to liposomes and emulsions, solid lipid nanoparticles (SLNs) are a kind of lipid carrier, but they differ from NLCs in that their particle matrices are solid rather than liquid (Pardeike et al. 2009). Because of their many benefits above standard formulation, they work as carriers for ingredients and hydrate the skin very well (Wissing et al., 2003). SLNs have a single shell layer and an oily or lipoidal center. Because of their diminutive size, SLNs make extensive contact with the stratum corneum, facilitating the absorption of active ingredients through the skin (Couteau et al. 2015). They have the potential to act as a UV blocker and can also help to increase the skin's water content. While nanostructured lipid carriers (NLCs) have a greater capacity for transporting bioactives, SLNs have greater chemical stability. They are simple to scale up, and they use tried-and-true manufacturing techniques. As a carrier delivery agent, they exhibit improved skin penetration, biocompatibility, and stability (Hilton, L. Cosmetic Dermatology, Aesthetic Technology, a Report). Sunscreens and moisturizers both include SLNs and NLCs, which are used to prevent skin damage. Highly tolerated carrier systems for cutaneous applications include SLNs and NLCs, both of which are amenable to both laboratory and industrial-scale manufacturing (Pardeike et al. 2009).

Nanocapsules

Polymeric NM capsules are encased in an oily or watery phase, making them nanocapsules (Wissing et al. 2003). Cosmetics employ nanocapsules for ingredient protection, odor control, and compatibility difficulties (Nafisi et al. 2017). Suspensions of polymeric nanocapsules may be used as a topical application or added to semisolid formulations. A substance's ability to permeate the skin may be adjusted by changing the proportion of polymer to surfactant in its production (Poletto et al., 2011). Through nanoprecipitation, we were able to construct stabilized poly-L-lactic acid nanocapsules with a diameter of around 115 nm, which enabled us to entrap fragrance molecules in a polymeric nano-carrier and therefore accomplish a prolonged release of scent (Hosseinkhani et al. 2015). The use of biocompatible nanocapsules to encase molecules has the potential to revolutionize the future of deodorant. To include vitamins and extracts in semisolid formulations like creams, scientists have recently produced innovative stimuli-responsive nanocapsules. After being applied to the skin, these formulations waited for indications such as a shift in pH or the presence of enzymes produced by wounded skin to release the active contents of the nanocapsules.

While perfluorocarbon emulsions are unstable, a new study has shown that encapsulating perfluorodecalin (an oxygen carrier) in a silica nanocapsule provides a fresh approach to the topical treatment of aging skin. Overall, the mixture outperformed emulsions in terms of both delivery and stability (Svarc et al. 2018). By combining carrot oil with Pluronic polymer, Barbosa et al. were able to create nanocapsules that contained the sunscreen component benzophenone-3 (-caprolactone). In addition to enhancing the benzophenone's stability in the topical formulation, these nanocapsules showed synergistic SPF activity with a non-irritant profile (Barbosa et al. 2019).

The Dendrimerics

Dendrimers are very flexible because of their spherical construction, which consists of a core around which symmetric units are formed (Santos et al. 2019). Dendrimers are a kind of polymer that, because of its stability, may aid in the delivery of active substances through the skin. Dendrimers may be found in antiperspirant and shampoo formulations. Symmetric dendrimer branches exhibit surface activity because of the interplay between their hydrophobic edge section and their hydrophilic center. Thanks to their favorable characteristics—including monodispersity, polyvalence, and stability—they are excellent drug delivery vehicles (Nafisi et al. 2017). Scaling up and commercializing a product based on the dendrimer structure allowed for greater loading and skin penetration of resveratrol (which has antioxidant and anti-aging properties) (Pentek et al. 2017).

Nanoemulsions

Colloidal dispersions containing droplets of sizes from a few nanometers to 200 nm are known as nanoemulsions (Tadros, T. 2004). Increasing patent protection actions using nanoemulsions demonstrate increasing business interest (Sonneville-Aubrun et al. 2018). The optical, stability, rheological, and ingredient transport qualities of microemulsions are better than those of traditional emulsions because of the droplets' tiny size. Nanoemulsions are typically stable and see-through. The hydroglycolic extract of Opuntia ficus-indica (L.) Mill was used to create an oil-in-water (O/W) nanoemulsion with desirable thermal stability and moisturizing performance (Ribeiro et al., 2015). The findings revealed that O/W nanoemulsions containing 1% of Opuntia ficus-indica (L.) Mill extract exhibited adequate stability for at least 2 months. The formulation's moisturizing effectiveness was further shown by its ability to raise the stratum corneum's water content. According to a study by Musazzi et al. (2019), nanoemulsions may dramatically affect the permeation profiles of molecules based on their physicochemical features; in particular, O/W nanoemulsions greatly enhance the permeation profiles of apolar components compared to traditional emulsions.

Buckyballs (Buckminsterfullerene/C60)

Atoms of odd-numbered carbon are grouped in rings to form buckyballs, which have the same overall shape as certain types of footballs but a nanometer-scale diameter. This compound is also known by its alternate chemical name, Buckminster fullerene. Moisturizers with these ingredients provide anti-oxidant benefits by scavenging free radicals, and their smoothing effects stimulate skin renewal. Because of their high hydrophobicity, they weren't used for a while, but surfactants helped them overcome this barrier (Duarah et al., 2016). Anti-aging cosmeceutical Zelens Fullerene C-60 night cream contains buckyball and has been shown to have potent antioxidant effects (Lohani et al. 2014).

Micelles

Micelles, which are fewer than 100 nm in size, have the potential to be employed as nanosized materials in the formulation of pharmaceuticals or cosmetic treatments that are poorly soluble in colloidal solutions. Micelles are composed of a variety of subunits and develop when a surfactant's critical micellar concentration is reached and the surfactant self-arranges in an aqueous environment. Micelles contain many of the same characteristics as liposomes, with the exception of an inside hydrophilic liquid compartment. Instead, they have a hydrophobic core and a hydrophilic exterior, making them effective in solubilizing poorly soluble cosmetic ingredients.

Nanocrystals

These nanoparticles are crystalline or ground down to the nanoscale to be very pure active materials. These crystal aggregates, which may vary in size from 10 to 400 nm and are composed of hundreds to thousands of atoms, are employed to transport insoluble actives (Keck et al. 2006). Juvedical®, made by cosmetics firm Juvena in 2000, is a nanocrystal formulation of rutin, an antioxidant with a wide range of potential uses. Compared to the water-soluble version, rutin glycoside, this formulation of the original rutin molecule is 500 times more bioactive, according to research. The UV-blocking properties of rutin nanocrystals were used to do this comparison with rutin glycoside (Petersen, R. 2010). Nanocrystal solution, which was less soluble than glycoside or powder, had a much greater concentration of actives formed as nanocrystals in the skin, which was particularly impressive (Lohani et al. 2014). Comparing the commercially available form with the powder, Pyo et al. (2016) found that the rutin nanocrystals in gel had the strongest neutralizing (antioxidant) action. The most recent research on the anti-pollution substance SymUrban, conducted by Köpke et al., found that its solubility and penetration profiles significantly increased when it was in nanocrystal form. These nanocrystals seemed to be an effective delivery mechanism for SymUrban, as their use boosted the material's dermal bioavailability despite the material's weakly soluble active component (Feltin, U.S. Patent, 2015).

Nano-Organic (Tris-Biphenyl Triazine)

A unique, strong, and photostable filter, tris-biphenyl triazine, is employed exclusively in sunscreen formulations (Coutea et al. 2015). It is often used in sunscreen formulations because its nano form acts as a broad-spectrum UV protector. It has excellent photostability and is recognized as a UV protector in Europe. BASF SE makes use of it, and it goes by the trade name TINOSORB® A2B. Another UV protectant that has been given the green light for use in cosmetics sold in the European Union is methylene bisbenzotriazolyl tetramethyl butylphenol (nano), or MBBT, which may be used at concentrations of up to 10% weight per weight in topical applications. If applied to healthy, unbroken skin, MBBT poses no risk to humans, according to an analysis by the Scientific Committee on Consumer Safety (SCCS). Yet, it has the ability to bioaccumulate in some tissues, which has prompted worries about its deleterious effects (Fytianos et al. 2020).

Nanoparticles of Micelles

These have been extensively employed in the cosmetics sector and are largely regarded as one of the most effective nanotechnology-based particles. They provide a solid and adaptable foundation for incorporating a large variety of lipophilic active substances with a wide range of physicochemical qualities into cosmetic products. These particles are superior to other nanocarriers because of their small size, high encapsulation effectiveness, and low production cost (Sonneville-Aubrun et al., 2004; Lee et al., 2010). They are used in place of traditional cleansers as a means of effectively removing oil and debris from the skin without disrupting the skin's protective barrier (Dhapte-Pawar et al., 2020; Morales et al., 2009). Using nanoparticles, transdermal medication delivery has undergone a dramatic shift (TDD). Micellar nanoparticle emulsions provide promising potential as a topical route to systemic medication delivery. The method enables a strong medication concentration to penetrate the skin, yielding a pharmacological formulation with the same advantages as TDD but being safer and more tolerable. Various cosmetics companies tout their micellar nanotechnology-based facial cleansers as the most effective on the market. Micellar shampoos were developed in one of the studies by Ziba et al., who found that they could better emulsify fatty deposits and had a greater viscosity overall.

Hydrogels

As a consequence of chemical or physical cross-links, these 3D hydrophilic polymer networks swell in water or biological fluids without disintegrating. By adapting to the changes they foresee, they may protect their property from any potential harm (Singh et al. 2016).

Distinct Classes

In the cosmetics sector, nanoparticles may be classified as either I biodegradable (like liposomes and chitosan) or II nonbiodegradable (Aranaz et al. 2018). (e.g., polystyrene, ZnO, and silica-based nanoparticles). The cosmeceutical industry has shown considerable interest in chitin and its deacetylated derivative, chitosan, owing to their distinctive biological and technical features (Morganti et al. 2008). One type of nanocrystal that is both harmless and useful is a chitin nanofibril, which is extracted from the exoskeleton of crustaceans by removing the carbonate and protein components (SCENIHR, Risk Assessment of Products of Nanotechnologies). Emulsions containing chitin nanofibrils may cause the development of a hygroscopic molecular layer, which reduces the rate at which water evaporates from the skin and helps keep it moist.

APPROACHES TO CHARACTERIZATION FOR RISK EVALUATION

The Scientific Committee on Emerging and Newly Identified Health Risks (SCENIHR) offers advice on how to conduct research on nanomaterials and how to conduct risk assessments to determine whether their use poses any danger to human health or the environment. According to the SCCP (Scientific Committee on Consumer Products), nanomaterials have unique properties, thereby requiring novel testing methods to identify their possible damage causes. For the purpose of determining whether or not nanomaterials are safe to use, the following factors are mostly considered:

Physical and Chemical Properties

Chemical properties such as structural formula or molecular structure, composition of nanomaterials, phase identity, surface chemistry, hydrophilicity or lipophilicity, and solubility must be analyzed alongside physical properties such as size, shape, specific surface area, agglomeration state, size distribution, and solubility (SCENIHR, A Report).

Intuitive Mathematical Simulation

These prediction models may be anything from a set of basic, empirical methods to a set of complicated mathematical equations, the solution to which may involve the knowledge or estimate of factors that are unavailable for direct experimentation. However, these models cannot be relied upon to forecast what would happen when such things approached the skin since they do not incorporate data pertaining to macromolecular compounds or particle shapes. (Report of the ICCR Joint)

Utilizing Microscopic Methods

Microscopic inspection of the skin after treatment provides more meaningful data for the in vitro research. Visualizing the tissue to which an active has been administered may give significant knowledge even if exact quantification is not achievable. (US Food and Drug Administration, Cosmetics)

Laser scanning confocal microscopy (LSCM)
High-resolution transmission electron microscopy (HRTEM)
Particle-induced X-ray emission (PIXE)
Radio-labeling with the positron emitter

CONCERNS (SAFETY AND HEALTH)

The usage of nanoparticles in cosmetics has been met with mixed reactions because of their unusual features. It has both positive and negative effects on the human body, with the former including the possibility of enhancement. Because particles' known characteristics may quickly change when shrunk to nanosize, this raises worries about their impact on human health and safety. Also, there is still a lot of mystery surrounding nanomaterial toxicity.

Therefore, the biggest worry regarding using cosmetics containing nanoparticles is whether or not any of the particles may enter the bloodstream during either the manufacturing process or the actual usage of the product. What potential damage might there be if they did? In the United States, the issue of nanomaterial safety sprang to the forefront when a group working for the Food and Drug Administration (FDA) made proposals for regulatory consideration, focusing on the security and purity of cosmetics (Baroli et al. 2007). The Food and Drug Administration needs two primary pieces of data (approaches) to evaluate the safety of nanoparticles in cosmetics.

One is specific to the material's characteristics, in particular the nanoform it takes. Because nanoparticles may change the physicochemical characteristics of components and final cosmetic products, it is important to have a thorough understanding of their qualities, their biological interactions, and the characterization of any related contaminants. Experts in the field of health safety may use this kind of up-to-date and precise data to better handle situations of poisoning, contamination, and allergies.

The toxicological information about the nanosized product is the second kind of data. This may be determined by testing cosmetic chemicals for a variety of adverse effects, including photoirritation and photoallergy, as well as acute, chronic, and subchronic toxicity to the central nervous system. effects on human health from repeated or prolonged use of certain substances or from repeated or prolonged exposure to a class of substances. Genotoxicity, foetal toxicity, carcinogenicity assessment, and reproductive health effects data will also be required.

There are primarily two groups of people that are vulnerable to nanomaterial exposure:

Industrial Producers: The inhalation route accounts for the majority of workers' exposure to nanoparticles in cosmetics factories. This is particularly true if the production region does not have or adhere to adequate safety precautions, since numerous microscopic nanoparticles will be released into the air throughout the manufacturing process.

Market Participants: Main users of the product and those (secondary users) exposed via contact or relationship with primary users are both considered customers (or primary users). Nanoparticles in cosmetics and personal care items may be transferred to unintended recipients via casual contact, such as an embrace or kiss.

Nanomaterials can enter the body in three different ways: through the lungs when consumers inhale sprayed products (perfumes, fragrances); through the mouth when they ingest lip and near-mouth products (lip gloss, lip paint, mouth fresheners); and through the skin when they apply skin care products. Nanoparticles have a strong propensity to interact with the membrane of the respiratory epithelium, and their small size allows them to enter the alveoli, the functional unit of the respiratory tract. They are taken in by the olfactory epithelium and distributed to the brain and other organs through the nasal nerves and the bloodstream (Duarah et al. 2006). It has been claimed that only a negligible fraction of ingested nanoparticles make it into the bloodstream. Most of it is eliminated from the body through feces (Baroli et al., 2007).

But when nanoparticles enter the body via the skin, things become more complicated. Since the skin is the body's biggest organ, it is often exposed to cosmetics via this method. When whole, however, the outer layer's lipophilicity and strong cell cohesion make it an effective, apparently impenetrable barrier. While there have been a number of attempts to determine how deeply nanoparticles may penetrate intact skin, so far no conclusive evidence has been found. However, skin penetration has been seen under the following circumstances:

Nano-sized dendrimers of 2.9 nm but 3256 Da (Daltons) were discovered to not enter intact skin, while particles of 10 nm but of smaller weight in Da have been demonstrated to sometimes reach viable epidermal layers of the skin via the hair follicles and the stratum corneum (Pyo et al. 2016). When the skin's integrity has been compromised by external factors such as cuts, burns, skin illnesses, or scrapes, allowing for the "opening of the door" for particles, particularly nanoparticles in cosmetic products, to enter the skin. Sometimes the damage is brought about when the stratum corneum is stripped away by adhesives (like elastoplasters), allowing nanoparticles to penetrate the skin. Using a tape stripping approach, researchers were able to observe the penetration patterns of a rutin nanocrystal gel on pig skin, reporting that the nanocrystals reached the horny layer of the skin (Baroli et al. 2007).

In the same way that certain chemical agents may irreversibly disturb intact, less permeable skin once applied, mechanical bending of the skin can as well. These aid the skin's ability to take in the active ingredients of cosmetics. In order to penetrate the skin, several surfactants such as water, alcohol, and dimethyl sulfoxide have been used. Quantum dot skin penetration was purportedly compared by Zhang and Monteiro-Riviere to that of quantum dots in 75% v/v glycerol. Glycerol at 75% volume in the preparation was more effective in penetrating (Wissing et al. 2003). Many of these penetration boosters may be found in common beauty products. According to a review conducted by "Friends of the Earth," 70% of the cosmetics analysed had chemicals known to serve as "penetration enhancers," increasing the likelihood that nanoparticles in cosmetics would be absorbed into the skin upon application (Gajbhiye et al. 2016).

Most diseases travel from one part of the body to another via the limbs. Hair follicles and their accompanying sebaceous glands, as well as sweat pores, may provide a route for nano-sized lipids, niosomes, and nanoparticles to penetrate deeper into the skin's viable layers (Wissing et al. 2003). Nevertheless, since the appendages cover such a small area of skin and some of them are already full of oil or perspiration flowing out in the opposite direction of the cosmetic component positioned for delivery into the skin, the amount that can be absorbed into the skin may be rather tiny.

ENVIRONMENTAL IMPACTS & ASSESSMENTS

Nanoparticles may have diverse effects on the environment depending on their composition, size, stability, mobility in different media (such as water and air), and how they are employed in the workplace. While such fundamental information on the behavior and toxicity of nanoparticles is necessary for risk assessment, it is not sufficient for a realistic assessment. Instead, data on the expected concentration of nanoparticles in environmental systems would be required, and there is currently no accurate information on such concentrations. Resources, environmental routes, and uses of nanoparticles, as well as nanoparticle-sensitive plants and animals, must be identified as a first step in conducting an environmental risk assessment of nanoparticles (Reijnders, L. 2006).

In Open Space

Nanoparticles are released into the environment via a variety of natural and anthropogenic mechanisms. Outdoor work conditions are common for numerous professions, including construction, gas and petroleum transmission pipelines, police work, farming, and others. Very little study has been done on the consequences of nanoparticle exposure on such individuals, but what little there is implies that they are at a higher risk of developing health problems as a result of their employment. Nanoparticles may be able to go from inside to outside under particular circumstances. Nanoparticles that make it past a filtering system may still endanger employees in the outside areas if they are dispersed there through ventilation ducts. Nanoparticles, with their unique physical and chemical qualities, may easily penetrate and spread across indoor and outdoor workplaces. Some of the responses they set off in human cells may be harmful to the body biochemically. Researchers are now assessing worker exposure by quantifying nanoparticle concentrations in indoor and outdoor work environments (Spielvogel et al. 2009).

In Plants

In a study looking at how aluminum oxide nanoparticles affect plant root development, researchers found that uncoated alumina nanoparticles somewhat slowed root development, but phenanthrene coating had no effect. Alumina's toxicity is significantly influenced by its surface characteristics (Yang et al. 2005). Smearing spinach seeds with titanium dioxide nanoparticles or spraying spinach leaves with these nanoparticles has been proven in several studies to promote spinach growth. Unlike larger samples of titanium dioxide, which do not affect enzymatic activity, nitrate absorption, or the rate at which inorganic nitrogen is converted to organic nitrogen, titanium dioxide nanoparticles do. Some evidence shows that inorganic oxide nanoparticles may interact with plant cells, including those of green algae, which share many characteristics with plant cell walls (Hund-Rinke et al. 2006).

In order to have an effect on plant roots, nanoparticles must first penetrate the root cell wall and then be absorbed into the root cell interior. Nanoparticles may also be taken up by the membranes of neighboring cells if they find their way into the intercellular gap. Root bark intercellular space is permeable to negatively charged species due to the negatively charged cell surfaces of plants. This is a pathway for nanoparticles to infiltrate the plant's woody tissue (Nowack et al 2006).

Absorption Under Environmental Conditions

Nanoparticle toxicity investigations have mostly been conducted on mammalian cells. In these tests, the nanoparticles were immersed in protein- and other-rich cell culture mediums. Since prokaryotes lack a method for transporting colloidal particles through the cell wall, their cellular surfaces serve as a protective barrier, preventing the absorption of many forms of nanoparticles. Nevertheless, eukaryotes, including early unicellular creatures and multicellular organisms, are in a distinct scenario because they include processes for the entrance of nanoparticles and microparticles, namely endocytosis and phagocytosis. Protozoa, for instance, may take in carbon nanotubes, which then aggregate in the mitochondria of the ingested cells (Zhu et al. 2006).

CONCLUSION

There are several "nano-enhanced" skin formulations on the market, and the number is growing as more people realise the purported benefits of using nanoparticles in cosmetics. But, we cannot rest easy merely because these skin treatments have economic potential; further research into their health effects is necessary. The growing body of information on the toxicity of nanoparticles, however, belies its importance. When it comes to new and current nanosized chemicals used in cosmetics, regulatory authorities will need to do more than just provide suitable standards and monitoring; they will also need to screen these substances proactively to decide acceptance or rejection. Cosmetics producers should be held to a higher level with regards to correct labelling and post-marketing surveillance by such organisations. There can be no worldwide coverage without the international cooperation of regulatory bodies in exchanging information on cosmetic components and their effect on health. Whenever a consumer has an adverse response to a nanocosmetic product, they must notify the relevant authorities immediately. So, we are all responsible for our own parts. Many signs point to the continued importance of cosmetics using nanoparticles, but this should not be prioritised above concerns about their potential health effects. The improved qualities of nanoparticles have sparked interest in R&D into new cosmetics thanks to the cosmetics industry's current focus on nanotechnology applications. More importantly for nanocosmetics, the cosmetics industry has significantly more relaxed regulatory standards than the pharmaceutical industry.

REFERENCES

Akturk, O., Kismet, K., Yasti, A. C., Kuru, S., Duymus, M. E., Kaya, F., Caydere, M., Hucumenoglu, S., & Keskin, D. (2016). Collagen/gold nanoparticle nanocomposites: A potential skin wound healing biomaterial. *Journal of Biomaterials Applications*, *31*(2), 283–301. doi:10.1177/0885328216644536 PMID:27095659

Aranaz, I., Mengíbar, M., Harris, R., Paños, I., Miralles, B., Acosta, N., Galed, G., & Heras, Á. (2018). Cosmetics and cosmeceutical applications of chitin, chitosan and their derivatives. *Polymers*, *10*(2), 213. doi:10.3390/polym10020213 PMID:30966249

Barbosa, T. C., Nascimento, L. É. D., Bani, C., Almeida, T., Nery, M., Santos, R. S., Menezes, L. R. O., Zielińska, A., Fernandes, A. R., Cardoso, J. C., Jäger, A., Jäger, E., Sanchez-Lopez, E., Nalone, L., Souto, E. B., & Severino, P. (2019). Development, cytotoxicity and eye irritation profile of a new sunscreen formulation based on benzophenone-3-poly (ε-caprolactone) nanocapsules. *Toxics*, *7*(4), 51. doi:10.3390/toxics7040051 PMID:31546707

Baroli, B., Ennas, M. G., Loffredo, F., Isola, M., Pinna, R., & López-Quintela, M. A. (2007). Penetration of metallic nanoparticles in human full-thickness skin. *The Journal of Investigative Dermatology*, *127*(7), 1701–1712. doi:10.1038j.jid.5700733 PMID:17380118

Castranova, V. (2011). Overview of Current Toxicological Knowledge of Engineered Nanoparticles. *Journal of Occupational and Environmental Medicine*, *53*(Supplement 6S), S14–S17. doi:10.1097/JOM.0b013e31821b1e5a PMID:21606847

Cevc, G., & Blume, G. (1992). Lipid vesicles penetrate into intact skin owing to the transdermal osmotic gradients and hydration force. *Biochimica et Biophysica Acta (BBA)-. Biomembranes*, *1104*(1), 226–232. doi:10.1016/0005-2736(92)90154-E PMID:1550849

Chaudhri, S. K., & Jain, N. K. (2009). History of cosmetics. [AJP]. *Asian Journal of Pharmaceutics*, *3*(3).

Chaudhry, Q.Scientific Committee on Consumer. (2016). Chaudhry Q. Opinion of the scientific committee on consumer safety (SCCS)—Second revision of the opinion on carbon black, nano-form, in cosmetic products. *Regulatory Toxicology and Pharmacology*, *79*, 103–104. doi:10.1016/j.yrtph.2016.02.021 PMID:26946407

Coelho, C. C., Grenho, L., Gomes, P. S., Quadros, P. A., & Fernandes, M. H. (2019). Nano-hydroxyapatite in oral care cosmetics: Characterization and cytotoxicity assessment. *Scientific Reports*, *9*(1), 11050. doi:10.103841598-019-47491-z PMID:31363145

Cornier, J., Keck, C., & Voorde, M. (2019). *Nanocosmetics from Ideas to Products: From Ideas to Products*. Springer. doi:10.1007/978-3-030-16573-4

Couteau, C., Paparis, E., Chauvet, C., & Coiffard, L. (2015). Tris-biphenyl triazine, a new ultraviolet filter studied in terms of photoprotective efficacy. *International Journal of Pharmaceutics*, *487*(1-2), 120–123. doi:10.1016/j.ijpharm.2015.03.077 PMID:25843762

Couteau, C., Paparis, E., Chauvet, C., & Coiffard, L. (2015). Tris-biphenyl triazine, a new ultraviolet filter studied in terms of photoprotective efficacy. *International Journal of Pharmaceutics*, *487*(1-2), 120–123. doi:10.1016/j.ijpharm.2015.03.077 PMID:25843762

Dhapte-Pawar, V., Kadam, S., Saptarsi, S., & Kenjale, P. P. (2020). Nanocosmeceuticals: facets and aspects. Fut Sci OA 6: FSO613.

Doak, S. (2016). Opinion of the Scientific Committee on consumer safety (SCCS)-Opinion on the use of 2, 2′-methylene-bis-(6-(2H-benzotriazol-2-yl)-4-(1, 1, 3, 3-tetramethylbutyl) phenol)(nano)–S79–In cosmetic products. *Regulatory Toxicology and Pharmacology*, *76*, 215–216. doi:10.1016/j.yrtph.2016.01.002 PMID:26776753

Duarah, S., Pujari, K., Durai, R. D., & Narayanan, V. H. B. (2016). Nanotechnology-based cosmeceuticals: A review. *Int J Appl Pharm*, *8*(1), 8–12.

Feltin, C., & Brun, G. (2015). Cosmetic Composition Based on Supramolecular Polymer and An Absorbent Filler. U.S. Patent US9000051B2.

Filipe, P., Silva, J. N., Silva, R., De Castro, J. C., Gomes, M. M., Alves, L. C., & Pinheiro, T. (2009). Stratum corneum is an effective barrier to TiO2 and ZnO nanoparticle percutaneous absorption. *Skin Pharmacology and Physiology*, *22*(5), 266–275. doi:10.1159/000235554 PMID:19690452

Fytianos, G., Rahdar, A., & Kyzas, G. Z. (2020). Nanomaterials in cosmetics: Recent updates. *Nanomaterials (Basel, Switzerland)*, *10*(5), 979. doi:10.3390/nano10050979 PMID:32443655

Gajbhiye, S., & Sakharwade, S. (2016). Silver nanoparticles in cosmetics. *Journal of Cosmetics. Dermatological Sciences and Applications*, *6*(1), 48–53.

Gautam, A., Singh, D., & Vijayaraghavan, R. (2011). Dermal exposure of nanoparticles: An understanding. *Journal of Cell and Tissue Research*, *11*(1), 2703–2708.

Gupta, V., Mohapatra, S., Mishra, H., Farooq, U., Kumar, K., Ansari, M. J., Aldawsari, M., Alalaiwe, A., Mirza, M., & Iqbal, Z. (2022). Nanotechnology in cosmetics and cosmeceuticals—A review of latest advancements. *Gels (Basel, Switzerland)*, *8*(3), 173. doi:10.3390/gels8030173 PMID:35323286

Hilton, L. (n.d.). Cosmetic Dermatology, Aesthetic Technology. *Dermatology Times*. http://www.dermatologytimes.com

Cosmetics Info. (n.d.). History of Cosmetics from Ancient Times. *Cosmetics Info*. https://www.cosmeticsinfo.org/Ancient-history-cosmetics

Hosseinkhani, B., Callewaert, C., Vanbeveren, N., & Boon, N. (2015). Novel biocompatible nanocapsules for slow release of fragrances on the human skin. *New Biotechnology*, *32*(1), 40–46. doi:10.1016/j.nbt.2014.09.001 PMID:25224920

Hougeir, F. G., & Kircik, L. (2012). A review of delivery systems in cosmetics. *Dermatologic Therapy*, *25*(3), 234–237. doi:10.1111/j.1529-8019.2012.01501.x PMID:22913440

Hund-Rinke, K., & Simon, M. (2006). Ecotoxic effect of photocatalytic active nanoparticles (TiO2) on algae and daphnids (8 pp). *Environmental Science and Pollution Research International*, *13*(4), 225–232. doi:10.1065/espr2006.06.311 PMID:16910119

Irshad, A., Zahid, M., Husnain, T., Rao, A. Q., Sarwar, N., & Hussain, I. (2020). A proactive model on innovative biomedical applications of gold nanoparticles. *Applied Nanoscience*, *10*(8), 2453–2465. doi:10.100713204-019-01165-4

Is It Safe to Use Cosmetics Containing Silica in Nanoform? (n.d.). Europea. https://ec.europa.eu/health/sites/health/files/scientific_committees/docs/citizens_nanosilica_en.pdf

Joseph, J., & BN, V. H. (2018). Experimental optimization of Lornoxicam liposomes for sustained topical delivery. *European Journal of Pharmaceutical Sciences*, *112*, 38–51. doi:10.1016/j.ejps.2017.10.032 PMID:29111151

Katz, L. M., Dewan, K., & Bronaugh, R. L. (2015). Nanotechnology in cosmetics. *Food and Chemical Toxicology, 85*, 127–137. doi:10.1016/j.fct.2015.06.020 PMID:26159063

Kaul, S., Gulati, N., Verma, D., Mukherjee, S., & Nagaich, U. (2018). Role of nanotechnology in cosmeceuticals: A review of recent advances. *Journal of Pharmaceutics, 2018*(34202043), 1–19. doi:10.1155/2018/3420204 PMID:29785318

Keck, C. M., & Müller, R. H. (2006). Drug nanocrystals of poorly soluble drugs produced by high pressure homogenisation. *European Journal of Pharmaceutics and Biopharmaceutics, 62*(1), 3–16. doi:10.1016/j.ejpb.2005.05.009 PMID:16129588

Kokura, S., Handa, O., Takagi, T., Ishikawa, T., Naito, Y., & Yoshikawa, T. (2010). Silver nanoparticles as a safe preservative for use in cosmetics. *Nanomedicine; Nanotechnology, Biology, and Medicine, 6*(4), 570–574. doi:10.1016/j.nano.2009.12.002 PMID:20060498

Lee, R., Shenoy, D., & Sheel, R. (2010). Micellar Nanoparticles: Applications for Topical and Passive Transdermal Drug Delivery. In Handbook of Non-Invasive Drug Delivery Systems. William Andrew.

Logothethetidis, S. (2012). *Nanoscience and Technology* (1st ed.). Springer.

Lohani, A., Verma, A., Joshi, H., Yadav, N., & Karki, N. (2014). Nanotechnology-based cosmeceuticals. *International Scholarly Research Notices, 2014*, 1–14. doi:10.1155/2014/843687 PMID:24963412

Lu, P. J., Huang, S. C., Chen, Y. P., Chiueh, L. C., & Shih, D. Y. C. (2015). Analysis of titanium dioxide and zinc oxide nanoparticles in cosmetics. *Yao Wu Shi Pin Fen Xi, 23*(3), 587–594. PMID:28911719

Mebert, A. M., Baglole, C. J., Desimone, M. F., & Maysinger, D. (2017). Nanoengineered silica: Properties, applications and toxicity. *Food and Chemical Toxicology, 109*, 753–770. doi:10.1016/j.fct.2017.05.054 PMID:28578101

Mohammed, Y. H., Holmes, A., Haridass, I. N., Sanchez, W. Y., Studier, H., Grice, J. E., Benson, H. A. E., & Roberts, M. S. (2019). Support for the safe use of zinc oxide nanoparticle sunscreens: Lack of skin penetration or cellular toxicity after repeated application in volunteers. *The Journal of Investigative Dermatology, 139*(2), 308–315. doi:10.1016/j.jid.2018.08.024 PMID:30448212

Monsé, C., Hagemeyer, O., Raulf, M., Jettkant, B., van Kampen, V., Kendzia, B., Gering, V., Kappert, G., Weiss, T., Ulrich, N., Marek, E.-M., Bünger, J., Brüning, T., & Merget, R. (2018). Concentration-dependent systemic response after inhalation of nano-sized zinc oxide particles in human volunteers. *Particle and Fibre Toxicology*, *15*(1), 1–11. doi:10.118612989-018-0246-4 PMID:29429408

Morales, M. E., Gallardo, V., Clares, B., García, M. B., & Ruiz, M. A. (2009). Study and description of hydrogels and organogels as vehicles for cosmetic active ingredients. *Journal of Cosmetic Science*, *60*(6), 627–636. PMID:20038351

Morganti, P., & Morganti, G. (2008). Chitin nanofibrils for advanced cosmeceuticals. *Clinics in Dermatology*, *26*(4), 334–340. doi:10.1016/j.clindermatol.2008.01.003 PMID:18691512

Musazzi, U. M., Franzè, S., Minghetti, P., & Casiraghi, A. (2018). Emulsion versus nanoemulsion: How much is the formulative shift critical for a cosmetic product? *Drug Delivery and Translational Research*, *8*(2), 414–421. doi:10.100713346-017-0390-7 PMID:28508377

Nafisi, S., & Maibach, H. I. (2017). Nanotechnology in cosmetics. *Cosmetic science and technology: theoretical principles and applications, 337.*

Nicole, F. (2019) A Brief History of Makeup. *Her Campus*. https://www.hercampus.com

Nigam, P. K. (2009). Adverse reactions to cosmetics and methods of testing. *Indian Journal of Dermatology, Venereology and Leprology*, *75*(1), 10–19. doi:10.4103/0378-6323.45214 PMID:19172025

Nowack, B., Schulin, R., & Robinson, B. H. (2006). Critical assessment of chelant-enhanced metal phytoextraction. *Environmental Science & Technology*, *40*(17), 5225–5232. doi:10.1021/es0604919 PMID:16999093

Pardeike, J., Hommoss, A., & Müller, R. H. (2009). Lipid nanoparticles (SLN, NLC) in cosmetic and pharmaceutical dermal products. *International Journal of Pharmaceutics*, *366*(1-2), 170–184. doi:10.1016/j.ijpharm.2008.10.003 PMID:18992314

Pentek, T., Newenhouse, E., O'Brien, B., & Chauhan, A. S. (2017). Development of a topical resveratrol formulation for commercial applications using dendrimer nanotechnology. *Molecules (Basel, Switzerland)*, *22*(1), 137. doi:10.3390/molecules22010137 PMID:28098828

Petersen, R. (2010). *Nanocrystals for Use in Topical Cosmetic Formulations and Method of Production Thereof.* US Patent US 20100047297A1.

Poland, C. A., Larsen, P. B., Read, S. A. K., Varet, J., Hankin, S. M., & Lam, H. R. (2016). *Assessment if Nano-Enabled Technologies in Cosmetics*. The Danish Environmental Protection Agency.

Poletto, F. S., Beck, R. C. R., Guterres, S. S., & Pohlmann, A. R. (2011). Polymeric nanocapsules: Concepts and applications. In R. Beck, S. Guterres, & A. Pohlmann (Eds.), *Nanocosmetics and Nanomedicines*. Springer. doi:10.1007/978-3-642-19792-5_3

Pyo, S. M., Meinke, M., Keck, C. M., & Müller, R. H. (2016). Rutin—Increased antioxidant activity and skin penetration by nanocrystal technology (smartCrystals). *Cosmetics*, *3*(1), 9. doi:10.3390/cosmetics3010009

Ramis, J. M., Coelho, C. C., Córdoba, A., Quadros, P. A., & Monjo, M. (2018). Safety assessment of nano-hydroxyapatite as an oral care ingredient according to the EU Cosmetics Regulation. *Cosmetics*, *5*(3), 53. doi:10.3390/cosmetics5030053

Reijnders, L. (2006). Cleaner nanotechnology and hazard reduction of manufactured nanoparticles. *Journal of Cleaner Production*, *14*(2), 124–133. doi:10.1016/j.jclepro.2005.03.018

Ribeiro, R. C. D. A., Barreto, S. M. D. A. G., Ostrosky, E. A., Rocha-Filho, P. A., Veríssimo, L. M., & Ferrari, M. (2015). Production and characterization of cosmetic nanoemulsions containing Opuntia ficus-indica (L.) Mill extract as moisturizing agent. *Molecules (Basel, Switzerland)*, *20*(2), 2492–2509. doi:10.3390/molecules20022492 PMID:25648593

Sahud, K. (2014). *Carbon black particle exhibits size dependent toxicity in human monocytes*, 827019.

Santos, A. C., Morais, F., Simões, A., Pereira, I., Sequeira, J. A., Pereira-Silva, M., Veiga, F., & Ribeiro, A. (2019). Nanotechnology for the development of new cosmetic formulations. *Expert Opinion on Drug Delivery*, *16*(4), 313–330. doi:10.1080/17425247.2019.1585426 PMID:30793641

SCCP (Scientific Committee on Consumer Products), Preliminary Opinion On Safety of nanomaterials in cosmetic products.

SCENIHR. (n.d.). *Opinion on The Appropriateness of The Risk Assessment Methodology in Accordance with The Technical Guidance Documents for New and Existing Substances for Assessing The Risks Of Nanomaterials*. SCENIHR. http://www.ec.europa.eu/health/ph_risk/committees/04_scenihr /docs/scenihr_o_023.pdf

Scientific Committee on Consumer Safety. (2019). *Scientific Committee on Consumer Safety Melatonin*. Scientific Committee on Consumer Safety.

Sharma, P. (2008). *Cosmetic Formulation Manufacture and Quality Control* (4th ed.). Vandanna Publication.

Simonetti, N., Simonetti, G., Bougnol, F., & Scalzo, M. (1992). Electrochemical Ag+ for preservative use. *Applied and Environmental Microbiology*, *58*(12), 3834–3836. doi:10.1128/aem.58.12.3834-3836.1992 PMID:1476427

Singh, R., Tiwari, S., & Tawaniya, J. (2013). Review on nanotechnology with several aspects. *Int. J. Res. Computer Eng. Electron*, *2*, 1–8.

Singh, T. G., & Sharma, N. (2016). Nanobiomaterials in cosmetics: Current status and future prospects. In A. M. Grumezescu (Ed.), *Nanobiomaterials in Galenic Formulations and Cosmetics* (pp. 149–174). William Andrew Publishing. doi:10.1016/B978-0-323-42868-2.00007-3

Sivanskar, M., & Kumar, B. P. (2010). Role of Nanoparticle in Drug Delivery System. *International Journal of Research in Pharmaceutical and Biomedical Sciences*, *1*, 41–66.

Sonneville-Aubrun, O., Simonnet, J. T., & L'alloret, F. (2004). Nanoemulsions: A new vehicle for skincare products. *Advances in Colloid and Interface Science*, *108*, 145–149. doi:10.1016/j.cis.2003.10.026 PMID:15072937

Sonneville-Aubrun, O., Yukuyama, M. N., & Pizzino, A. (2018). Application of nanoemulsions in cosmetics. In *Nanoemulsions* (pp. 435–475). Academic Press. doi:10.1016/B978-0-12-811838-2.00014-X

Spielvogel, J., Guo, X., Pesch, M., Keck, L., & Hagler, R. (2009). A New Real-time Exposure Monitor for Measuring Airborne Nanoparticles. *GRIMM Aerosol Techni GmbH & Co, Bayern, Germany*. Stephanie, S.G. History of Makeup. https://www.webmd.com/beauty/history-makeup

Svarc, F., Ranocchia, R. P., Perullini, A. M., Jobbagy, M., & Aldabe, S. A. (2018). *A New Route to Obtain Perfluorodecalin Nanocapsules as An Oxygen Carrier in Cosmetic Formulations*.

Tadros, T. (2004). Application of rheology for assessment and prediction of the long-term physical stability of emulsions. *Advances in Colloid and Interface Science*, *108*, 227–258. doi:10.1016/j.cis.2003.10.025 PMID:15072944

Thornfeldt, C. (2005). Cosmeceuticals containing herbs: Fact, fiction, and future. *Dermatologic Surgery*, *31*, 873–881. doi:10.1111/j.1524-4725.2005.31734 PMID:16029681

Triazine, B. (n.d.). TINOSORB® A 2 B. *Career Creations*. https://www.carecreations.basf.com/product-formulations/prod uct-highlights/product-highlights-detail/TINOSORB-A2B/304781 25

U. S. Food Drug Administration. (2018). *Is It a Cosmetic, a Drug, or Both? (Or is it soap?)*. FDA. https://www.fda.gov/cosmetics/guidanceregulation/lawsregulat ions/ucm074201.htm

US Food and Drug Administration. (n.d.). *Cosmeceuticals*. FDA. https://www.fda.gov

Watchtower Bible and Tract Society of Pennsylvania (1995) Fashion—Ancient Greek Style. Awake (March 8), 21-24.

Watchtower Bible and Tract Society of Pennsylvania. (2012). *The Use of Cosmetics in Bible Times*. The Watchtower Public Edition (December 1).

Wissing, S. A., & Müller, R. H. (2003). The influence of solid lipid nanoparticles on skin hydration and viscoelasticity–in vivo study. *European Journal of Pharmaceutics and Biopharmaceutics*, *56*(1), 67–72. doi:10.1016/S0939-6411(03)00040-7 PMID:12837483

Xu, X., Costa, A. P., Khan, M. A., & Burgess, D. J. (2012). Application of quality by design to formulation and processing of protein liposomes. *International Journal of Pharmaceutics*, *434*(1-2), 349–359. doi:10.1016/j.ijpharm.2012.06.002 PMID:22683453

Yang, L., & Watts, D. J. (2005). Particle surface characteristics may play an important role in phytotoxicity of alumina nanoparticles. *Toxicology Letters*, *158*(2), 122–132. doi:10.1016/j.toxlet.2005.03.003 PMID:16039401

Zhu, Y., Zhao, Q., Li, Y., Cai, X., & Li, W. (2006). The interaction and toxicity of multi-walled carbon nanotubes with Stylonychia mytilus. *Journal of Nanoscience and Nanotechnology*, *6*(5), 1357–1364. doi:10.1166/jnn.2006.194 PMID:16792365

Chapter 6
NanoTechnology in Cosmetics and Cosmeceuticals:
Future Direction and Safety

N. Suthanthira Vanitha
https://orcid.org/0000-0002-9579-7651
Muthayammal Engineering College, India

K. Radhika
https://orcid.org/0000-0003-4849-8680
Muthayammal Engineering College, India

G. Sudarmozhi
Muthayammal Engineering College, India

G. Kavitha
https://orcid.org/0000-0002-9667-6122
Muthayammal Engineering College, India

M. Shenbagapriya
Muthayammal Engineering College, India

ABSTRACT

Nanotechnology has the potential to generate innovation in formulations and delivery systems. Due to its rapid technological development, it has widely been exploited for diagnostic and therapeutic purposes. Cosmetic formulations incorporating nanotechnology seems to be a most promising research area. The role of nanoparticles in cosmetic applications has shown promising signs to overcome the problems associated with conventional cosmetics. Insertion of nanomaterials helps to improve product efficacy and consumer satisfaction which leads to the replacement of many traditional cosmeceuticals with Nanocosmeceuticals. Nanotoxicological studies on Nanocosmeceuticals have elevated concerns in terms of health hazards due to their potential skin penetration resulting in toxic effects. This chapter discusses various nanotechnology approaches employed in the delivery of cosmetics as well as cosmeceutical products. Further, it depicts the benefits, potential health, future

DOI: 10.4018/978-1-6684-9135-5.ch006

direction, safety, and an overview of Nanocosmetics and Nanocosmeceuticals applications in cosmetic industries.

1. INTRODUCTION

Nanotechnology plays a broad application in many fields such as Agricultural Field, Food Technology, Healthcare, Textiles and Fashion Technology etc., It comprises the design, characterization, manufacturing and application of materials, devices and systems at the nanoscale level (1–100 nm). Nanotechnology is recognized as one of the revolutionizing technologies in the Cosmetics and Cosmeceutical industries because of its enhanced properties which includes color, transparency, solubility, chemical reactivity, attracting the personal care and cosmetics sectors to nanomaterials. With increasing customers demand led to the development of formulations in cosmetics with better performance, retentions, attractive appearance and more importantly safety .At present, nanomaterials occupies the center of attention in this area and offer better merits over conventional cosmetic products. Further, the amalgamation of nanomaterials has significantly contributed to the global growth in the market share of pharmaceuticals and cosmetics. Basically the concept of nanomaterials like gold and silver are used in cosmetics for many years, but the applications are intensified in recent years (Wux et al.,2009 & Kaul et al.,2018)

Cosmetics preparations are used externally and formulated from its own or mixture of substances obtained from either natural or artificial sources.US Food and Drug Administration (USFDA) report says, cosmetics formulation intended to be applied to any human body for cleansing, beautifying and prettiness. The soap is plainly barred from this class that encompasses any material proposed as a component of a cosmetic item.As per the Federal Food Drug and Cosmetic the word as "Cosmeceutical" is merely for industrial purposes for referring therapeutic action of cosmetic products. In other way cosmetics as "any substance intended to be applied in the various outer parts of the human body such as, hair, nails, lips and external genital organs etc.,. It exclusively used for the purpose of cleaning, perfuming, appearance and correcting body odours, protecting them with good condition (Tripathy et al.,2015).

Currently, Cosmeceuticals considered as one of the rapidly growing sector in nanocosmeceutical research, exploration and application of the personal care industry and the market for individual consideration is massively expanding. Cosmeceuticals demonstrate therapeutically active ingredients that explicitly possess remedial effects on surface applications. These cosmetic products are utilized for the treatment of diverse conditions like, scratched hair, wrinkles, aging, dryness, acne, hyper pigmentation, etc. which has quantifiable restorative effects on the skin and hair. It act as a connection between drugs and beauty care products, assure an progress in appearance (Fatima et al., 2021).

The progression of nanotechnology in research and development is increasing through delivery of innovative solutions. In order to defeat firm problems related with conventional products, applications of nanotechnology is growing towards cosmeceutical segments of the personal care industry over the years. Cosmetics are enormous in the 21th century due to the developments, incorporation of the new technologies and cosmetic formulations (Sharma, 2012 & Dureja et al.,2005). Cosmeceuticals are the cosmetic products which includes biologically dynamic ingredients Cosmeceuticals are active between personal care products and pharmaceuticals products. The therapeutic efficacy on the skin, drugs and diversified formulations are used ensured for the treatment of various conditions like, hair damage, skin, face wrinkles, aging, dark spots, uneven complexion and hyper pigmentation, etc (Srinivas, 2016).Cosmeceuticals has high efficiency with superior sensorial properties and stable than the conventional cosmetics. Majority of the nanoparticles are fit for both lipophilic and hydrophilic drug delivery system. Nanomaterials are functional in the preparation of creams like, antiwrinkle moisturizing, skin whitening, hair repairing shampoos, conditioners and hair serums (Mu et al., 2010 & Nohynek et al.,2007).

Some of the ultrafine nanomaterials are carbon nanotubes, carbon based fullerenes, TiO2, copper nanoparticles and silver nanoparticles is toxic to human tissues and cells. Titanium dioxide is found in sunscreens is demonstrated to cause damage to DNA, RNA and fats within cells. For approving the regulation of nanocosmeceuticals no stringent scrutiny has been imposed by the regulatory agencies .Nanocosmeceuticals are sometimes harmful to environment as well and clinical trials are not mandatory for the approval of nanocosmeceuticals, thus raising a anxiety of toxicity after use.

The integration of a variety of nanomaterials during the enlargement of cosmetics and cosmeceutical products outcome in Nanocosmetics/Nanocosmeceuticals. By employing nanotechnology materials are manipulated at the atomic level to a great prospective in the cosmeceuticals field creates new avenues for the cosmetics industry. In order to improve UV protection, skin penetration, color, fragrance, quality, anti-aging effect cosmetic manufacturers employs nanosized ingredients. Prolongation of action is achieved either by controlling the delivery of active ingredients, causing site-specificity, enhancing the drug-loading capacity. All these factors, necessitates the clinical trials to address the safety concerns among consumers.

2. CLASSIFICATION OF NANOMATERIALS IN COSMETICS PREPARATION

Nanomaterials are considered asone dimension in the nano range with significant physicochemical properties. Nanomaterials Cosmetics has proven better while compared to microscale cosmetics. The large surface area of these particles is responsible for their efficient transportation, absorption, bioavailability, transparency and the sustained effect of the product. The following materials are commonly used in the cosmetic industry.

2.1. Inorganic Nanoparticles

These Particles are more hydrophilic and biocompatible, high stable, safer as compared to natural nanoparticles. Generally, nanoparticles are derived from inorganic components (Ag, Au, Ti, etc.), while the natural ones are from polymers. Figure 2 shows the statistics of various inorganic nanoparticles in cosmetic and cosmeceutical formulations. Some of the Inorganic Particles are discussed below and listed in table 1.

2.1.1 Titanium Dioxide and Zinc Oxide

TiO2 and Zinc oxide (ZnO) are the main compounds used as nanoparticles in UV filters. The nanoparticles contains two kinds of oxides extensively in sunscreens to prevent UV damage to the skin. In order to protect the skin from the hazardous impacts of solar radiation the sunscreens protects through shielding (Khan et al.,2016 & Poletto et al.,2011). Research says that these ivy nanoparticles are highly efficient and four times better than metallic oxide nanoparticles in blocking ultraviolet rays (Tripathy et al.,2015). The material ZnO is considered as efficient for obstructing UVA while TiO2 is superior for the UVB range with a size ranges of 20 nm. They show better scattering and produce a superior restorative or protective effect but inhalation of a large amount of these nanoparticles seems to be harmful (Peterson,2010).The merits of nanoparticles are more efficient than larger particle sized counterpart inabsorbing UV light, Sunscreen formulations with nanoparticles of TiO2 are less greasy, transparent,Less smelly and aesthetically appealing whereas ZnO holds a safe entity in cosmeceuticals.

2.1.2. Gold Nanoparticles

Gold and silver nanoparticles are widely utilized in cosmetic formulations such as antiperspirants, anti-aging creams face masks and display antibacterial as well as antifungal properties (Arora et al.,2012).At present, gold is included into diverse skincare items, such as salves, creams and skincare treatments.Gold nanoparticles has different shapes like,nanospheres, nanorods, nanoclusters, nanostars, nanoshells, nanocubes, and nanotriangles that determines their cell uptake and optical behaviors. Usually gold in skincare products is also called as colloidal gold or nanogold and color varies from red to purple, based on the total surface area size (Karim et al.,2010).Due to its stability and biocompatibility properties make more appropriate for skincare and cosmetics Kaul et al.,2018) .It plays a sizeable role in setting up skin damage and improving skin surface, grace and flexibility.

2.1.3 Silver Nanoparticles

Silver nanoparticles are used for inhibitors of various microorganisms. In order to control bacterial development in cosmetics formulations, silver-based mixtures are employed (Yapar et al.,2012 & Kokura et al., 2010).The management of silver in cosmetics are problematic, as silver precipitates easily in silver-based mixtures which overcome by the consumption of silver nanoparticles. Research says, the use of silver nanoparticles in cosmetics makes the formulation stable, limit sedimentation for longer duration (Simonetti et al., 1992). In addition, due to sufficient protection against microbes and their growth is impossible to enter the human skin (Kopke & Pyo.,2020).

2.1.4 Nano-Hydroxyapatite

The function of Nano-hydroxyapatite in cosmetic products specifically intended for oral preparations for treating extreme dental sensitivity and polish remineralization of the teeth.These are integrated into oral formulations, dentifrices and mouthwashes, owing to their remineralization and desensitization properties. All these preparations provides an substitute to fluoride toothpaste.

2.1.5 Buckyballs

Fullerene is a three-dimensional spherical compound that encompass a carbon ring with an odd number of carbon atoms is called “buckyballs” or buckminsterfullerene C60. Carbon fullerene is expansively worn in cosmetics and cosmeceuticals because of its antioxidative properties. Due to their potent scavenging ability of free radical oxygen species, Fullerenes are commonly employed in skin-rejuvenating cosmeceutical formulations thus serving to minimize the effects of UV damage, such as hyper pigmentation and wrinkles. Fullerenes alone limited applications because of their hydrophobic nature, high aqueous solubility, pharmaceutical applications and very expensive face creams.

2.1.6 Tris-Biphenyl Triazine

Tris-biphenyl triazine or Nano Organic, TINOSORB® A2B by BASF SE is a innovative, dominant, and photostable filter exclusively plays in sunscreen formulations . In its, it functions in broad-spectrum UV of nano form protectant thus frequentlyapplied in preparations of sunscreen. It is an approved UV protectant that offers significant photostability. Methylene bis-benzo triazolyl tetramethyl butylphenol(nano), or MBBT, is another approved UV protectant at percentages of up to 10% . As per the Scientific Committee on Consumer Safety study, MBBT is out of a danger to people when applied to solid, unbroken skin. However, it has elevate concerns related to possible hurtful impacts and has the potential to accumulate in selected tissues (Alka et al.,2014).

2.1.7 Carbon Black

Carbon black, CI 77266, is known to be a important ingredient in cosmetic formulations which is frequently make use of colorant in eye and skin cosmetic products as a colorant at a maximum of 10%.With the evaluation of carbon black nanoparticles confirm that they exhibit a higher propensity for causing cytotoxicity, aggravation and changes in phagocytosis in human monocytes. As per the Research, cosmetic items has no danger of being breathed.

2.1.8 Silica

Silica nanoparticles are in nature of hydrophilic surfaces favoring extended distribution and low manufacturing costs, interest towards these materials has increased, particularly in the cosmetic sector. Nanosilica is to improve the competence, surface and period of actual usability of cosmetic items.It supports to improve the manifestation and appropriation of shades in lipsticks and Silica nanoparticles keep colors with size range from 5 to 100 nm present as nanodispersions that deliver both hydrophilic and lipophilic entities as per targets by encapsulation . Normally it initiate in leave-on and washoff cosmetic items for hair, skin, lips, face and nails .Conversely, the realistic silica-based nanoparticles are uncertain and raise apprehension safety, but size and surface factors changes ought to be measured while surveying its toxicity (Akturk et al.,2016).The commercial exercise of silica nanoparticles in beautifying agents is still ambiguous, necessitate long-term trials.

Table 1. Classification of nanomaterials in cosmetics preparation

Classification of Cosmeceuticals	Type of Nanomaterial	Merits	Demerits
Lipstick	Silica (Sio_2)	Decreased Manufacturing Cost, Hydrophilic	Pulmonary Toxicity
Sunscreen	Tris-Biphenyl Triazine (Nano-Organic)	Act as Photostable filter	Harmful to Aquatic Environment
Face Cream	Bukminster-Fullerene/ C60 (Buckyballs)	Prevents skin problems Thermostability Antioxidant	Damage of Brain Tissues,Pulmonary Toxicity
Sunscreen	TiO_2,ZN (Inorganic Particles)	Biocompatible Stable,Safe & Hydrophilic	Pulmonary Toxicity
Toothpaste	Nano Hydroxyapatite	Teeth remineralization,Dental desensitizer	Very Brittle
Facemask and Mascara	Carbon Black	High Chemical & thermal Stability,Less weight	Changes Phagocytic property of Macrophages
Antiaging Cream and Facemask	Silver and Gold Nanoparticles	Antibacterial & Antifungal activity Chemical stability	Pulmonary Toxicity Cells and DNA damage

3. NANOFORMULATIONS/NANOCARRIERS IN COSMETICS INDUSTRY

Presently, Nanotechnology offers a innovative solutions to several problems in the field of cosmetics that results in novel formulations named as nanocosmeceuticals with better solubility, transparency, chemical reactivity and stability. Nanomaterials in cosmetics are increasing, various unique properties, affecting suitable type of cosmetic products. The types of nanomaterials that are used include nanocapsules, dendrimers, nanoemulsions, nanoliposomes, nano-hydroxyapatite materials are used in cosmetics. It has various nano formulations in drug delivery of cosmeceuticals (Nasir,2010 & Fox,1998) are depicted in Figure 1.

3.1 NanoLiposomes

Liposomes consists of vesicular structure aqueous core enclosed by a hydrophobic lipid bilayer for the preparation of cosmeceuticals and delivery of hydrophobic as well as hydrophilic compounds (Tripura et al.,2017).Their size varies from 20 nm to several micrometers for the multilamellar or unilamellar structure. Liposomes are large where one liposomes aligned inside another liposome with ability to penetrate into narrow blood vessels/skin whereas nano liposomes hasthe ability for better

Figure 1. Nanocarriers for cosmeceuticals

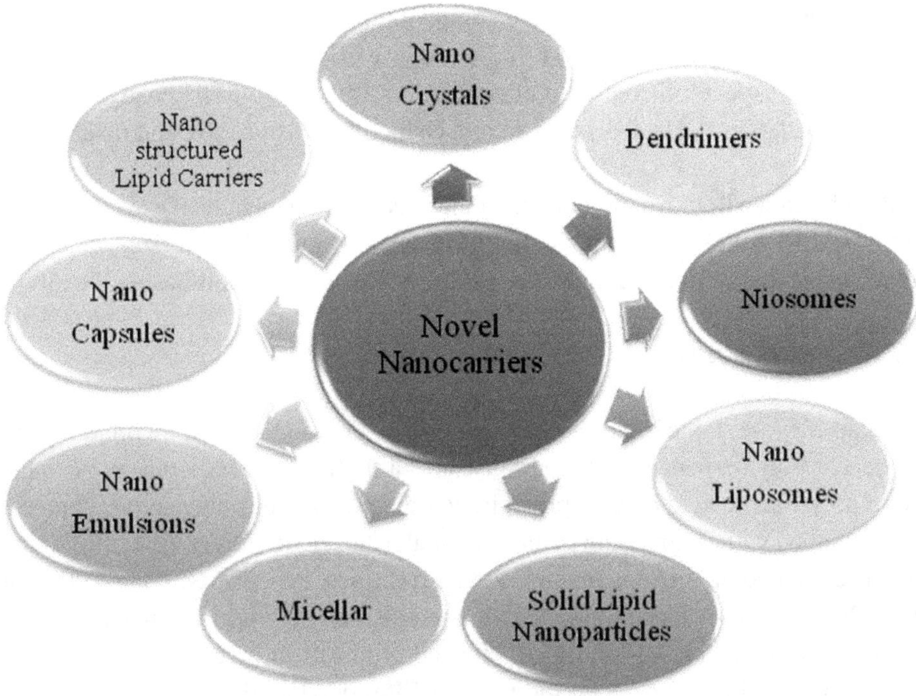

penetration due to its biodegradable and biocompatible. Cosmetic formulations integrating liposomes has high stability in skin which are not easily washed off. Phospho lipid is act as the key component of liposome lipid bilayer is recognized as safe ingredients to minimize the risk for adverse effects. Liposome encapsulates the drug and releases active ingredients in a forbidden manner for protecting the drug from metabolic degradation (Hope & Kitson,1993).Antioxidants like, carotenoids, CoQ10 and lycopene vitamins A, E and K are integrated into liposomes to amplify their physical and chemical stability while isolated in water.

Nanoliposomes in cosmetics enhance the hydration in skin due to its smaller size particles, build smooth skin, elastic and transport active moieties into the skin deeper layers, systemic circulation, transdermal drug delivery system (TDDS). Further, Liposomes are urbanized for the delivery of fragrances, botanicals, anhydrous formulations vitamins, antiperspirants, body sprays, deodorants, lipsticks, antiaging creams, deep moisturizing cream, sunscreen, beauty creams and hair loss treatments. Experimental study says that flexible liposomes support in decreasing wrinkle and show effects by increasing skin smoothness(Bhupendra et al.,2013).Nanoliposome encapsulated cream is capable of penetrating into deeper layers that resulting in sensible moisturizing capacity and anti-aging activity (Kocic et al.,2020). It includes Positive aspects and Negative aspects like,

Positive Aspects: High stability, Biocompatible and Biodegradable, High efficacy, Less toxicity Easy penetration and Site avoidance effect.

Negative Aspects: Increased production cost, Less solubility, Leakage of drug, irregular oxidation and hydrolysis reaction, Osmotically responsive and Insufficient stability

Applications: Removes wrinkles and dark spots has radiance effect with sunscreen, Ant wrinkle, firming, Nourishes skin, prevents photo aging, Moisturizer, Increases lips volume, fills wrinkles contour, moisturizes the skin and outlines the lips

3.2 Nanocapsules

Nanocapsules are stable in an aqueous medium, biodegradable and biocompatible in nature.The polymeric nanoparticles consists of a shell and a hollow space. With the use of poly-l-lactic acid by nanoprecipitation dimensions 115 nm can be achieved. It is entrapped with fragrant molecules that make the continuous release of perfume which are biocompatible in nature used in various deodorants.It has different methods such as surfactant-assisted, miniemulsion, polymerization techniques, chemical vapour condensation process for the synthesis. It is found that the use of nanocapsules reduces the penetration of UV filter octyl methoxycinnamate in pig skin while compared with conventional emulsions. Due to the inherent instability of perfluorocarbon emulsions, a novel method for topical therapy of aged skin was developed by integrating perfluorodecalin (an oxygen carrier) inside a silica nanocapsule core. In the case of anti-aging cosmetic formulations nanocapsules are used as a carrier for the antioxidant components (Katz et al.,2015).

Applications

- Protect vital actives that are confined by an oily or water state
- Delivery of 100% botanically pure vitamin E overcome conditions of lip bleeding and feathering that results from fine lines and wrinkles
- Nanocapsules are used to encapsulate several components
- Alteration of nanocapsules are achieved through the nature of polymer and surfactants
- Used as a carrier for the antioxidant components in anti-aging cosmetic formulations

3.3 Nanoemulsions

Nanoemulsions are considered as one of the most sophisticated nanomaterials for cosmetic production. Nanoemulsions consists of nanoparticles in the range of 50 to 1000 nm with a mixture of oil and water. It includes various cosmetics properties like, progress in shelf life products and textures, reduces the delicacy, oily nature of hair and safety. Nanoemulsions are relatively stable systems of fine texture whose structure is manipulated based on the preparation method that possess good skin penetration and hydration power. The presence of characteristic properties such as high surface area, minimal viscosity, stability, solubility employs the more practice of nanoemulsions. The synthesis involves a two-step process namely macroemulsion and nanoemulsion classified as low and high energy methods. The synthesis methods as, Microfluidization, high-pressure homogenization, evaporative ripening method, bubble bursting at an oil or water interface are used (Hu et al.,2012, Sonneville et al.,2004).

Applications

- Widely used in sunscreen, deodorant, nail enamels, body lotions, hair conditioners and shampoos and hair serums .
- Highly used in cosmetic products
- Supportive in the enhanced skin penetrations
- Improved skin hydrations due to small dimensions particles
- Skin has better infiltrations of active substances and is more fluid with a glossy coating
- Sprayable form attractive choice for cosmetics and drug delivery applications

Merits of Nanoemulsion: As Formulated into foams, creams, sprays and liquids, High absorption rate,Nontoxic & Nonirritant, Less problems like, sedimentation, coalescence, creaming and flocculation and Fit for medication delivery.

3.4 Nanocrystals

Nanocrystals are particles made of pure active materials that condensed to nanoscale through crystallization methods. The crystal size ranges 10 - 400nm that combine into a cluster of atoms and for the delivery of weakly soluble actives (Keck et al.,2006). It mainly integrate bioactive compounds and aid to progress their dissolution rate. The penetration profiles and solubility are observed to be extremely increased in its nanocrystal form. The rutin nanocrystal and rutin glycoside are compared based on its sun protective factor. Nanocrystals are remarkable due to its nano-crystal suspension with less soluble concentration of actives formulated as nanocrystals in the skin is identified as high compared with the glycoside . They permits safe and effective channel through skin.

3.5 Dendrimers

Dendrimers are unimolecular, monodisperse, micellar nanostructures are largerly branced symmetrical structure, star-shaped large molecules and a lot of functional end groups around the edges. It is typically built from a core with number of successive series of branches are engrafted in an arborescent manner and frequently adopts a spherical three-dimensional morphology (Fruchon et al.,2017).Dendrimers generation is based on the overall quantity of branch series. The size is small and range of 2–20 nm that contains large number of external groups suitable for multifunctionalization with three components: a central core, an interior dendritic structure a surface outside with functional surface groups, including the branches (Klajnert, 2001). Its stability, monodispersity, and polyvalence combine to make it the perfect carrier for precise, selective drug delivery. Dendrimers has macromolecular architecture formulated by nanotechnology based cosmeceuticals for diverse applications like, hair care, skin care and nail care.. Shampoos, sunscreen, hair-styling gels, and anti-acne treatments are only a few examples of the cosmetics that can benefit from dendrimers (Dhapte Pawar et al.,2020).

Merits are: Easy to functionalize due to structure, Degradationncan be controlled structure, Biocompatible,Molecular weight & size can be controlled

Applications

- Molecular weight and size determination, energy harvesting, photoactivity, drug delivery, gene transfection, catalysis and energy harvesting
- Science and Technology related to nanoscales, as well as rheology modification

3.6 Niosome

Niosomes consists of nano-vesicles formed of self-assembly non-ionic surfactants either incorporation of cholesterol or lipids. The unilamellar vesicles in the range of (0.025–0.05 μm) and multilamellar vesicles has more than>0.05μm containing aqueous solutions. Cholesterol, nonionic surfactants such as spans, tweens, brijs, alkyl amides, sorbitan ester, crown ester, polyoxyethylene alkyl ether, and steroid-linked surfactants are major components of niosomes in preparation (Khan et al.,2016). The synthesis methods of niosomes are ether injection method, film method, sonication method, reverse-phase evaporation and heating method. In addition to cosmetic applications, these were exploit in various other sectors like, pharmaceuticals and food due its higher bioavailability of actives which is considered in drug delivery vehicles. Proniosomes are used for improved drug delivery vesicles.The moisturizers, skin whitening creams, anti-wrinkle creams, shampoos and conditioners also been formulated (Yeo et al.,2018 & Ge et al.,2019). The properties of Niosomes are similar to those of liposomes except for better stability and flexibility.

Positive Aspects

Controlled drug delivery, Stable and Osmotically active, enhanced dermal penetration and oral bioavailability, No immunogenic, Nontoxic, Biocompatible, and Biodegradable.

Negative Aspects

May exhibit fusion, leaching, hydrolysis of entrapped drug which limits the shelf life, inadequate drug loading capacity, specialized equipment necessary for manufacture, entrapped drug, Physically instable, Time consuming techniques need for formulation, aggregation and expensive

Applications

- Used for parenteral and oral topical routes
- Surfactant handling and storage don't need any particular circumstances
- Better drug therapeutic performance

3.7 Micellar Nanoparticles

These type of Nanoparticles are identified as most effective nanotechnology-based particles in the cosmetic industry due to its robustness and versatile platform to encompass wide-ranging lipophilic active ingredients possessing diverse physicochemical properties in cosmetic formulations. The key features of these particles are smaller in size, better encapsulation efficiency and reasonable manufacturing cost that makes them more efficient (Lee et al.,2010). Generally, it is employed in skin cleansing products for removal of oil and dirt from the skin without affecting barrier integrity. It led to a revolution in transdermal drug delivery (TDD) and for systemic drug delivery through topical application which allow a high concentration of the drug to permeate the skin, creating a drug formulation attaining the similar benefits as TDD. For Facial cleansing formulations micellar are assured as successful products by different cosmetic brands.

3.8 Solid Lipid Nanoparticles

The Solid Lipid Nanoparticles (SLN) has unique oily or lipid materials outer shell core of sub-micrometer in dimensions. It is polymeric in nature includes diverse types of fatty acids as stearic acid or a mixture of fatty acids (Dingler,2002,Joohh et al.,2009,Muller et al., 2007 and 2016,Pardeike et al.,2009). The skin structure is composed of the outer epidermis which cover up the layer of the dermis that confines blood and lymph vessels which allow painless absorption of these nanoparticles through topical applications. Some of the techniques based on microemulsions, double emulsion formation, phase inversion induced by temperature, membrane contactor and cold homogenization. The formulations are characterized by a solid-state lipid matrix with the nano size range. Due to its miniaturization, formulations allows straight entry into the corneum layer supports to enhance the infiltration of active ingredients into the skin (Pardeike et al.,2009 & Singh et al.,2021). In addition, SLN being hydrophobic in nature, the particles are capable to protect the skin from dehydration and maintain the skin moisture. When compared to conventional carriers such as nanoemulsions, liposomes. lipid nanoparticles are promising carrier systems.Both SLN and Nanostructured lipid carriers (NLCs)has shown many beneficial features such as the controlled release of actives, occlusion and increased skin hydration effect for cosmetic applications. The synthesis methods are solvent emulsification-evaporation technique, solvent emulsification diffusion technique, high shear homogenization and ultrasonication technique .

3.8.1 Positive Aspects

It has better skin penetration, non-toxicity and effective carrier, to improve the penetration of active compounds into the stratum corneum, Better encapsulation, SLNs showed enhanced UVB protection action, crystalline nature of SLNs leads to less drug encapsulation efficiency, short shelf life along with slower drug release rates improved bioavailability, greater stability and good sunscreen-boosting action, Easy upgradability and Application versatility

3.8.2 Negative Aspects

Poor drug loading capacity, Low hydrophilic drugs, loading capacity due to partitioning effects,Unpredictable gelation tendency, High water content, Burst release can take place,Less particle concentration and expulsion of drug during storage

Applications

- Mostly used in sunscreen, anti-acne, and anti-aging ingredients (Singh et al.,2021,Wissing & Muller,2003,Lee et al.,2020)
- Properties used to treat irritated skin and dermatitis
- Utilized for the controlled administration of cosmetic agents over an extended period of time
- Film formation, aids in the re-enforcement and restoration of the skin barrier

3.9 Nanostructured Lipid Carriers (NLC)

In recent years, nanostructured lipid carriers are gaining significance and has numerous benefits in drug delivery. Lipid carriers with a nanostructure made up of emulsifying agents, water, solid lipids, and liquid lipids. The shortcomings associated with SLN is overcome with the development of NLC .Here the preparation of the carriers is through blending of solid lipids with spatially incompatible liquid lipid leading to amorphous solids in preferable ratio at body temperature [51]. The structure of the NLC is developed mainly based on three types of formulation composition and production parameters that includes imperfect type, amorphous type and multiple type with the size of the particle ranges from 10 to 1000nm (Shailesh et al.,2016 & purohit et al.,2016).Due to the lower risk of systemic side effects there is an better scientific and commercial attention for NLC during the past few years. It has higher drug-loading capacity for entrapped bioactive compound because of the distorted structure which supports in creating more space. NLC have modulated drug delivery profile, that is, a biphasic drug release pattern; in which the drug is released gradually at a constant rate (Westesen et al.,1997).

3.9.1 Positive Aspects

Reduced polymorphic transition, little crystalline index, enhanced encapsulation efficiency, drug loading, physical stability, improved chemical stability, bioavailability, and the controlled release of encapsulated components, Increased particle concentration and no expulsion of drug during storage, Very low biodegradable and physiological lipids formulation toxicity, enhanced UV protection system with reduced side effects

3.9.2 Negative Aspects

Lipids in the solvents are not sufficiently soluble, solvent residues are harmful and cause diluted NLC particles.

Applications

- Pharmaceutical and food industries
- Illnesses include cancer, alopecia, Parkinson's disease, diabetes, hyperlipidemia and epilepsy
- Hormone deficiency, topical inflammation, ocular, hepatic and fungal diseases

4. NANOCOSMECEUTICALS CLASSES

Cosmeceuticals are reflected as the fastest rising segment of personal care industry. A surplus of nanocosmeceuticals are assimilated in nail, hair, lip and skin care. Some of the nanocosmeceuticals classes are depicted in Figure 2.

Figure 2. Classes of nanocosmeceuticals

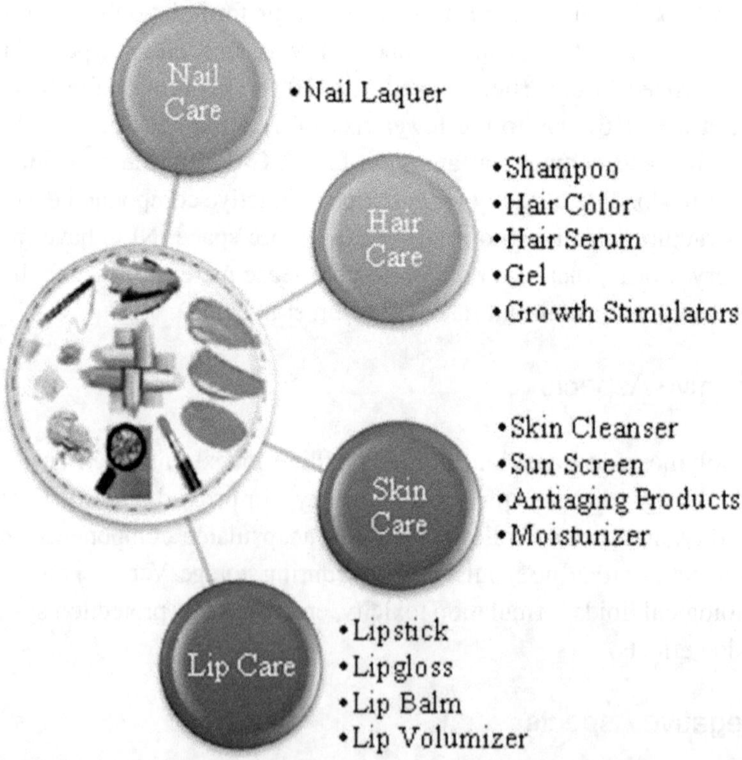

4.1 Hair Care

Hair concern includes various hair nanocosmeceutical products shampoos, conditioning agents, hair growth stimulants, Coloring and styling products. Nanoparticles of distinctive size and intrinsic properties are achieved in hair follicle, shaft targeting and increased quantity of active ingredient. Nanoparticles count present in shampoos contains seal moisture by optimizing resident contact time with scalp and hair follicles within the cuticles for protective film (Rosen et al.,2015). Conditioning agents upholds the special function to pass on softness, shine, silkiness, gloss and boost hair disentangling. The innovative nanocarriers like niosomes, microemulsion, nanoemulsion, nanospheres, and liposomes has exclusive function of repairing damaged cuticles, restoring texture, hair nongreasy, shiny and less brittle (Hu et al.,2012).

4.2 Skin Care

Skincare products in cosmeceuticals restructure the skin texture and performance through stimulating the growth of collagen by fighting harmful effect of free radicals. The formulation makes the skin healthier by sustain the structure of keratin superior. Nanomaterials like zinc oxide and titanium dioxide are most efficient minerals shield the skin by penetrating into the cavernous layers of skin. Further formulate the product less greasy, smelly and transparent (Glaser, 2004) in moisturizing formulations.Several nanoparticles namely, SLNs, nanoemulsions, liposomes and niosomes are extensively used which forms thin film of humectants and retain moisture for prolonged span. In addition, nanocapsules, liposomes, nanosome and nanospheres which assimilating and marked benefits for the skin collagen renewal, skin rejuvenation, firming and lifting the skin are broadly in antiaging nanocosmeceutical products (Smijs & Pavel,2011).

4.3 Nail Care

Cosmeceuticals based on nano nail care products seems to be superior when compared to traditional products. The nanomaterial nail paints has better toughness, sudden dryness, stability, chip resistance, trouble-free elasticity. The novel approach in nanoparticles as amalgamating silver and metal oxide antifungal properties nature in nail polish for the treatment of toe nails due to fungal infections (Pereira et al.,2014).

4.4. Lip Care

The products in nanocosmeceuticals involve lipstick, lip balm, lip gloss, and lip volumizer. Several nanoparticles in nanocosmeceuticals are coalesced into lip gloss and lipstick to soften the lips by impeding transepidermal water loss (Pereira et al.,2014). It prevents the pigments to migrate from the lips and keeping the colour of lips for a longer time. Lip volumizer containing liposomes raises the lip volume, hydrates and outlines the lips and fills wrinkles in the lip contour.

5. SAFETY ASPECTS OF NANOMATERIALS IN COSMETICS

In order to aid the cosmetics industry and academicians, researchers in identifying and investigating the safety aspects of nanomaterials in cosmetics it includes several important factors like, Physicochemical characteristics, Agglomeration and size, toxicity, testing and final product, Impurities, Potential routes and aggregation, Dosimetry for in vitro studies and toxicological nanomaterial ingredients, dermal penetration, potential inhalation, irritation (skin and eye), sensitization studies and mutagenicity/genotoxicity studies. The safety of product is obtained by analyzing the physicochemical properties and toxicological nature of each ingredient projected in cosmetic. In addition, Safety approaches to Nanomaterials in Cosmetics as Risk Assessment Pattern, identification of physicochemical parameters for the quantify the raw materials Assessment includes the investigation of systemic exposure, dermal, respiratory, systemic Absorption and Effects of the formulation. The guidance of safety assessment in nanomaterials cosmetic products holds the elements as general considerations, material characterization, exposure assessment, hazard identification, response characterization, and risk assessment.Major benefits of nanoparticles in cosmetics, act on the top of the skin, used in tiny measures, improve the texture, boost absorption rate, develope solubility, increase the surface area and fuse improved shelf life

5.1 Human Safety Concern in Cosmetics

The nanomaterials in cosmetic products has been attributed to its multi-faceted nature because of its biocompatible nature and biodegradation. Conversely, toxicity factors subsist with the human body using nanomaterials in large quantities in the cosmetic industry (Katz et al.,2015 & Muller et al.,2011). Traces of different metal nanoparticles such as copper, iron, cobalt, nickel,chromium, cadmium, lead, arsenic, mercury, etc noticed in cosmetics that cause a severe threat to human health.It happened due to tiny size, penetrate through the dermal tissue leading severe response

Figure 3. Nanomaterial safety assessment

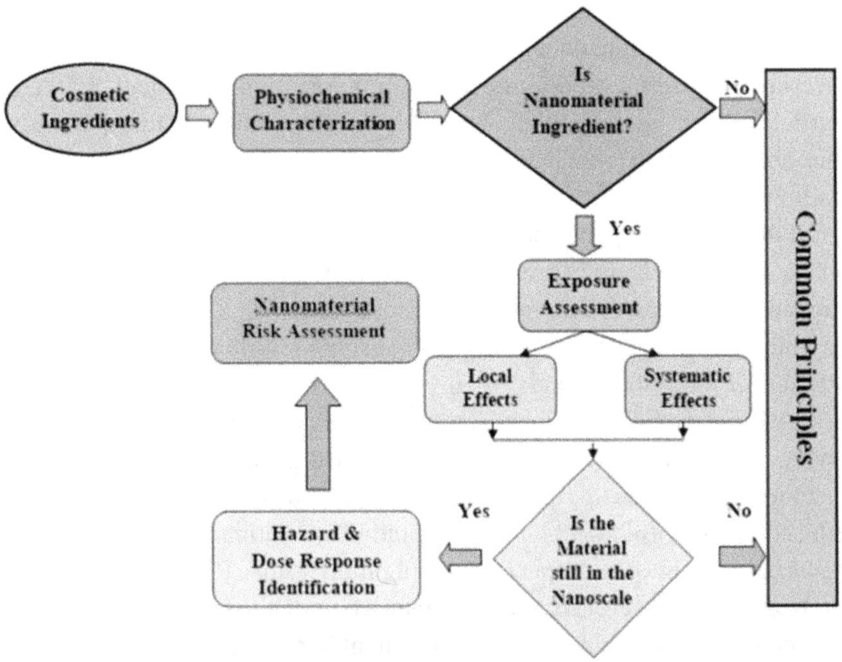

after its accumulation (Borowska,2015). Also the effect of polyvinyl pyrrolidone-functionalized silver nanoparticles is evaluated, observed that decreases cell viability, metabolic rate it severely impairs its migration and proliferation.

6. CHALLENGES OF NANOCOSMETICS

Nanoparticles in healthcare has found to be serious health risks to humans due to their toxicity, quantity, route and time of exposure. The factor includes shape, surface structure, surface charge, chemical composition and solubility. Nanoparticles are capable of moving easily inside the human body due to small size, shape, able to cross membranes, gain access to cells, tissues and organs which is not accessible to larger-sized particles and even enter into cells, causes more cell death (Li et al.,2003). The fundamental substances properties are changed substances at nanoscale level. The chemical reactivity and biological activity are repeatedly higher while compared to those of larger-sized particles, which is higher surface-area-to-volume ratios (Gunasekaran et al.,2022).In addition, nanoparticles having higher chemical reactivity increases the production of reactive oxygen with free radicals (Brown et al.,2002).

The key mechanisms of toxicity initiation leads to inflammation, oxidative stress and damage to membranes, proteins and DNA. Nanomaterials induce toxicity in various structure of human pulmonary, reticuloendothelial, neurological, cardiovascular systems, endocrine disrupting or immunological effects.The nanoparticles enter into the body by three different routes,namely, inhalation, ingestion and through the skin.The broadly recognized route of exposure to airborne nanoparticles is inhalation as per the National Institute of Occupational Health and Safety (NIOHS).when applying nanomaterials products to customers happen breathe in problem. For example, sunscreen sprays with TiO2 nanoscale cause the inhalation of nanomaterials, if passes through the nasal nerves to reach the brain and sensory system, enter the blood and various organs, causing life-threatening adverse effects (Schulte et al.,2008,Ostiguy et al.,2010).

Further, the nanomaterials ingestion take place accidentally and transfer from the hand to the mouth or ingested intentionally.During ingestion a sensible amount of the nanoparticles taken up from the body and pass to vital organs, tissues causing side effects and harmful effects. Practical study says within 24 hours of exposure nanomaterials enter into the deeper layers of pig skin (Ryman Rasmussen et al.,2006). Neonatal toxicity cause severe harm in pregnant women and used for various purposes placenta, endometrium, yolk sac or fetus resultant in oxidative stress and irritation. Further, nanoparticles may induce cytokine production in pregnant women These problems leads to placental damage, delayed neonatal development, fetal deformities, neurotoxicity and reproductive dysfunction in infants.

7. FUTURE DIRECTIONS

In a modern scenario, the center of attention is the applications of nanomaterials but the side effects on health is taken into consideration. The toxic effects such as skin inflammation, skin cancer and genotoxicity arise that need to be thoroughly deliberated. The cosmetics industry was the first to implement nanotechnological principles in product development. More than one thousand registered nanotechnology-based products on the global market in 2009, around 13% were classified as cosmetic products. In the year 2000 and 2010 the use of nanotechnology in dermatological, dental and haircare products intended for improving the outward show of the customer. More than a few other innovations, nanotechnology has the potential to significantly alter dental medicine, healthcare and human life. However, they even have the potential to evoke merits like improved health, higher use of natural resources, and reduced environmental pollution.Latest invention in delivery systems by scientists are currently being applied in the cosmeceuticals process. the conventional drug delivery systems find alternate by the novel systems.

Today, applications of nanotechnology in every facet of life, sunscreens, deodorants to smartphones and medicine. Some of the nanocarriers are liposomes, niosomes, NLC, SLNs, gold nanoparticles, nanoemulsion and nanosomes are used in various cosmeceuticals.The cosmetic formulation is considered as the hottest and promising technology available. Manufacturers of cosmetics use chemicals of nanoscale dimensions to deliver improved UV prevention, deeper skin penetration, long-lasting effects, enhanced colour, finish quality and many other benefits. Many researches carried out to determine the toxicity and safety concern of the nanomaterials related to health and developing green cosmetics from renewable plant and bacterial sources with decreased toxicity, transparent manufacturing processes, biodegradable formulations. Nanoproducts are fabricated by considering their value and health of the customers are raised. For clinical trials approval of cosmeceuticals are not mandatory. Finally, rigorous laws imposed on the regulation and safety of usuage cosmeceuticals and nanoparticles.

8. CONCLUSION

This chapter discussed about the overview of Nanocosmetics and Nanocosmeceuticals applications in cosmetic industries. It depicts the Classification of Nanomaterials in Cosmetics Preparation, nanoformulations/nanocarriers in cosmetics industry, benefits, potential health, future direction and safety . Further, it confers on general classes of Nanocosmeceuticals and challenges . In order to meet the customers demand which leads to the development of formulations in cosmetics with better performance.

REFERENCES

Akturk, O., Kismet, K., Yasti, A. C., Kuru, S., Duymus, M. E., Kaya, F., Caydere, M., Hucumenoglu, S., & Keskin, D. (2016). Collagen/gold nanoparticle nanocomposites: A potential skin wound healing biomaterial. *Journal of Biomaterials Applications*, *31*(2), 283–301. doi:10.1177/0885328216644536 PMID:27095659

Alka, L., Anurag, V., Himanshi, J., Niti, Y., & Neha, K. (2014). Nanotechnology-Based Cosmeceuticals. *ISRN Dermatology*, *2014*, 1–14. doi:10.1155/2014/843687 PMID:24963412

Arora, N., Agarwal, S., & Murthy, R. S. R. (2012). Latest technology advances in cosmaceuticals. *International Journal of Pharmaceutical Sciences and Drug Research*, *4*(3), 168–182.

Borowska, S., & Brzóska, M. M. (2015). Metals in cosmetics: Implications for human health. *Journal of Applied Toxicology*, *35*(6), 551–572. doi:10.1002/jat.3129 PMID:25809475

Brown, J. S., Zeman, K. L., & Bennett, W. D. (2002). Ultrafine particle deposition and clearance in the healthy and obstructed lung. *American Journal of Respiratory and Critical Care Medicine*, *166*(9), 1240–1247. doi:10.1164/rccm.200205-399OC PMID:12403694

Dhapte-Pawar, V.; Kadam, S.; Saptarsi, S.; Kenjale, P.P.(2020), Nanocosmeceuticals: Facets and aspects. *Futur. Sci. OA*.

Dingler, A., & Gohla, S. (2002). Production of solid lipid nanoparticles (SLN): Scaling up feasibilities. *Journal of Microencapsulation*, *19*(1), 11–16. doi:10.1080/02652040010018056 PMID:11811752

Dureja, H., Kaushik, D., Gupta, M., Kumar, V., & Lather, V. (2005). Cosmeceuticals: An emerging concept. *Indian Journal of Pharmacology*, *37*(3), 155–159. doi:10.4103/0253-7613.16211

Fatima, M., Monawwar, S., Mohapatra, S., Alex, T. S., Ahmed, A., Taleuzzaman, M., Ali, A., Ansari, M. J., Mirza, M. A., & Iqbal, Z. (2021). In silico drug screening based development of novel formulations for onychomycosis management. *Gels (Basel, Switzerland)*, *2021*(7), 221. doi:10.3390/gels7040221 PMID:34842710

Fox, C. (1998). *Cosmetic and Pharmaceutical Vehicles: Skin Care, Hair Care, Makeup and Sunscreens* (Vol. 113). Cosmetics & Toiletries.

Fruchon, S., & Poupot, R. (2017). Pro-inflammatory versus anti-inflammatory effects of dendrimers: The two faces of immunomodulatory nanoparticles. *Nanomaterials (Basel, Switzerland)*, *7*(9), 251. doi:10.3390/nano7090251 PMID:28862693

Ge, X., Wei, M., He, S., & Yuan, W. E. (2019). Advances of non-ionic surfactant vesicles (niosomes) and their application in drug delivery. *Pharmaceutics*, *11*(2), 55. doi:10.3390/pharmaceutics11020055 PMID:30700021

Glaser, D. A. (2004). Anti-aging products and cosmeceuticals. *Facial Plastic Surgery Clinics of North America*, *12*(3), 363–372. doi:10.1016/j.fsc.2004.03.004 PMID:15261173

Gunasekaran, K., Suthanthira Vanitha, N., Radhika, K., & Suresh, P. (2022). Nanomedicine in healthcare: Impact and challenges for future generation. Research Gate.

Hope, M. J., & Kitson, C. N. (1993). Liposomes: A perspective for dermatologists. *Dermatologic Clinics*, *11*(1), 143–154. doi:10.1016/S0733-8635(18)30291-2 PMID:8435909

Hu, Z., Liao, M., & Chen, Y. (2012). A novel preparation method for silicone oil nanoemulsions and its application for coating hair with silicone. *International Journal of Nanomedicine*, *7*, 5719–5724. PMID:23166436

Joo, H. H., Lee, H. Y., & Kim, J. C. (2009). Stability, release property and skin penetration of stearic acid nanoparticles. *Molecular Crystals and Liquid Crystals (Philadelphia, Pa.)*, *508*(1), 137–149. doi:10.1080/15421400903060466

Karim, K. M. (2010). A future of targeted drug delivery systems. *Journal of Advanced Pharmaceutical Technology & Research*, *1*(4), 374–380. doi:10.4103/0110-5558.76435 PMID:22247876

Katz, L. M., Dewan, K., & Bronaugh, R. L. (2015). Nanotechnology in cosmetics. *Food and Chemical Toxicology*, *85*, 127–137. doi:10.1016/j.fct.2015.06.020 PMID:26159063

Kaul, S., Gulati, N., Verma, D., Mukherjee, S., & Nagaich, U. (2018). Role of Nanotechnology in Cosmeceuticals: *A Review of Recent Advances. Journal of Pharmaceutics*, *2018*, 3420204. doi:10.1155/2018/3420204 PMID:29785318

Keck, C. M., & Muller, R. H. (2006). Drug Nanocrystals of Poorly Soluble Drugs Produced by High Pressure Homogeny. *European Journal of Pharmaceutics and Biopharmaceutics*, *62*(1), 3–16. doi:10.1016/j.ejpb.2005.05.009 PMID:16129588

Khan, R., & Irchhaiya, R. (2016). Niosomes: A Potential Tool for Novel Drug Delivery. *Journal of Pharmaceutical Investigation*, *46*(3), 195–204. doi:10.100740005-016-0249-9

Klajnert, B., & Bryszewska, M. (2001). Dendrimers: Properties and Applications. *Acta Biochimica Polonica*, *48*(1), 199–208. doi:10.18388/abp.2001_5127 PMID:11440170

Kocic, H., Stankovic, M., Tirant, M., Lotti, T., & Arsic, I. (2020). Favorable effect of creams with skimmed donkey milk encapsulated in nanoliposomes on skin physiology. *Dermatology and Therapy*, *2020*(33), e13511. PMID:32372458

Kokura, S., Handa, O., Takagi, T., Ishikawa, T., Naito, Y., & Yoshikawa, T. (2010). Silver nanoparticles as a safe preservative for use in cosmetics. *Nanomedicine; Nanotechnology, Biology, and Medicine*, *6*(4), 570–574. doi:10.1016/j.nano.2009.12.002 PMID:20060498

Köpke, D., & Pyo, S. M. (2020). Symurban nanocrystals for advanced anti-pollution skincare. *Cosmetics*, *2020*(7), 17. doi:10.3390/cosmetics7010017

Lee, R., Shenoy, D., & Sheel, R. (2010). Micellar Nanoparticles: Applications for Topical and Passive Transdermal Drug Delivery. In *Handbook of Non-Invasive Drug Delivery Systems*. William Andrew. doi:10.1016/B978-0-8155-2025-2.10002-2

Lee, Y. J., & Nam, G. W. (2020). Sunscreen boosting effect by solid lipid nanoparticles-loaded fucoxanthin formulation. *Cosmetics*, *2020*(7), 14. doi:10.3390/cosmetics7010014

Li, N., Sioutas, C., Cho, A., Schmitz, D., Misra, C., Sempf, J., Wang, M., Oberley, T., Froines, J., & Nel, A. (2003). Ultrafine particulate pollutants induce oxidative stress and mitochondrial damage. *Environmental Health Perspectives*, *111*(4), 455–460. doi:10.1289/ehp.6000 PMID:12676598

Mu, L., & Sprando, R. L. (2010). Application of nanotechnology in cosmetics. *Pharmaceutical Research*, *27*(8), 1746–1749. doi:10.100711095-010-0139-1 PMID:20407919

Muller, H. R., Shegokar, R., & Keck, M. C. (2011). 20 years of lipid nanoparticles (SLN & NLC): Present state of development & industrial applications. *Current Drug Discovery Technologies*, *8*, 207–227. doi:10.2174/157016311796799062 PMID:21291409

Müller, R. H., Alexiev, U., Sinambela, P., & Keck, C. M. (2016). *Nanostructured lipid carriers (NLC): The second generation of solid lipid nanoparticles Percutaneous Penetration Enhancers chemical Methods in Penetration Enhancement*. Springer., doi:10.1007/978-3-662-47862-2_11

Müller, R. H., Petersen, R. D., Hommoss, A., & Pardeike, J. (2007). Nanostructured lipid carriers (NLC) in cosmetic dermal products. *Advanced Drug Delivery Reviews*, *59*(6), 522–530. doi:10.1016/j.addr.2007.04.012 PMID:17602783

Nasir, A. (2010). Nanotechnology and dermatology: Part II-risks of nanotechnology. *Clinics in Dermatology*, *28*(5), 581–588. doi:10.1016/j.clindermatol.2009.06.006 PMID:20797523

Nohynek, G. J., Lademann, J., Ribaud, C., & Roberts, M. S. (2007). Grey Goo on the skin? Nanotechnology, cosmetic and sunscreen safety. *Critical Reviews in Toxicology*, *37*(3), 251–277. doi:10.1080/10408440601177780 PMID:17453934

Ostiguy, C., Roberge, B., Woods, C., & Soucy, B. (2010), Engineered Nanoparticles: Current Knowledge about OHS Risks and Prevention Measures. Studies and Research Projects. Robert-Sauvé Research Institute for Occupational Health and Safety (IRSST).

Pardeike, J., Hommoss, A., & Müller, R. H. (2009). Lipid nanoparticles (SLN, NLC) in cosmetic and pharmaceutical dermal products. *International Journal of Pharmaceutics, 366*(1-2), 170–184. doi:10.1016/j.ijpharm.2008.10.003 PMID:18992314

Pereira, L., Dias, N., Carvalho, J., Fernandes, S., Santos, C., & Lima, N. (2014). Synthesis, characterization and antifungal activity of chemically and fungal-produced silver nanoparticles against Trichophyton rubrum. *Journal of Applied Microbiology, 117*(6), 1601–1613. doi:10.1111/jam.12652 PMID:25234047

Petersen, R. (2010). *Nanocrystals for Use in Topical Cosmetic Formulations and Method of Production Thereof.* US Patent US 20100047297A1.

Poletto, F. S., Beck, R. C. R., Guterres, S. S., & Pohlmann, A. R. (2011). Polymeric Nanocapsule: Concepts and Applications. In R. Beck, S. Guterres, & A. Pohlmann (Eds.), *Nanocosmetics and Nanomedicines: New Approaches for Skin Care* (pp. 47–51). Springer. doi:10.1007/978-3-642-19792-5_3

Purohit, D. K., Nandgude, T. D., & Poddar, S. S. (2016). Nano-lipid carriers for topical application: Current scenario. *Asian Journal of Pharmaceutics, 9*(5), 544–553.

Bhupendra, G., Prajapati Niklesh, K., Manan, M., Rakesh, P. P. (2013). Topical Liposomes in Drug Delivery. *International Journal of Pavement Research and Technology, 4*(1), 39–44.

Rosen, J., Landriscina, A., & Friedman, A. (2015). Nanotechnology Based Cosmetics for Hair Care. *Cosmetics, 2*(4), 211–224. doi:10.3390/cosmetics2030211

Ryman-Rasmussen, J. P., Riviere, J. E., & Monteiro-Riviere, N. A. (2006). Penetration of intact skin by quantum dots with diverse physicochemical properties. *Toxicological Sciences, 91*(1), 159–165. doi:10.1093/toxsci/kfj122 PMID:16443688

Schulte, P., Geraci, C., Zumwalde, R., Hoover, M., & Kuempe, E. (2008). Occupational risk management of engineered nanoparticles. *Journal of Occupational and Environmental Hygiene, 5*(4), 239–249. doi:10.1080/15459620801907840 PMID:18260001

Shailesh, L., Patwekar Snehal, R., Ashwini, P., Manoj, S., & Arvind, B. (2016). Nanostructured lipid carriers in stability improvement forcosmetic nanoparticles. *International Journal of Pharmacy and Pharmaceutical Research, 6*(1), 168–180.

Sharma, S. (2012). Cosmeceuticals and herbal drugs: Practical uses. *International Journal of Pharmaceutical Sciences and Research*, *3*(1), 59–65.

Simonetti, N., Simonetti, G., Bougnol, F., & Scalzo, M. (1992). Electrochemical Ag+ for preservative use. *Applied and Environmental Microbiology*, *58*(12), 3834–3836. doi:10.1128/aem.58.12.3834-3836.1992 PMID:1476427

Smijs, T. G., & Pavel, S. (2011). Titanium dioxide and zinc oxide nanoparticles in sunscreens: Focus on their safety and efectiveness. *Nanotechnology, Science and Applications*, *4*(1), 95–112. doi:10.2147/NSA.S19419 PMID:24198489

Srinivas, K. (2016). The current role of nanomaterials in cosmetics. *Journal of Chemical and Pharmaceutical Research*, *8*(5), 906–914.

Tripathy, S., & Dureja, H. (2015). Cosmetics: Regulatory Scenario in USA, EU and India. *J. Pharm. Technol. Res. Manag*, *3*(2), 127–139. doi:10.15415/jptrm.2015.32010

Tripura, P., & Anushree, H. (2017). Novel delivery systems: Current trend in cosmetic industry. *European Journal Pharmaceutical and Medical Research*, *4*(8), 617–627.

Westesen, K., Bunjes, H., & Koch, M. H. J. (1997). Physicochemical characterization of lipid nanoparticles and evaluation of their drug loading capacity and sustained release potential. *Journal of Controlled Release*, *48*(2-3), 223–236. doi:10.1016/S0168-3659(97)00046-1

Wissing, S. A., & Müller, R. H. (2003). Cosmetic applications for solid lipid nanoparticles (SLN). *International Journal of Pharmaceutics*, *254*(1), 65–68. doi:10.1016/S0378-5173(02)00684-1 PMID:12615411

Wu, X., & Guy, R. H. (2009). Applications of nanoparticles in topical drug delivery and in cosmetics. *Journal of Drug Delivery Science and Technology*, *19*(6), 371–384. doi:10.1016/S1773-2247(09)50080-9

Yapar, E. A., & Inal, O. (2012). Nanomatrials and cosmetics. *Journal of Pharmacy of Istanbul University*, *42*(1), 43–70.

Yeo, P. L., Lim, C. L., Chye, S. M., Ling, A. P., & Kand Koh, R. Y. (2018). Niosomes: A review of their structure, properties, methods of preparation,and medical applications. *Asian Biomedicine*, *11*(4), 301–314. doi:10.1515/abm-2018-0002

APPENDIX: ABBREVIATIONS

USFDA: US Food and Drug Administration
NLCs: Nanostructured Lipid Carriers
DNA: DeoxyriboNucleic Acid
RNA: RiboNucleic Acid
ZnO: Zinc Oxide
TiO2: Titanium Dioxide
TDD: Transdermal Drug Delivery
SLN: Solid Lipid Nanoparticles
UV: Ultraviolet
NIOHS: National Institute of Occupational Health and Safety
MBBT: Methylene Bis-Benzo Triazolyl Tetramethyl Butylphenol

Chapter 7
Internet of Things and Artificial Intelligence-Integrated Nano-Materials Research

Nalla. Bhanu Teja
Aditya College of Engineering, India

M. Sudhakar
Sri Sairam Engineering College, India

G. Rathinasabapathi
https://orcid.org/0000-0003-2619-466X
Panimalar Engineering College, India

M. Anusuya
Indra Ganesan College of Engineering, India

Mimi Akash Pateria
https://orcid.org/0000-0002-8283-1769
Shri Shankaracharya Technical Campus, India

ABSTRACT

This chapter presents a system design and architecture that incorporates AI, IoT, and ML-based solutions for data collection, analysis, and interpretation in the nano-materials industry. It consists of three components: data acquisition, data analytics, and decision-making. The system incorporates robust security measures, such as authentication, access control, encryption, and Industry 4.0 principles, to ensure the security and privacy of data. The proposed system was used to collect, analyze, and interpret data from the production, resulting in improved quality, increased efficiency, and reduced waste. AI, IoT, and ML-based solutions have the potential to revolutionize nano-materials research and production, improving productivity, reducing costs, and improving quality.

DOI: 10.4018/978-1-6684-9135-5.ch007

INTRODUCTION

Nano-materials are materials that have unique physical, chemical, and mechanical properties due to their nanoscale size. These materials are typically composed of particles that are smaller than 100 nanometers in size, and they can be engineered to have specific properties for a variety of applications. Nano-materials can be made from a range of materials, including metals, polymers, ceramics, and composites. They have a large surface area to volume ratio, which can lead to increased reactivity, strength, and other unique properties compared to their bulk counterparts(Wu et al., 2020). Some of the properties that are unique to nano-materials include:

- **Optical properties:** Nano-materials can exhibit unique optical properties, such as color changes and fluorescence, due to the quantum confinement of electrons in their structure.
- **Mechanical properties:** Nano-materials can have high strength and toughness due to their small size and unique structure.
- **Electrical properties:** Nano-materials can have different electrical properties than their bulk counterparts due to the quantum confinement of electrons.
- **Chemical properties:** Nano-materials can have increased reactivity due to their large surface area, making them useful for catalysis and other chemical reactions.

Due to their unique properties, nano-materials have a wide range of applications in fields such as electronics, energy, medicine, and environmental science. Some examples of nano-materials include carbon nanotubes, quantum dots, nanoparticles, and nanofibers. However, the use of nano-materials also raises concerns about their potential environmental and health impacts, and research is ongoing to understand these issues and ensure safe use and disposal of these materials.

The Internet of Things (IoT) is a network of connected devices that can communicate with each other and share data. These devices can include sensors, smartphones, wearable devices, and other objects that are embedded with electronics, software, and network connectivity. IoT technology enables devices to gather and transmit data to cloud-based servers or other devices in real-time, allowing for remote monitoring and control of physical systems(Jeevanantham et al., 2023; Sahal et al., n.d.).

Artificial Intelligence (AI) is a branch of computer science that focuses on creating intelligent machines that can perform tasks that typically require human intelligence, such as learning, reasoning, problem-solving, and decision-making. AI technology can be divided into two categories: narrow or weak AI and general or strong AI. Narrow AI is designed to perform a specific task, such as image recognition or speech-to-text conversion. General AI, on the other hand, aims to replicate human-level intelligence across a wide range of tasks.

Optimization technologies refer to methods and algorithms used to find the best solution to a given problem, often involving maximizing or minimizing an objective function subject to certain constraints. Optimization techniques are widely used in a variety of fields, including engineering, economics, finance, and computer science. Optimization techniques can include linear programming, nonlinear programming, integer programming, dynamic programming, and heuristic algorithms(Gardašević et al., 2017).

Together, IoT, AI, and Optimization technologies can be used to create powerful solutions for a variety of applications. For example, IoT can be used to collect data from sensors in real-time, AI can be used to analyze the data and make predictions, and Optimization can be used to find the best course of action based on the analysis. This approach can be applied to many fields, including transportation, energy management, healthcare, and environmental monitoring, among others(Senthil et al., 2023; Vanitha et al., 2023).

IoT-based solutions can be applied in various stages of nano-materials research and production, from material synthesis to characterization and analysis. Here are some examples of IoT-based solutions for nano-materials research and production:

- **Smart synthesis:** IoT sensors can be used to monitor various parameters during the synthesis process, such as temperature, pressure, and gas flow. This data can be collected and analyzed in real-time to optimize the synthesis process and ensure consistent product quality.
- **Real-time monitoring:** IoT sensors can be used to monitor various parameters during the production process, such as temperature, humidity, and pressure. This data can be collected and analyzed in real-time to optimize the production process and ensure consistent product quality.
- **Quality control:** IoT sensors can be used to monitor product quality during production and after it's been produced. This data can be analyzed to identify any quality issues and improve the Hence quality of the product.
- **Supply chain management:** IoT sensors can be used to track and monitor raw materials and finished products as they move through the supply chain. This data can be used to optimize logistics and reduce waste and costs.
- **Predictive maintenance:** IoT sensors can be used to monitor equipment health and predict maintenance needs. This can help prevent equipment failures and improve production efficiency.
- **Remote monitoring:** IoT solutions can enable remote monitoring of nano-materials production facilities, allowing for real-time monitoring and control of the production process from anywhere in the world.

Thus, IoT-based solutions can help improve the efficiency, consistency, and quality of nano-materials production while also reducing costs and waste. The use of IoT in this field is still in its early stages, but it has the potential to revolutionize the way nano-materials are produced and used in a wide range of applications.

AI-BASED SOLUTIONS FOR NANO-MATERIALS RESEARCH AND PRODUCTION

AI-based solutions can be applied in various stages of nano-materials research and production, from material synthesis to characterization and analysis. Here are some examples of AI-based solutions for nano-materials research and production(Singh et al., 2020):

- **Predictive modeling:** AI algorithms can be trained to predict the properties and behaviors of nano-materials based on their chemical composition, structure, and synthesis conditions. This can help guide the development of new materials with desired properties and reduce the need for trial-and-error experiments.
- **Intelligent optimization:** AI algorithms can be used to optimize the synthesis and production processes for nano-materials, taking into account various parameters such as temperature, pressure, and reactant concentrations. This can help improve efficiency, reduce waste, and improve product quality.
- **Image analysis:** AI algorithms can be used to analyze images of nano-materials obtained through microscopy and other imaging techniques, allowing for automated characterization and analysis. This can help speed up the characterization process and reduce the potential for human error.
- **Virtual screening:** AI algorithms can be used to screen large databases of potential nano-materials for specific properties or applications. This can help identify promising candidates for further study and reduce the need for expensive and time-consuming experiments.
- **Intelligent process control:** AI algorithms can be used to monitor and control the production process in real-time, adjusting various parameters to maintain optimal conditions and improve product quality.

Hence, AI-based solutions have the potential to revolutionize nano-materials research and production by enabling faster, more efficient, and more accurate materials development and production. The use of AI in this field is still in its early stages, but it is expected to play an increasingly important role in the development of new materials for a wide range of applications.

OPTIMIZATION TECHNIQUES FOR NANO-MATERIALS RESEARCH AND PRODUCTION

Optimization techniques can be applied to various stages of nano-materials research and production, from material synthesis to characterization and analysis. Here are some examples of optimization techniques for nano-materials research and production(Yang et al., 2021)(Boopathi, Arigela, et al., 2023; Jeevanantham et al., 2023; Saravanan et al., 2022):

- **Design of experiments (DoE):** DoE is a statistical method that can be used to systematically evaluate the effect of multiple variables on a process or outcome. In nano-materials research and production, DoE can be used to optimize the synthesis process by identifying the optimal combination of reactant concentrations, temperature, and other parameters.
- **Response surface methodology (RSM):** RSM is a statistical method that can be used to model the relationship between multiple variables and a response or outcome. In nano-materials research and production, RSM can be used to optimize the synthesis process by identifying the optimal conditions for maximum yield or desired properties.
- **Genetic algorithms (GA):** GA is a type of optimization algorithm that is inspired by natural selection. In nano-materials research and production, GA can be used to search for the optimal combination of synthesis parameters by simulating the process of natural selection(Boopathi, 2022e; Boopathi & Sivakumar, 2013, 2016).
- **Artificial neural networks (ANN):** ANN is a type of machine learning algorithm that can be used to model complex relationships between input and output variables. In nano-materials research and production, ANN can be used to optimize the synthesis process by predicting the outcome of different synthesis conditions.
- **Multi-objective optimization:** Multi-objective optimization is a technique that can be used to optimize multiple objectives simultaneously. In nano-materials research and production, multi-objective optimization can be used to optimize the synthesis process for multiple criteria such as yield, purity, and desired properties.

Hence, optimization techniques can help improve the efficiency, consistency, and quality of nano-materials production while also reducing costs and waste. The use of optimization techniques in this field is still in its early stages, but it has the potential to revolutionize the way nano-materials are produced and used in a wide range of applications.

SYSTEM DESIGN AND ARCHITECTURE

The system design and architecture for IoT-based, AI-based, and optimization-based nano-materials research and production can vary depending on the specific application and requirements (Figure 1). However, here are some general considerations for each technology(Daveiga & Ferreira, 2005):

Figure 1. AI, IoT techniques: General implementation procedure

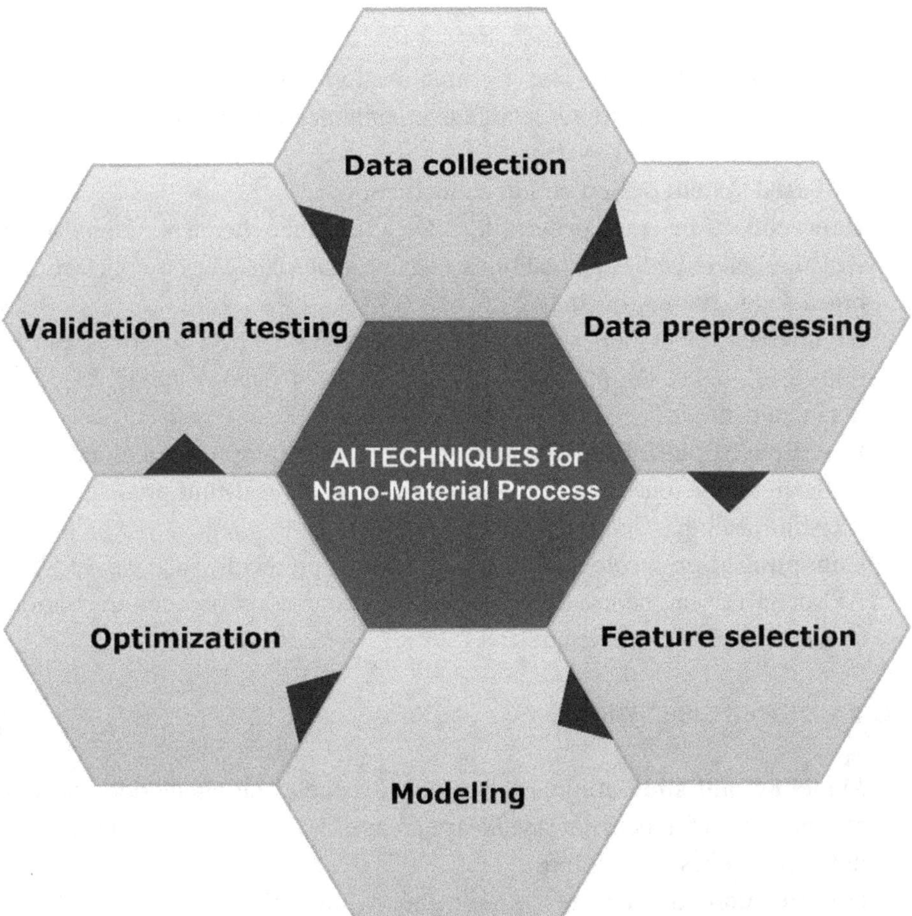

IoT-Based System Design and Architecture

- **Sensor network:** A network of sensors can be deployed to collect data on various parameters such as temperature, pressure, and reactant concentrations during the synthesis process.
- **Data acquisition and processing:** The data collected from the sensors can be processed and analyzed in real-time to monitor the synthesis process and identify any issues.
- **Cloud-based platform:** The data can be sent to a cloud-based platform for storage, analysis, and visualization. This can enable remote monitoring and control of the synthesis process.
- **Communication protocols:** Communication protocols such as MQTT or HTTP can be used to facilitate communication between the various components of the IoT system.
- **AI-based system design and architecture:**
- **Data collection and processing:** Data on the chemical composition, structure, and synthesis conditions of nano-materials can be collected and processed to train AI models.
- **AI models:** AI models such as neural networks or decision trees can be trained to predict the properties and behaviors of nano-materials based on the input data.
- **Cloud-based platform:** The AI models can be deployed on a cloud-based platform for inference and analysis. This can enable real-time predictions and decision-making.
- **Communication protocols:** Communication protocols such as RESTful API or gRPC can be used to facilitate communication between the various components of the AI system.

Optimization-Based System Design and Architecture

- **Modeling and simulation:** Mathematical models and simulation software can be used to model the synthesis process and predict the outcome of different synthesis conditions.
- **Optimization algorithms:** Optimization algorithms such as genetic algorithms or particle swarm optimization can be used to search for the optimal combination of synthesis parameters.
- **Cloud-based platform:** The optimization algorithms can be deployed on a cloud-based platform for optimization and analysis. This can enable real-time optimization and decision-making.

- **Communication protocols:** Communication protocols such as MQTT or HTTP can be used to facilitate communication between the various components of the optimization system.

Hence, the system design and architecture for IoT-based, AI-based, and optimization-based nano-materials research and production can be complex and require integration of multiple technologies. However, with careful planning and design, these technologies can enable faster, more efficient, and more accurate materials development and production.

Sensors and Devices

There are various sensors and devices that can be used for IoT-based nano-materials research and production. Here are some examples:

- **Temperature sensors:** Temperature sensors can be used to monitor the temperature during the synthesis process. They can help ensure that the temperature is maintained within the desired range to achieve the desired properties of the nano-material.
- **Pressure sensors:** Pressure sensors can be used to monitor the pressure during the synthesis process. They can help ensure that the pressure is maintained within the desired range to achieve the desired properties of the nano-material.
- **pH sensors:** pH sensors can be used to monitor the pH during the synthesis process. They can help ensure that the pH is maintained within the desired range to achieve the desired properties of the nano-material.
- **Spectrometers:** Spectrometers can be used to analyze the composition and structure of nano-materials. They can provide information on the size, shape, and composition of nano-particles.
- **Microscopes:** Microscopes can be used to visualize the nano-materials and analyze their morphology. They can provide information on the size, shape, and distribution of nano-particles.
- **Gas sensors:** Gas sensors can be used to monitor the presence of gases such as oxygen, nitrogen, and hydrogen during the synthesis process. They can help ensure that the reaction environment is optimized for the desired properties of the nano-material.
- **Flow meters:** Flow meters can be used to measure the flow rate of reactants during the synthesis process. They can help ensure that the reactants are supplied at the desired rate to achieve the desired properties of the nano-material.

- **Actuators:** Actuators can be used to control the synthesis process by adjusting parameters such as temperature, pressure, and flow rate. They can help ensure that the synthesis process is optimized for the desired properties of the nano-material.

Hence, the sensors and devices used for IoT-based nano-materials research and production can provide real-time data and control to optimize the synthesis process and achieve the desired properties of the nano-material.

Data Collection and Analysis

Data collection and analysis are critical components for effective nano-materials research and production using IoT and AI. Here are some key steps involved in this process:

- **Sensor data collection:** IoT sensors are used to collect data on various parameters such as temperature, pressure, pH, gas concentration, and flow rate during the synthesis process. This data is collected in real-time and stored in a database.
- **Data preprocessing:** The raw data collected from the sensors may contain noise or outliers, which need to be removed. Data preprocessing techniques such as filtering, normalization, and outlier detection are used to clean the data.
- **Data integration:** Data from different sensors and sources are integrated to create a unified dataset. This dataset can be used for further analysis and modeling.
- **Data analysis:** Data analysis techniques such as statistical analysis, machine learning, and deep learning are used to analyze the dataset. These techniques can help identify patterns and relationships in the data and extract insights.
- **Modeling and prediction:** Machine learning and deep learning models are used to predict the properties of nano-materials based on the data collected from the sensors. These models can help optimize the synthesis process and achieve the desired properties of the nano-material.
- **Feedback and control:** The predictions and insights from the data analysis and modeling can be used to adjust the synthesis process in real-time. This feedback and control loop can help improve the quality and efficiency of the synthesis process.

Hence, data collection and analysis are critical for effective nano-materials research and production using IoT and AI. They can help optimize the synthesis process, improve the quality of the nano-materials, and reduce the production costs.

Deep Learning Models

Deep Learning is a powerful tool for AI-based nano-materials research and production. Here are some of the popular deep learning models used in this domain(Courville, 2016):

- **Convolutional Neural Networks (CNNs):** CNNs are widely used in image processing and analysis of micrographs to extract morphological features of nano-materials, such as size and shape. They can help identify and classify nano-materials based on their physical characteristics.
- **Recurrent Neural Networks (RNNs):** RNNs are used for time-series analysis and sequence modeling, which are important in tracking and optimizing the synthesis process. They can help predict the behavior of the synthesis process over time and optimize the process for the desired properties of the nano-material.
- **Autoencoders:** Autoencoders are used for unsupervised learning and feature extraction. They can help identify the most important features of the nano-materials from large and complex datasets, and reduce the dimensionality of the data.
- **Generative Adversarial Networks (GANs):** GANs are used for generating new samples of nano-materials based on the learned distribution of the data. They can help explore the potential range of the synthesis process and identify new and unique nano-materials.
- **Deep Reinforcement Learning:** Deep Reinforcement Learning is used to optimize the synthesis process by learning from experience. It involves an agent that interacts with the environment and receives rewards based on its actions. The agent learns to optimize the synthesis process by maximizing the reward over time.

Hence, deep learning models can help extract insights and patterns from large and complex datasets, optimize the synthesis process, and generate new and unique nano-materials for a wide range of applications.

Integration With Existing Nano-Materials Production Systems

Integrating IoT and AI-based solutions with existing nano-materials production systems can be challenging but is important for maximizing the benefits of these technologies. Here are some key considerations for integration:

- **Compatibility:** IoT and AI-based solutions should be compatible with existing systems, including hardware and software. This may require customization or integration with existing systems.
- **Data integration:** Data from existing systems should be integrated with the IoT and AI-based solutions to provide a comprehensive view of the production process. This may involve data mapping and data standardization.
- **Scalability:** IoT and AI-based solutions should be scalable to accommodate the size and complexity of the production system. This may require the use of cloud-based systems or distributed computing.
- **Security:** IoT and AI-based solutions should be secure and protect the data from unauthorized access. This may require the use of encryption, authentication, and access control.
- **User interface:** The user interface of IoT and AI-based solutions should be designed to be user-friendly and intuitive for the operators and engineers of the production system.
- **Training and support:** The operators and engineers of the production system should be trained and provided with ongoing support to effectively use the IoT and AI-based solutions.

Hence, integration with existing nano-materials production systems requires careful planning and consideration of various factors. By effectively integrating IoT and AI-based solutions, it is possible to improve the efficiency, quality, and safety of the production process, leading to significant benefits for the industry.

Security and Privacy

Security and privacy are critical considerations in the implementation of IoT, AI, and Optimization-based solutions in Nano-materials production. Here are some important security and privacy considerations:

- **Data Security:** Data generated by IoT devices, sensors, and AI algorithms should be secured and protected from unauthorized access. This may require the use of encryption, access control, and data backups.

- **Network Security:** The network infrastructure used in IoT and AI-based solutions should be secure and protected from cyber-attacks. This may require the use of firewalls, intrusion detection systems, and network segmentation.
- **Device Security:** IoT devices and sensors used in Nano-materials production should be secured and protected from physical and cyber-attacks. This may require the use of tamper-resistant and tamper-evident devices and secure booting mechanisms.
- **Privacy:** The use of IoT and AI-based solutions in Nano-materials production may involve the collection of personal and sensitive data. Appropriate measures should be taken to ensure privacy, such as anonymization and de-identification.
- **Regulatory Compliance:** The use of IoT and AI-based solutions in Nano-materials production may be subject to regulatory requirements, such as data protection and privacy laws. These requirements should be considered in the design and implementation of the solutions.
- **User Awareness:** The operators and engineers of the Nano-materials production system should be trained and made aware of the security and privacy considerations associated with the use of IoT and AI-based solutions. Hence, security and privacy considerations are essential for the successful implementation of IoT, AI, and Optimization-based solutions in Nano-materials production. By incorporating appropriate security and privacy measures, it is possible to ensure the safety and integrity of the production process and maintain the trust of stakeholders.

Evaluation and Comparisons

When evaluating and comparing IoT, AI, and Optimization-based solutions with traditional Nano-materials production methods, several factors can be considered:

- **Efficiency:** IoT and AI-based solutions can significantly improve the efficiency of Nano-materials production by automating and optimizing various aspects of the production process.
- **Quality:** The use of IoT and AI-based solutions can improve the quality of Nano-materials production by providing real-time monitoring and control of the production process, reducing errors and defects.
- **Cost:** The implementation of IoT and AI-based solutions can involve significant upfront costs, such as hardware, software, and training. However, in the long run, these solutions can result in cost savings through improved efficiency and reduced waste.

- **Safety:** IoT and AI-based solutions can enhance safety in Nano-materials production by reducing the risk of accidents and exposure to hazardous materials.
- **Environmental impact:** IoT and AI-based solutions can help reduce the environmental impact of Nano-materials production by optimizing the use of resources and reducing waste.
- **Flexibility:** IoT and AI-based solutions can provide greater flexibility in Nano-materials production by allowing for real-time adjustments to the production process.

Hence, the use of IoT, AI, and Optimization-based solutions in Nano-materials production can offer significant advantages over traditional methods. While the upfront costs may be higher, the potential for improved efficiency, quality, safety, and environmental impact can lead to significant benefits for the industry.

Future Directions and Potential Improvements

The future of Nano-materials research and production using IoT, AI, and Optimization is promising, and there are several potential improvements and directions that can be explored:

- **Integration of Big Data:** The use of Big Data analytics can provide valuable insights into Nano-materials production by analyzing large datasets generated by IoT devices and sensors.
- **Machine Learning for real-time decision-making:** Machine learning algorithms can be used to analyze data in real-time, allowing for faster decision-making in Nano-materials production.
- **Advanced control systems:** Advanced control systems using AI and optimization can be used to further improve the efficiency and quality of Nano-materials production.
- **Cybersecurity:** Advancements in cybersecurity technologies can help ensure the security and privacy of IoT and AI-based Nano-materials production systems.
- **Smart sensors and devices:** Advances in sensor and device technologies can enhance the accuracy and reliability of data collection in Nano-materials production.
- **Augmented Reality (AR) and Virtual Reality (VR) technologies:** AR and VR technologies can be used to enhance training, simulation, and visualization in Nano-materials production.

- **Collaborative Robotics (Cobots):** Cobots can be used to work alongside human operators, enhancing safety and improving efficiency in Nano-materials production.
- **Autonomous systems:** Autonomous systems can be used to automate various aspects of Nano-materials production, reducing the need for human intervention and improving efficiency.

Hence, the future of Nano-materials research and production using IoT, AI, and Optimization is exciting, and there are several potential improvements and directions that can be explored to further improve the efficiency, quality, and safety of the production process.

EMERGING TRENDS IN NANOPARTICLES AND NANOFLUID INDUSTRY 4.0

Advanced materials for Nano-particles and Nano-fluid: Development and application of advanced materials such as graphene, carbon nanotubes, and metal-organic frameworks in Industry 4.0 Nano-particles and Nano-fluid production(Choi, 1998).

- **Hybrid Nanomaterials:** Integration of different types of Nano-particles and Nano-fluid to form hybrid nanomaterials with improved properties.
- **Additive manufacturing for Nano-particles and Nano-fluid:** The use of 3D printing technology in the production of Nano-particles and Nano-fluid.
- **Quantum computing for Nano-particles and Nano-fluid:** The application of quantum computing in Nano-particles and Nano-fluid research and development for improved efficiency and accuracy.
- **Smart sensors and devices for Nano-particles and Nano-fluid:** The development of smart sensors and devices for monitoring and controlling the production process of Nano-particles and Nano-fluid.
- **Internet of Things (IoT) for Nano-particles and Nano-fluid:** The use of IoT in the production process for real-time monitoring, data collection, and analysis for improved efficiency and accuracy.
- **Big Data and Machine Learning for Nano-particles and Nano-fluid:** The use of Big Data and Machine Learning for data analysis, predictive modeling, and decision-making in Nano-particles and Nano-fluid production.
- **Augmented Reality (AR) and Virtual Reality (VR) for Nano-particles and Nano-fluid:** The use of AR and VR technology in the design and simulation of Nano-particles and Nano-fluid production processes.

- **Cobots and Autonomous Systems for Nano-particles and Nano-fluid:** The use of collaborative robots and autonomous systems in the production process of Nano-particles and Nano-fluid for improved efficiency and safety.
- **Sustainable and Green Nano-particles and Nano-fluid:** The development of sustainable and environmentally friendly methods for the production of Nano-particles and Nano-fluid in Industry 4.0.

Applications of Additive Manufacturing

Additive manufacturing (AM) techniques have shown great potential for the production of Nano-particles and Nano-fluids with improved properties. Here are some possible applications of additive manufacturing for Nano-particles and Nano-fluids(Boopathi, Khare, et al., 2023; Mohanty et al., 2023; Palaniappan et al., 2023):

- **Functionalized Nano-particles and Nano-fluids:** AM can be used to create functionalized Nano-particles and Nano-fluids with specific surface properties, such as increased surface area, enhanced catalytic activity, or improved stability.
- **Complex Nano-structures:** AM can produce complex Nano-structures that are difficult or impossible to create using traditional methods. These structures can have applications in various fields, such as biomedical engineering, electronics, and energy storage.
- **Customized Nano-particles and Nano-fluids:** AM can be used to create customized Nano-particles and Nano-fluids tailored to specific applications. This can result in better performance and increased efficiency in various industries, including aerospace, automotive, and manufacturing.
- **Nano-composites:** AM can produce Nano-composites by combining Nano-particles and Nano-fluids with other materials to create materials with unique properties. These composites can have applications in various fields, such as biomedical engineering, electronics, and energy storage.
- **Microfluidic devices:** AM can be used to create microfluidic devices for the controlled production and manipulation of Nano-particles and Nano-fluids. This can have applications in drug delivery, tissue engineering, and lab-on-a-chip devices.
- **Nanomaterial coatings:** AM can produce coatings of Nano-particles and Nano-fluids with enhanced surface properties, such as improved adhesion, wear resistance, or anti-corrosion. These coatings can have applications in various fields, such as aerospace, automotive, and manufacturing.

Biomedical implants and devices: AM can create biomedical implants and devices with Nano-particles and Nano-fluids for improved performance and biocompatibility. These implants and devices can have applications in orthopedics, dentistry, and drug delivery.

Hence, additive manufacturing has the potential to revolutionize the production of Nano-particles and Nano-fluids and enable the development of novel materials with enhanced properties and applications.

PROCEDURE FOR NANO-MATERIALS MANUFACTURING USING IOT

The manufacturing of Nano-materials using IoT involves the integration of various devices and sensors that collect real-time data from the manufacturing process to optimize production and improve product quality(Fuskele & Sarviya, 2017). The following are some working principles of Nano-materials manufacturing using IoT(Arunprasad RV, 2019; Harikaran et al., 2023; Kumara et al., 2023; Samikannu et al., 2023; Sampath et al., 2022; Vanitha et al., 2023; Vennila et al., 2023):

- **Smart sensors:** Smart sensors are used to monitor the manufacturing process and collect data on factors such as temperature, pressure, humidity, and particle size distribution. The data collected from these sensors can be used to optimize the manufacturing process and ensure the quality of the final product.
- **Machine learning algorithms:** Machine learning algorithms are used to analyze the data collected from the sensors to optimize the manufacturing process. The algorithms can be used to predict the quality of the final product based on real-time data, identify potential production issues, and make recommendations to improve the process.
- **Automated production:** IoT devices can be used to automate the production of Nano-materials, reducing the need for manual labor and increasing the speed and efficiency of the manufacturing process. Automated systems can adjust the manufacturing process in real-time based on the data collected from sensors to ensure optimal production.
- **Cloud-based data management:** IoT devices can be connected to a cloud-based data management system that can store and analyze data from multiple manufacturing processes. This allows for centralized monitoring and analysis of the manufacturing process, enabling manufacturers to identify trends, track performance, and optimize production on a larger scale.

- **Quality control:** IoT devices can be used to monitor the quality of the final product, ensuring that it meets the required specifications. Data collected from sensors can be used to identify any defects or inconsistencies in the manufacturing process, allowing for real-time adjustments to be made to ensure optimal quality.

Hence, the use of IoT in Nano-materials manufacturing allows for a more efficient, automated, and data-driven approach to production. By collecting real-time data and using machine learning algorithms to analyze it, manufacturers can optimize the manufacturing process to ensure the quality of the final product while reducing costs and increasing efficiency.

PROCEDURE FOR NANO-MATERIALS MANUFACTURING USING AI TECHNIQUES

The manufacturing of Nano-materials using AI techniques involves the use of various components and processes that enable the optimization of the production process(Boopathi, Khare, et al., 2023; Boopathi, Siva Kumar, et al., 2023; Boopathi, Venkatesan, et al., 2023; Domakonda et al., 2023; Janardhana et al., 2023; S. et al., 2022; Senthil et al., 2023). The following are the working principles of Nano-materials manufacturing using AI techniques and the function of the components in the schematic diagram:

- **Data collection:** The first step in Nano-materials manufacturing using AI techniques is data collection. This involves the use of various sensors and devices to collect real-time data on factors such as temperature, pressure, humidity, and particle size distribution.
- **Data preprocessing:** The collected data is then preprocessed to remove any noise and inconsistencies. This involves the use of various techniques such as data cleaning, normalization, and feature selection.
- **Machine learning models:** Machine learning models are then used to analyze the preprocessed data to optimize the manufacturing process. This involves the use of various techniques such as classification, regression, clustering, and reinforcement learning.
- **Decision-making:** The output of the machine learning models is used to make decisions about the manufacturing process. This involves the use of various techniques such as decision trees, random forests, and neural networks.
- **Feedback loop:** A feedback loop is established to continuously improve the manufacturing process based on the data collected from the sensors. This

Figure 2. General steps for ML: nano materials process

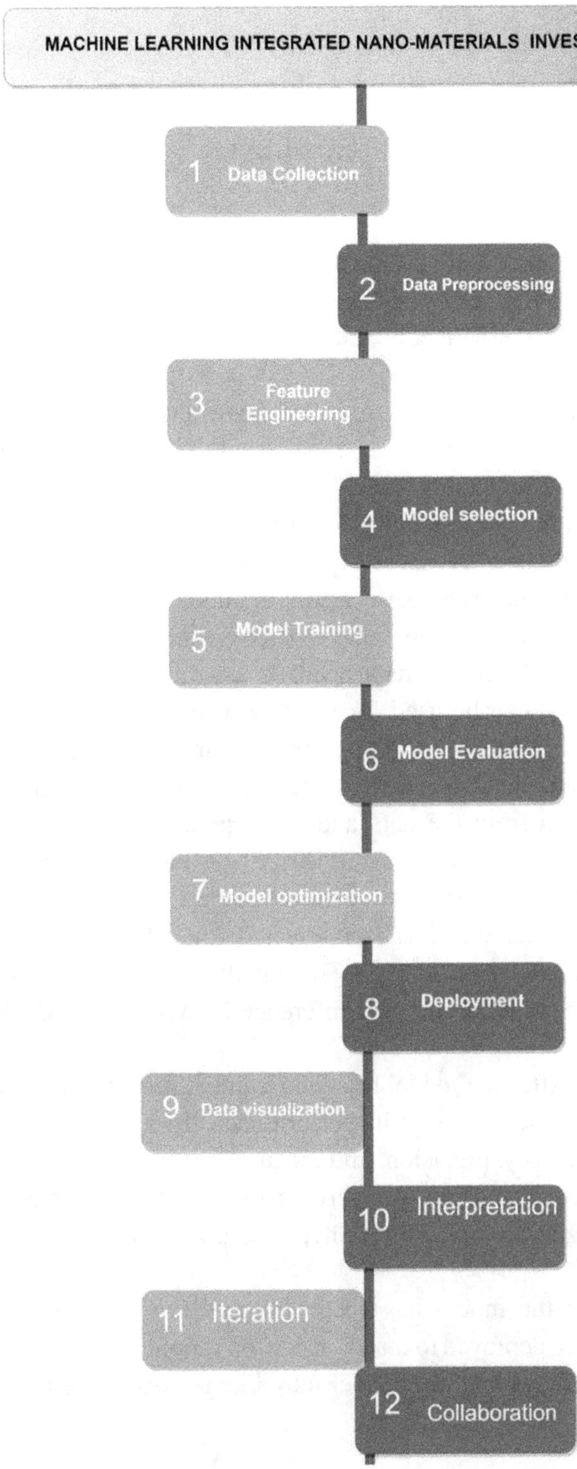

involves the use of various techniques such as online learning and model updating.

MACHINE LEARNING INTEGRATED NANO-MATERIALS INVESTIGATIONS

Here is a general procedure and flow chart for investigating nano-materials using machine learning(Chen et al., 2011; Malakar et al., 2022)(Babu et al., 2023; Boopathi, Arigela, et al., 2023; Jeevanantham et al., 2023) (Figure 2):

- Data collection: The first step is to collect data on the nano-materials of interest. This may involve conducting experiments, simulations, or analyzing existing data sets. The data collected should include information on the material's composition, structure, and properties.
- Data preprocessing: Once the data has been collected, it needs to be cleaned and preprocessed to remove any noise, errors, or outliers. This step may involve techniques such as normalization, feature scaling, and data imputation.
- Feature engineering: Next, the data needs to be transformed into a set of relevant features that can be used for machine learning. This step involves selecting and extracting the most informative features from the raw data.
- Model selection: The next step is to select an appropriate machine learning model that can learn from the data and make predictions about the nano-materials. This may involve using a variety of models, such as neural networks, decision trees, or support vector machines.
- Model training: Once a model has been selected, it needs to be trained on the data using a set of training examples. This involves iteratively adjusting the model's parameters to minimize the difference between its predictions and the actual values.
- Model evaluation: After the model has been trained, it needs to be evaluated on a set of test examples to assess its performance. This involves calculating metrics such as accuracy, precision, and recall.
- Model optimization: If the model's performance is not satisfactory, it may need to be optimized by adjusting its hyperparameters or using a different model altogether.
- Deployment: Once the model has been optimized and its performance is satisfactory, it can be deployed to make predictions about new nano-materials. This may involve integrating the model into a larger system or using it as a standalone tool.

- Data visualization: Data visualization is an essential step to help understand the data, identify trends, and communicate the results of the analysis effectively. This step involves creating visual representations of the data, such as scatter plots, heat maps, and histograms.
- Interpretation: Once the model has made predictions, it's essential to interpret the results and understand what they mean in the context of the nano-materials. This may involve analyzing the model's predictions in combination with domain-specific knowledge and expertise.
- Iteration: The process of investigating nano-materials using machine learning is iterative, and the steps outlined above may need to be repeated several times to refine the analysis and improve the model's performance.
- Collaboration: Investigating nano-materials using machine learning often involves interdisciplinary teams, including material scientists, data scientists, and domain experts. Collaboration and communication between these different groups are essential for successful analysis.

In summary, investigating nano-materials using machine learning involves collecting and preprocessing data, engineering relevant features, selecting and training appropriate models, evaluating and optimizing the models, visualizing the results, interpreting the findings, and collaborating with interdisciplinary teams. The process is iterative, and the steps may need to be repeated several times to refine the analysis and improve the model's performance.

CASE STUDY: THE OPTIMIZATION OF NANO-PARTICLE QUALITY

It involves the use of various parameters such as particle size, shape, and surface area. One way to optimize these parameters is through the use of the NSGA-II algorithm in Matlab(Patil et al., 2023). The following is a flow chart for the parameter optimization process (Figure 3):

- **Define the objective function:** The first step is to define the objective function that needs to be optimized. In this case, the objective function is to optimize the quality of Nano-particles based on parameters such as particle size, shape, and surface area(Boopathi et al., 2021; Boopathi, 2022f, 2022d, 2022a; Boopathi, Haribalaji, et al., 2022; Boopathi, Sureskumar, et al., 2022; Boopathi, Thillaivanan, et al., 2022).

- **Define the decision variables:** The decision variables are the parameters that will be optimized using the NSGA-II algorithm. These include parameters such as particle size, shape, and surface area.
- **Set up the NSGA-II algorithm:** The NSGA-II algorithm is set up in Matlab, which involves specifying the number of generations, population size, and other algorithm-specific parameters(Boopathi, 2021, 2022b, 2022c, 2022e).
- **Run the NSGA-II algorithm:** The NSGA-II algorithm is run in Matlab, which involves the evaluation of the objective function for each individual in the population and the selection of the best individuals based on their fitness.
- **Evaluate the results:** The results of the NSGA-II algorithm are evaluated to determine the optimal values for the decision variables.
- **Validate the results:** The optimal values for the decision variables are validated to ensure that they are consistent with the desired quality of Nano-particles.

Chemical Vapor Deposition (CVD) is a commonly used technique for depositing thin films of Nano-materials. CVD involves the reaction of precursor gases on a substrate surface at elevated temperatures to produce the desired Nano-materials.

Figure 3. Flow chart NSGA-II optimization procedure for nano material process

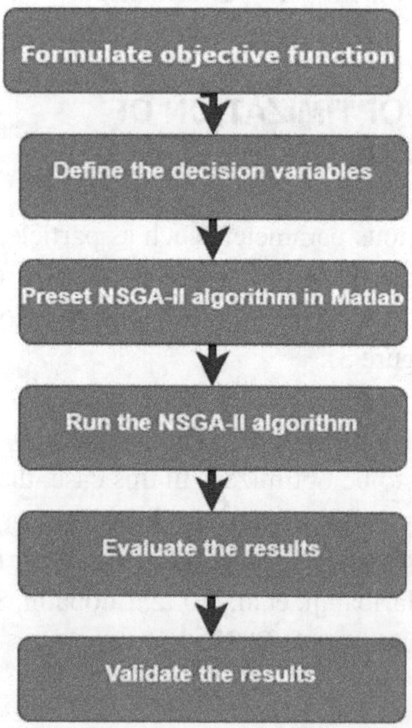

IoT technology can be used to control and optimize the CVD process, resulting in more efficient and consistent production of Nano-materials. The following are the constructional features and working principle of a typical CVD system.

Constructional Features

- **Reaction Chamber:** The reaction chamber is typically made of quartz or ceramic material and is designed to withstand high temperatures and corrosive gases.
- **Heating Element:** The heating element is used to heat the substrate and precursor gases to the desired temperature for the CVD process.
- **Gas Inlet and Outlet:** The gas inlet and outlet are used to introduce precursor gases into the reaction chamber and to remove by-products from the chamber.
- **Vacuum Pump:** A vacuum pump is used to evacuate the chamber and create a low-pressure environment for the CVD process.
- **Substrate Holder:** The substrate holder is used to hold the substrate in place during the CVD process.

Working Principle

- The CVD process typically involves the following steps:
- The precursor gases are introduced into the reaction chamber.
- The substrate is heated to the desired temperature using the heating element.
- The precursor gases react on the substrate surface, depositing a thin film of Nano-materials.
- By-products are removed from the chamber using the gas outlet.
- The process is repeated until the desired thickness of the Nano-materials is achieved.

IoT technology can be used to monitor and control the Chemical Vapor Deposition (CVD) process. By integrating sensors, actuators, and other IoT devices into the CVD system, real-time data can be collected and analyzed to optimize the process and improve the quality of the produced materials.

For example, IoT sensors can be placed at various points in the CVD system to measure parameters such as temperature, pressure, gas flow rate, and deposition rate. This data can be collected and analyzed using IoT platforms and machine learning algorithms to predict and control the CVD process.

In addition, IoT-enabled actuators can be used to adjust the system parameters in real-time to maintain optimal conditions and prevent errors. This can lead to increased efficiency, reduced waste, and improved product quality.

SUMMARY

- The integration of AI, IoT, and ML in the research and production of nano-materials can improve efficiency, accuracy, and productivity. This paper presents a system design and architecture that uses AI, IoT, and ML-based solutions for data collection, analysis, and interpretation in the nano-materials industry. The proposed system comprises three main components: data acquisition, data analytics, and decision-making, incorporating Industry 4.0 principles. Robust security measures are implemented to ensure data privacy and security.
- A case study involving the production of carbon nanotubes (CNTs) demonstrates the effectiveness of the proposed system. The system was used to collect, analyze, and interpret data from the CNT production process, resulting in improved quality, increased efficiency, and reduced waste.
- Overall, the integration of AI, IoT, and ML-based solutions has significant potential to revolutionize the research and production of nano-materials, improving productivity, reducing costs, and enhancing the quality of nano-materials.

REFERENCES

Arunprasad, R. V. (2019). Alternate Refrigerants for Minimization Environmental Impacts: A Review. In *Advances in Engineering Technology* (p. 69). AkiNik Publications New Delhi. https://www.researchgate.net/publication/340939609

Babu, B. S., Kamalakannan, J., Meenatchi, N., M, S. K. S., S, K., & Boopathi, S. (2023). Economic impacts and reliability evaluation of battery by adopting Electric Vehicle. *IEEE Explore*, (pp. 1–6). IEEE. doi:10.1109/ICPECTS56089.2022.10046786

Boopathi, S. (2021). Improving of Green Sand-Mould Quality using Taguchi Technique. *Journal of Engineering Research*. doi:10.36909/jer.14079

Boopathi, S. (2022a). An experimental investigation of Quench Polish Quench (QPQ) coating on AISI 4150 steel. *Engineering Research Express*, *4*(4), 45009. doi:10.1088/2631-8695/ac9ddd

Boopathi, S. (2022b). An Extensive Review on Sustainable Developments of Dry and Near-Dry Electrical Discharge Machining Processes. *Journal of Manufacturing Science and Engineering*, *144*(5), 50801. doi:10.1115/1.4052527

Boopathi, S. (2022c). An investigation on gas emission concentration and relative emission rate of the near-dry wire-cut electrical discharge machining process. *Environmental Science and Pollution Research International*, *29*(57), 86237–86246. doi:10.100711356-021-17658-1 PMID:34837614

Boopathi, S. (2022d). Cryogenically treated and untreated stainless steel grade 317 in sustainable wire electrical discharge machining process: A comparative study. *Environmental Science and Pollution Research International*, 1–10. doi:10.100711356-022-22843-x PMID:36057706

Boopathi, S. (2022e). Experimental investigation and multi-objective optimization of cryogenic Friction-stir-welding of AA2014 and AZ31B alloys using MOORA technique. *Materials Today. Communications*, *33*, 104937. doi:10.1016/j.mtcomm.2022.104937

Boopathi, S. (2022f). Performance Improvement of Eco-Friendly Near-Dry Wire-Cut Electrical Discharge Machining Process Using Coconut Oil-Mist Dielectric Fluid. *Journal of Advanced Manufacturing Systems*, 1–20. doi:10.1142/S0219686723500178

Boopathi, S., Arigela, S. H., Raman, R., Indhumathi, C., Kavitha, V., & Bhatt, B. C. (2023). Prominent Rule Control-based Internet of Things: Poultry Farm Management System. *IEEE Explore*, (pp. 1–6). IEEE. doi:10.1109/ICPECTS56089.2022.10047039

Boopathi, S., Balasubramani, V., Kumar, R. S., & Singh, G. R. (2021). The influence of human hair on kenaf and Grewia fiber-based hybrid natural composite material: An experimental study. *Functional Composites and Structures*, *3*(4), 45011. doi:10.1088/2631-6331/ac3afc

Boopathi, S., Haribalaji, V., Mageswari, M., & Asif, M. M. (2022). Influences of Boron Carbide Particles on the Wear Rate and Tensile Strength of Aa2014 Surface Composite Fabricated By Friction-Stir Processing. *Materiali in Tehnologije*, *56*(3), 263–270. doi:10.17222/mit.2022.409

Boopathi, S., Khare, R., Jaya Christiyan, K. G., Muni, T. V., & Khare, S. (2023). Additive manufacturing developments in the medical engineering field. In Development, Properties, and Industrial Applications of 3D Printed Polymer Composites (pp. 86–106). IGI Global. doi:10.4018/978-1-6684-6009-2.ch006

Boopathi, S., Siva Kumar, P. K., & Meena, R. S. J., S. I., P., S. K., & Sudhakar, M. (2023). Sustainable Developments of Modern Soil-Less Agro-Cultivation Systems. In Human Agro-Energy Optimization for Business and Industry (pp. 69–87). IGI Global. doi:10.4018/978-1-6684-4118-3.ch004

Boopathi, S., & Sivakumar, K. (2013). Experimental investigation and parameter optimization of near-dry wire-cut electrical discharge machining using multi-objective evolutionary algorithm. *International Journal of Advanced Manufacturing Technology*, *67*(9–12), 2639–2655. doi:10.100700170-012-4680-4

Boopathi, S., & Sivakumar, K. (2016). Optimal parameter prediction of oxygen-mist near-dry Wire-cut EDM. *International Journal of Manufacturing Technology and Management*, *30*(3–4), 164–178. doi:10.1504/IJMTM.2016.077812

Boopathi, S., Sureskumar, M., Jeyakumar, M., Sanjeev Kumar, R., & Subbiah, R. (2022). Influences of Fabrication Parameters on Natural Fiber Reinforced Polymer Composite (NFRPC) Material: A Review. *Materials Science Forum*, *1075*, 115–124. doi:10.4028/p-095f0t

Boopathi, S., Thillaivanan, A., Azeem, M. A., Shanmugam, P., & Pramod, V. R. (2022). Experimental investigation on abrasive water jet machining of neem wood plastic composite. *Functional Composites and Structures*, *4*(2), 25001. doi:10.1088/2631-6331/ac6152

Boopathi, S., Venkatesan, G., & Anton Savio Lewise, K. (2023). Mechanical Properties Analysis of Kenaf–Grewia–Hair Fiber-Reinforced Composite. In *Lecture Notes in Mechanical Engineering* (pp. 101–110). Springer., doi:10.1007/978-981-16-9057-0_11

Chen, M., Ma, C. Y., Mahmud, T., Darr, J. A., & Wang, X. Z. (2011). Modelling and simulation of continuous hydrothermal flow synthesis process for nano-materials manufacture. *The Journal of Supercritical Fluids*, *59*, 131–139. doi:10.1016/j.supflu.2011.07.002

Choi, S. (1998). Nanofluid technology: current status and future research. In *Energy*. http://www.ncbi.nlm.nih.gov/pubmed/11048%5Cnhttp://www.osti.gov/energycitations/product.biblio.jsp?osti_id=11048

Courville, I. G., Y. B., & A. (2016). Deep learning. In *Nature, 29*. MIT press. http://deeplearning.net/

Daveiga, J., & Ferreira, P. (2005). Smart and Nano materials in architecture. *ACADIA 2005 Conference: Smart Architecture*, 58–67. 10.52842/conf.acadia.2005.058

Domakonda, V. K., Farooq, S., Chinthamreddy, S., Puviarasi, R., Sudhakar, M., & Boopathi, S. (2023). Sustainable Developments of Hybrid Floating Solar Power Plants. In *Human Agro-Energy Optimization for Business and Industry* (pp. 148–167). IGI Global. doi:10.4018/978-1-6684-4118-3.ch008

Fuskele, V., & Sarviya, R. M. (2017). Recent developments in Nanoparticles Synthesis, Preparation and Stability of Nanofluids. *Materials Today: Proceedings*, *4*(2), 4049–4060. doi:10.1016/j.matpr.2017.02.307

Gardašević, G., Veletić, M., Maletić, N., Vasiljević, D., Radusinović, I., Tomović, S., & Radonjić, M. (2017). The IoT Architectural Framework, Design Issues and Application Domains. *Wireless Personal Communications*, *92*(1), 127–148. doi:10.100711277-016-3842-3

Harikaran, M., Boopathi, S., Gokulakannan, S., & Poonguzhali, M. (2023). Study on the Source of E-Waste Management and Disposal Methods. In *Sustainable Approaches and Strategies for E-Waste Management and Utilization* (pp. 39–60). IGI Global. doi:10.4018/978-1-6684-7573-7.ch003

Janardhana, K., Singh, V., Singh, S. N., Babu, T. S. R., Bano, S., & Boopathi, S. (2023). Utilization Process for Electronic Waste in Eco-Friendly Concrete: Experimental Study. In Sustainable Approaches and Strategies for E-Waste Management and Utilization (pp. 204–223). IGI Global.

Jeevanantham, Y. A., A, S., V, V., J, S. I., Boopathi, S., & Kumar, D. P. (2023). Implementation of Internet-of Things (IoT) in Soil Irrigation System. *IEEE Explore*, 1–5. doi:10.1109/ICPECTS56089.2022.10047185

Kumara, V., Mohanaprakash, T. A., Fairooz, S., Jamal, K., Babu, T., & B., S. (2023). Experimental Study on a Reliable Smart Hydroponics System. In *Human Agro-Energy Optimization for Business and Industry* (pp. 27–45). IGI Global. doi:10.4018/978-1-6684-4118-3.ch002

Malakar, P., Thakur, M. S. H., Nahid, S. M., & Islam, M. M. (2022). Data-Driven Machine Learning to Predict Mechanical Properties of Monolayer Transition-Metal Dichalcogenides for Applications in Flexible Electronics. *ACS Applied Nano Materials*, *5*(11), 16489–16499. doi:10.1021/acsanm.2c03564

Mohanty, A., Venkateswaran, N., Ranjit, P. S., Tripathi, M. A., & Boopathi, S. (2023). Innovative Strategy for Profitable Automobile Industries: Working Capital Management. In Handbook of Research on Designing Sustainable Supply Chains to Achieve a Circular Economy (pp. 412–428). IGI Global.

Palaniappan, M., Tirlangi, S., Mohamed, M. J. S., Moorthy, R. M. S., Valeti, S. V., & Boopathi, S. (2023). Fused deposition modelling of polylactic acid (PLA)-based polymer composites: A case study. In Development, Properties, and Industrial Applications of 3D Printed Polymer Composites (pp. 66–85). IGI Global. doi:10.4018/978-1-6684-6009-2.ch005

Patil, P., Kardekar, N., & Kamble, D. (2023). Performance analysis and optimization of nano particle base Libr-H2o vapor absorption system by using Taguchi, Grey relational analysis and ANOVA. *Materials Today: Proceedings*, *72*, 1413–1419. doi:10.1016/j.matpr.2022.09.340

S., P. K., Sampath, B., R., S. K., Babu, B. H., & N., A. (2022). Hydroponics, Aeroponics, and Aquaponics Technologies in Modern Agricultural Cultivation. In *Trends, Paradigms, and Advances in Mechatronics Engineering* (pp. 223–241). IGI Global. doi:10.4018/978-1-6684-5887-7.ch012

Saha1, B. C., R, D., A, A., Thrinath, B. V. S., Boopathi, S., J. R., & Sudhakar, M. (n.d.). *IoT Based Smart Energy Meter For Smart Grid.*

Samikannu, R., Koshariya, A. K., Poornima, E., Ramesh, S., Kumar, A., & Boopathi, S. (2023). Sustainable Development in Modern Aquaponics Cultivation Systems Using IoT Technologies. In *Human Agro-Energy Optimization for Business and Industry* (pp. 105–127). IGI Global. doi:10.4018/978-1-6684-4118-3.ch006

Sampath, B. C. S., & Myilsamy, S. (2022). Application of TOPSIS Optimization Technique in the Micro-Machining Process. In Trends, Paradigms, and Advances in Mechatronics Engineering (pp. 162–187). IGI Global. doi:10.4018/978-1-6684-5887-7.ch009

Saravanan, A., Venkatasubramanian, R., Khare, R., Surakasi, R., Boopathi, S., Ray, S., & Sudhakar, M. (2022). *Policy Trends of Renewable Energy and Non Renewable Energy.* Research Gate.

Senthil, T. S. R. Ohmsakthi vel, Puviyarasan, M., Babu, S. R., Surakasi, R., & Sampath, B. (2023). Industrial Robot-Integrated Fused Deposition Modelling for the 3D Printing Process. In Development, Properties, and Industrial Applications of 3D Printed Polymer Composites (pp. 188–210). IGI Global. doi:10.4018/978-1-6684-6009-2.ch011

Singh, A. V., Rosenkranz, D., Ansari, M. H. D., Singh, R., Kanase, A., Singh, S. P., Johnston, B., Tentschert, J., Laux, P., & Luch, A. (2020). Artificial Intelligence and Machine Learning Empower Advanced Biomedical Material Design to Toxicity Prediction. *Advanced Intelligent Systems*, *2*(12), 2070125. doi:10.1002/aisy.202070125

Vanitha, S. K. R., & Boopathi, S. (2023). Artificial Intelligence Techniques in Water Purification and Utilization. In *Human Agro-Energy Optimization for Business and Industry* (pp. 202–218). IGI Global. doi:10.4018/978-1-6684-4118-3.ch010

Vennila, T., Karuna, M. S., Srivastava, B. K., Venugopal, J., Surakasi, R., & B., S. (2023). New Strategies in Treatment and Enzymatic Processes. In *Human Agro-Energy Optimization for Business and Industry* (pp. 219–240). IGI Global. doi:10.4018/978-1-6684-4118-3.ch011

Wu, Q., Miao, W. S., Du Zhang, Y., Gao, H. J., & Hui, D. (2020). Mechanical properties of nanomaterials: A review. *Nanotechnology Reviews*, *9*(1), 259–273. doi:10.1515/ntrev-2020-0021

Yang, B., Gao, L., Xue, M., Wang, H., Hou, Y., Luo, Y., Xiao, H., Hu, H., Cui, C., Wang, H., Zhang, J., Li, Y. F., Xie, G., Tong, X., & Xie, Y. (2021). Experimental and simulation research on the preparation of carbon nano-materials by chemical vapor deposition. *Materials (Basel)*, *14*(23), 7356. doi:10.3390/ma14237356 PMID:34885507

Chapter 8
Applications of Nano-Fluids in the Various Engineering Fields

Parth Gaud
Parul Institute of Engineering and Technology, Parul University, India

C. Sailaja
Bangalore College of Engineering and Technology, India

Kode Jaya Prakash
https://orcid.org/0000-0002-4947-7745
VNR Vignana Jyothi Institute of Engineering and Technology, India

Yagya Dutta Dwivedi
https://orcid.org/0000-0002-5793-9364
Institute of Aeronautical Engineering, India

L. Karthick
https://orcid.org/0000-0002-1403-8179
Hindusthan College of Engineering and Technology, India

ABSTRACT

In this chapter, nano-fluids are a type of fluid with suspended nano-sized particles that are useful in various fields, such as manufacturing, medical, 3D printing, tissue engineering, and fluid effects. Nano-fluids can be used as coolants and lubricants to improve machining accuracy and reduce tool wear, as well as additives in coatings and adhesives. Nano-fluids can be used as printing inks to create intricate and precise structures and enhance their properties. Nano-fluids can improve drug delivery and enhance visibility of tissues and organs. Nano-fluids are manufactured using two-phase and one-phase methods, involving the dispersion of nanoparticles in a base fluid.

DOI: 10.4018/978-1-6684-9135-5.ch008

INTRODUCTION

Nano-fluid is a type of fluid that contains suspended nanoparticles, typically with a diameter of less than hundred nanometres. These nanoparticles can be made of various materials, such as metals, oxides, and polymers. The addition of nanoparticles to the base fluid can enhance the fluid's properties, such as thermal conductivity, viscosity, and stability, and can result in improved performance in various applications. Nano-fluids have gained increasing attention in recent years due to their unique properties and potential applications in various fields, including engineering, medicine, and environmental science. In engineering applications, nano-fluids can be used to improve the heat transfer and lubrication in machining, thermal management, and energy systems. In medicine, nano-fluids can be used for drug delivery and tissue engineering. In environmental science, nano-fluids can be used for water treatment and pollution control. However, the use of nano-fluids also presents some challenges, such as the potential for particle aggregation, instability, and increased viscosity. Therefore, extensive research and development are required to optimize the properties and performance of nano-fluids for specific applications. Overall, nano-fluids are a promising area of research and development that has the potential to revolutionize various fields by providing improved performance and efficiency in various applications(Mebarek-Oudina & Chabani, 2022).

This article highlights the applications of nano-fluids in different engineering fields. In heat transfer, nano-fluids have higher thermal conductivity than conventional fluids, making them useful in cooling systems for electronic devices, heat exchangers, and other applications where efficient heat transfer is essential. In energy conversion, nano-fluids are explored for use in solar energy systems, fuel cells, and thermoelectric generators to increase their efficiency. They can be used as lubricants in mechanical systems, thus increasing the lifespan and efficiency of the system. Nano-fluids have potential applications in drug delivery, tissue engineering, and medical imaging in biomedical engineering. In environmental engineering, they can be used in water treatment systems to remove pollutants and contaminants from wastewater, and in air filters to improve air quality. In aerospace engineering, nano-fluids can improve the efficiency and reduce weight in rocket engines and spacecraft thermal management systems. Overall, nano-fluids have the potential to enhance the performance and efficiency of various engineering systems, making them a crucial area of research and development in many fields.

Nanofluids are a new class of heat transfer fluids that are gaining popularity in engineering applications. These fluids consist of nanoparticles suspended in a base fluid, which can be a liquid or a gas. The use of nanofluids has been found to significantly enhance the heat transfer properties of the base fluid. Here are some applications of nanofluids in engineering fields(Yang et al., 2019):

- **Thermal management:** Nanofluids are being used in electronics cooling, automotive radiators, and in other applications where heat management is critical. The use of nanofluids can improve heat transfer and reduce the size of heat exchangers.
- **Solar thermal energy:** Nanofluids have been found to be effective in increasing the efficiency of solar thermal systems. The use of nanofluids in solar thermal systems can help to increase the amount of heat transferred from the collector to the working fluid, thus increasing the efficiency of the system.
- **Aerospace and aviation:** Nanofluids have been used to improve the thermal management of aircraft engines and to reduce the weight and size of heat exchangers in spacecraft.
- **Biomedical applications:** Nanofluids have been used in various biomedical applications, such as targeted drug delivery and hyperthermia therapy for cancer treatment. The use of nanofluids in these applications can enhance the effectiveness of treatments while minimizing side effects.
- **Manufacturing:** Nanofluids can be used in manufacturing processes such as metalworking and cutting fluids. The use of nanofluids in these processes can improve the quality of the products and reduce the amount of waste generated.

Overall, the use of nanofluids has the potential to revolutionize many engineering fields by improving the efficiency and effectiveness of various processes.

Thermal management of nanofluids involves controlling the heat transfer properties of the fluids to optimize their performance in various applications. Here are some key factors that affect the thermal management of nanofluids(Rubbi et al., 2020):

- **Particle size:** The size of the nanoparticles in the nanofluid is an important factor that affects its heat transfer properties. Smaller particles tend to have a higher surface area to volume ratio, which can lead to more efficient heat transfer.
- **Particle concentration:** The concentration of nanoparticles in the nanofluid also plays a key role in its thermal management. Higher particle concentrations can enhance the heat transfer properties of the nanofluid, but can also increase the viscosity and potentially lead to clogging of the flow channels.
- **Base fluid properties:** The properties of the base fluid used in the nanofluid also affect its thermal management. Factors such as viscosity, density, and thermal conductivity can impact the heat transfer properties of the nanofluid.
- **Flow rate and turbulence:** The flow rate and turbulence of the nanofluid can affect its thermal management by influencing the distribution of the nanoparticles in the fluid and the rate of heat transfer.

- **Surface interactions:** The interaction between the nanoparticles and the surface of the container or heat transfer surfaces can also impact the thermal management of the nanofluid. Surface modifications can be used to improve the adhesion and dispersion of the nanoparticles in the fluid.

Overall, effective thermal management of nanofluids involves optimizing the particle size, concentration, base fluid properties, flow rate and turbulence, and surface interactions to enhance the heat transfer properties of the fluid and maximize its performance in various applications.

Nano-fluids have been found to be effective in enhancing the performance of solar thermal energy systems, which convert solar energy into heat for use in various applications. Here are some ways in which nano-fluids can be used in solar thermal energy(Yoo et al., 2007):

- **Improved heat transfer:** Nano-fluids can increase the heat transfer coefficient of the heat transfer fluid, which allows for more efficient heat transfer from the solar collector to the working fluid. This can improve the efficiency of the solar thermal system.
- **Increased energy absorption:** Nano-fluids can increase the solar energy absorption of the working fluid, which improves the heat transfer properties and increases the temperature of the fluid.
- **Reduced thermal losses:** Nano-fluids can reduce thermal losses due to convection and radiation, which can improve the overall efficiency of the solar thermal system.
- **Reduced corrosion and scaling:** The use of nano-fluids can reduce the risk of corrosion and scaling in the solar thermal system, which can lead to longer system life and lower maintenance costs.
- **Reduced system size:** The improved heat transfer properties of nano-fluids can allow for the design of smaller and more compact solar thermal systems, which can be more cost-effective and easier to install.

Overall, the use of nano-fluids in solar thermal energy systems can enhance the efficiency, performance, and lifespan of these systems, making them more viable for a wider range of applications.

Nanofluids have potential applications in aerospace and aviation industries where thermal management is critical for maintaining the performance and safety of various components. Here are some ways in which nanofluids can be used in aerospace and aviation(Moita et al., 2021; Yurtkuran, 2021):

- **Aircraft engine cooling:** Nanofluids can be used as cooling fluids in aircraft engines to improve the efficiency of the cooling system and increase the heat transfer rate. This can improve the performance and lifespan of the engine.
- **Heat exchangers:** Nanofluids can be used in heat exchangers in spacecraft and aircraft to increase the efficiency of heat transfer and reduce the weight and size of the heat exchangers.
- **Thermal barrier coatings:** Nanofluids can be used as thermal barrier coatings on the surface of engine components to protect them from high temperatures and extend their lifespan.
- **De-icing fluids:** Nanofluids can be used as de-icing fluids to remove ice from the wings and other surfaces of aircraft. The improved heat transfer properties of nanofluids can enable faster and more efficient de-icing.
- **Fuel additives:** Nanofluids can be added to fuel to improve its combustion efficiency and reduce emissions. This can help to make aircraft engines more environmentally friendly.

Overall, the use of nanofluids in aerospace and aviation can help to improve the efficiency, performance, and safety of various components and systems, making air travel more reliable and sustainable.

Nanofluids have several potential biomedical applications due to their unique physical and chemical properties. Here are some ways in which nanofluids can be used in biomedical applications(Hamad et al., 2021):

- **Drug delivery:** Nanofluids can be used as carriers for targeted drug delivery in various diseases, including cancer. The high surface area to volume ratio of nanoparticles can increase the surface contact area with cells, enabling more efficient drug delivery.
- **Tissue engineering:** Nanofluids can be used to create scaffolds for tissue engineering applications. The small size of nanoparticles can improve the mechanical and biological properties of the scaffold, allowing for better tissue regeneration.
- **Imaging agents:** Nanoparticles in nanofluids can be used as imaging agents for various imaging techniques, including magnetic resonance imaging (MRI), computed tomography (CT), and ultrasound. The nanoparticles can be functionalized to target specific tissues or cells for better diagnostic accuracy.
- **Biosensors:** Nanofluids can be used in biosensors for rapid and sensitive detection of biomolecules, including proteins and DNA. The small size of nanoparticles can increase the sensitivity and specificity of the biosensor, allowing for faster and more accurate detection.

- **Wound healing:** Nanofluids can be used to improve wound healing by promoting cell proliferation and differentiation. The high surface area to volume ratio of nanoparticles can increase the interaction between the fluid and cells, improving the healing process.

Overall, the use of nanofluids in biomedical applications can offer several advantages, including improved drug delivery, tissue engineering, imaging, biosensing, and wound healing. However, more research is needed to fully explore the potential of nanofluids in these applications and to address any potential safety concerns.

Nanofluids have several potential applications in manufacturing due to their unique properties, including improved heat transfer and enhanced mechanical properties. Here are some ways in which nanofluids can be used in manufacturing(Hegab et al., 2018; Smaisim et al., 2022):

- **Machining:** Nanofluids can be used as coolants and lubricants in machining operations to improve the cutting performance and extend the tool life. The high thermal conductivity and low viscosity of nanofluids can improve the heat transfer and reduce friction during machining.
- **Metalworking:** Nanofluids can be used in metalworking operations, such as forging, casting, and extrusion, to improve the mechanical properties of the final product. The addition of nanoparticles can improve the strength, ductility, and wear resistance of the metal.
- **Coatings:** Nanofluids can be used to create coatings for various applications, including anti-corrosion, wear-resistant, and thermal barrier coatings. The addition of nanoparticles can improve the adhesion, durability, and performance of the coating.
- **Additive manufacturing:** Nanofluids can be used in additive manufacturing processes, such as 3D printing, to improve the print quality and mechanical properties of the final product. The high thermal conductivity and low viscosity of nanofluids can improve the heat transfer and reduce the likelihood of defects during printing.
- **Electronics:** Nanofluids can be used in the manufacturing of electronic components, such as heat sinks, to improve the thermal management and reliability of the component. The high thermal conductivity of nanofluids can improve the heat dissipation and reduce the risk of overheating.

Overall, the use of nanofluids in manufacturing can offer several advantages, including improved heat transfer, enhanced mechanical properties, and improved performance of the final product. However, more research is needed to fully explore

Figure 1. General methods to make nano-fluids

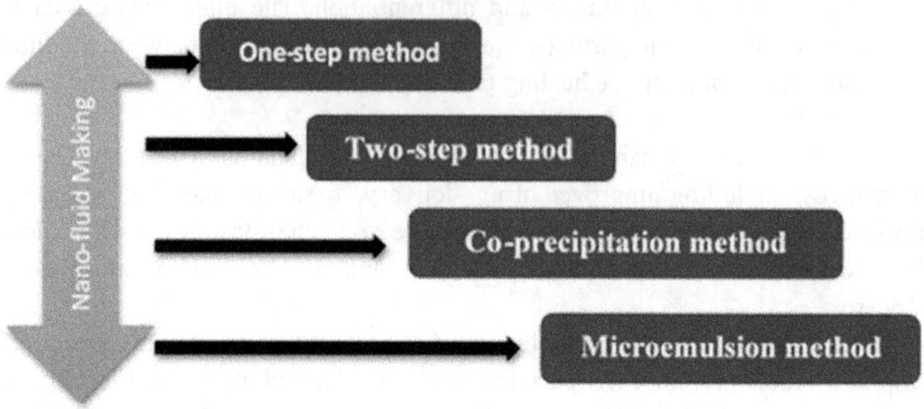

the potential of nanofluids in these applications and to address any potential safety concerns.

METHODS TO PREPARING NANO-FLUIDS

There are different methods to prepare nano-fluids depending on the type of nanoparticles and the base fluid used (Figure 1). Here are some common methods for preparing nano-fluids(Mebarek-Oudina & Chabani, 2022; Smaisim et al., 2022):

- **One-step method:** This method involves dispersing nanoparticles in the base fluid using a high-shear mixer or ultrasonic homogenizer. The nanoparticles are added directly to the base fluid, and the mixture is stirred for a specific time to ensure even dispersion.
- **Two-step method:** In this method, the nanoparticles are first dispersed in a solvent or surfactant solution to prevent agglomeration. The nanoparticle dispersion is then mixed with the base fluid using a high-shear mixer or ultrasonic homogenizer.
- **Co-precipitation method:** This method involves preparing nanoparticles and base fluid simultaneously by mixing precursor materials in a solution under controlled conditions. The nanoparticles are formed in the solution and dispersed in the base fluid.
- **Microemulsion method:** In this method, a microemulsion is prepared by mixing an oil phase, water phase, and surfactant. The nanoparticles are

then added to the microemulsion, and the mixture is stirred to form a stable nano-fluid.

It is important to note that the preparation of nano-fluids requires careful control of the process parameters, such as the concentration and size of nanoparticles, the type and concentration of surfactant, and the mixing conditions. Proper characterization of the nano-fluid, such as measuring the particle size distribution and zeta potential, is also important to ensure the quality and stability of the nano-fluid.

Figure 2. Types of nano-fluid in various engineering sectors

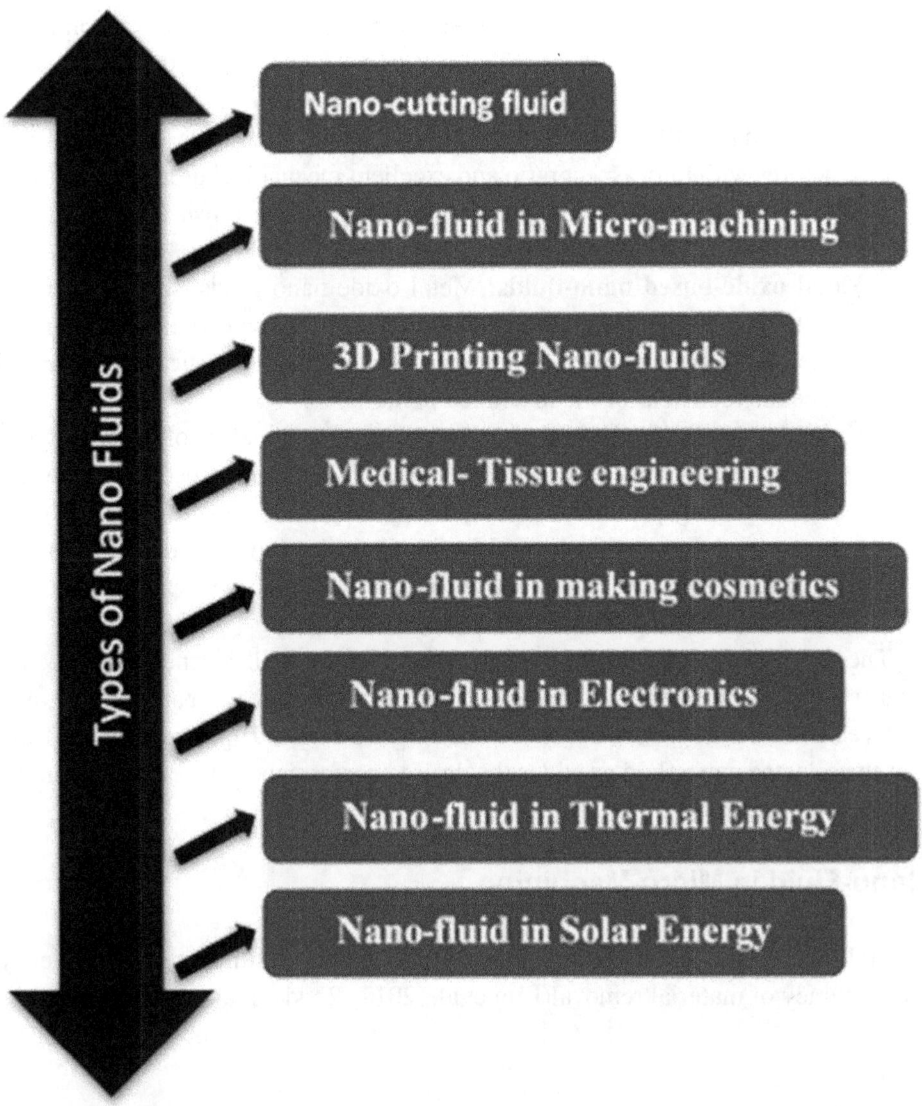

TYPES OF NANO-FLUIDS IN VARIOUS APPLICATIONS

The types Nano-fluids used in the various engineering applications are illustrated in the Figure 2.

Nano-Cutting Fluid

There are several types of metal cutting nano-fluids that can be used in metal cutting processes, each with its own unique properties and advantages(Hegab et al., 2018). Here are some examples:

- **Graphene-based nano-fluids:** Graphene is a two-dimensional material with high thermal conductivity and excellent mechanical properties. Graphene-based nano-fluids can improve the heat transfer and reduce the tool wear in metal cutting processes.
- **Carbon nanotube-based nano-fluids:** Carbon nanotubes are cylindrical structures with high aspect ratio and excellent mechanical properties. Carbon nanotube-based nano-fluids can improve the surface finish and reduce the tool wear in metal cutting processes.
- **Metal oxide-based nano-fluids:** Metal oxide nano-fluids, such as alumina and titania, have high thermal conductivity and can improve the heat transfer in metal cutting processes. They can also enhance the surface finish and reduce the tool wear of the machined parts.
- **Hybrid nano-fluids:** Hybrid nano-fluids are a combination of different types of nanoparticles and base fluids, designed to provide specific properties for metal cutting processes. For example, a hybrid nano-fluid consisting of graphene and diamond nanoparticles in oil can provide excellent lubrication and reduce the tool wear in metal cutting.

The choice of nano-fluid for metal cutting applications will depend on the specific requirements of the machining process, such as cutting speed, material type, and surface finish. It is important to conduct thorough testing and optimization to ensure that the selected nano-fluid provides the desired performance and efficiency in the metal cutting process.

Nano-Fluid in Micro-Machining

The following nano-fluids are used in the EDM process to improve the efficiency and accuracy of material removal(Lim et al., 2017; Reddy et al., 2020).

- Copper oxide nano-fluids are known for their high thermal conductivity and high electrical resistivity, making them an ideal choice for EDM applications.
- Graphene oxide nano-fluids have exceptional mechanical, electrical, and thermal properties, making them a promising choice for improving the efficiency and accuracy of the EDM process.
- Aluminum oxide nano-fluids are widely used in EDM due to their excellent dielectric properties and high thermal conductivity.
- Silicon carbide nano-fluids are known for their excellent hardness, wear resistance, and thermal stability, making them a popular choice for high-speed machining applications.
- Diamond nano-fluids are known for their exceptional hardness, thermal conductivity, and low friction coefficient, making them a promising choice for high-precision EDM applications.

Ultrasonic Machining (USM)

It is a non-traditional machining process that uses high-frequency ultrasonic vibrations to remove material from a workpiece. Here are some examples of nano-fluids that can be used in USM:

- **Alumina-based nano-fluids:** Alumina-based nano-fluids, such as alumina nanoparticles dispersed in water or oil, can improve the machining rate and surface quality in USM. Alumina-based nano-fluids can also reduce the tool wear and improve the stability of the machining process.
- **Titanium dioxide-based nano-fluids:** Titanium dioxide-based nano-fluids, such as titanium dioxide nanoparticles dispersed in water or oil, can improve the machining efficiency and surface finish in USM. Titanium dioxide-based nano-fluids can also reduce the tool wear and improve the machining accuracy of the process.
- **Diamond-based nano-fluids:** Diamond-based nano-fluids, such as diamond nanoparticles dispersed in water or oil, can improve the machining performance and surface quality in USM. Diamond-based nano-fluids can also reduce the tool wear and improve the stability of the machining process.
- **Carbon nanotube-based nano-fluids:** Carbon nanotube-based nano-fluids, such as carbon nanotubes dispersed in water or oil, can improve the machining rate and surface quality in USM. Carbon nanotube-based nano-fluids can also reduce the tool wear and improve the corrosion resistance of the machined part.

3D Printing Nano-Fluids

There are several types of nano-fluids that can be used in 3D printing. Here are some examples(Sharma et al., 2016):

- **Metal-based nano-fluids:** Metal-based nano-fluids, such as copper and silver, can improve the thermal conductivity and electrical conductivity of the 3D-printed part. Metal-based nano-fluids can also enhance the mechanical properties and reduce the deformation of the printed part.
- **Polymer-based nano-fluids:** Polymer-based nano-fluids, such as polyethylene glycol and polyvinylpyrrolidone, can improve the flowability and adhesion of the printing material. Polymer-based nano-fluids can also enhance the thermal stability and reduce the warping of the printed part.
- **Carbon nanotube-based nano-fluids:** Carbon nanotubes have high aspect ratio and excellent mechanical properties, which can improve the strength and stiffness of the 3D-printed part. Carbon nanotube-based nano-fluids can also enhance the thermal and electrical conductivity of the printed part.
- **Ceramic-based nano-fluids:** Ceramic-based nano-fluids, such as alumina and zirconia, can improve the hardness and wear resistance of the 3D-printed part. Ceramic-based nano-fluids can also enhance the thermal stability and reduce the shrinkage of the printed part.

The choice of nano-fluid for 3D printing will depend on the specific requirements of the printing process, such as material properties, printing resolution, and surface finish. It is important to conduct thorough testing and optimization to ensure that the selected nano-fluid provides the desired performance and quality in the 3D printing process.

Medical-Tissue Engineering

Nano-fluids have several potential applications in tissue engineering, including drug delivery, cell culture, and scaffolding. Here are some examples of nano-fluids used in tissue engineering:

- **Hydroxyapatite-based nano-fluids:** Hydroxyapatite is a biocompatible and bioactive material that is commonly used as a scaffold in tissue engineering. Hydroxyapatite-based nano-fluids can improve the mechanical properties and biocompatibility of the scaffold, as well as promote cell adhesion and proliferation.

- **Chitosan-based nano-fluids:** Chitosan is a biopolymer that has antimicrobial and wound-healing properties. Chitosan-based nano-fluids can be used as a drug delivery system, as well as a scaffold material in tissue engineering.
- **Polylactic acid-based nano-fluids:** Polylactic acid is a biodegradable polymer that has been used as a scaffold material in tissue engineering. Polylactic acid-based nano-fluids can improve the mechanical properties and biocompatibility of the scaffold, as well as promote cell growth and differentiation.
- **Alginate-based nano-fluids:** Alginate is a polysaccharide that is commonly used as a scaffold material in tissue engineering. Alginate-based nano-fluids can improve the mechanical properties and biocompatibility of the scaffold, as well as promote cell adhesion and proliferation.
- **Gelatin-based nano-fluids:** Gelatin is a protein that is commonly used as a scaffold material in tissue engineering. Gelatin-based nano-fluids can improve the mechanical properties and biocompatibility of the scaffold, as well as promote cell growth and differentiation.

The selection of a nano-fluid for tissue engineering will depend on the specific application and the properties required of the scaffold or drug delivery system.

Nano-Fluid in Making Cosmetics

Nano-fluids are used in cosmetics for various purposes, such as improving the texture and feel of the product, enhancing its efficacy, and providing sun protection. Some examples of nano-fluids used in cosmetics include:

- **Titanium dioxide nano-fluids:** Titanium dioxide is a common ingredient in sunscreens, and its nano-fluid form is used to improve the product's sun protection factor (SPF) and reduce the amount of white residue left on the skin.
- **Zinc oxide nano-fluids:** Zinc oxide is another common ingredient in sunscreens, and its nano-fluid form is used to improve the product's SPF and reduce the white cast on the skin.
- **Silica nano-fluids:** Silica is used in cosmetics to improve the texture and feel of the product, as well as to provide oil absorption and mattifying properties.
- **Gold nano-fluids:** Gold nanoparticles are used in cosmetics for their antioxidant and anti-inflammatory properties, as well as for their ability to improve skin elasticity and reduce the appearance of fine lines and wrinkles.

- **Silver nano-fluids:** Silver nanoparticles are used in cosmetics for their antimicrobial properties, as well as for their ability to soothe and calm the skin.

Nano-Fluid in Electronics

Nano-fluids are finding numerous applications in the field of electronics. Some common types of nano-fluids used in electronics include:

- **Carbon nanotube (CNT) nano-fluids:** These are used as thermal interface materials to improve the thermal management of electronic devices.
- **Metal oxide nano-fluids:** These are used as dielectric materials in capacitors and as insulators in electronic devices.
- **Graphene oxide nano-fluids:** These are used as conductive ink in printed electronics, and also as a transparent and flexible electrode material.
- **Silver nano-fluids:** These are used as conductive inks for printed electronics and as a surface coating material to improve the conductivity of electronic devices.
- **Copper nano-fluids:** These are used as an interconnect material in electronic devices due to their excellent electrical conductivity.
- **Dielectric nano-fluids:** These are used as coolant and insulating materials in electronic devices to improve their thermal management.
- **Magnetic nano-fluids:** These are used in magnetic storage devices, magnetic resonance imaging (MRI) machines, and other applications in the field of electronics.

Overall, nano-fluids are proving to be a promising area of research for improving the performance and reliability of electronic devices.

EFFECTS OF NANO-FLUIDS IN THE MANUFACTURING SECTORS

Nano-fluids have several potential applications in the machining process due to their unique properties, including improved heat transfer, enhanced lubricity, and reduced wear. Here are some ways in which nano-fluids can be used in the machining process:

- **Cutting fluids:** Nano-fluids can be used as cutting fluids to improve the cutting performance and extend the tool life. The high thermal conductivity and low viscosity of nano-fluids can improve the heat transfer and reduce friction during machining. The addition of nanoparticles can also improve the lubricity and reduce the wear of the cutting tool.

- **Grinding fluids:** Nano-fluids can be used as grinding fluids to improve the surface finish and reduce the grinding force. The high thermal conductivity and low viscosity of nano-fluids can improve the heat transfer and reduce the likelihood of thermal damage to the workpiece. The addition of nanoparticles can also improve the lubricity and reduce the wear of the grinding wheel.
- **Electrochemical machining:** Nano-fluids can be used as electrolytes in electrochemical machining processes to improve the material removal rate and surface finish. The high thermal conductivity and low viscosity of nano-fluids can improve the heat transfer and reduce the likelihood of thermal damage to the workpiece. The addition of nanoparticles can also improve the electrolyte conductivity and reduce the anode passivation.
- **Electrical discharge machining:** Nano-fluids can be used as dielectric fluids in electrical discharge machining processes to improve the machining accuracy and surface finish. The high thermal conductivity and low viscosity of nano-fluids can improve the heat transfer and reduce the likelihood of thermal damage to the workpiece. The addition of nanoparticles can also improve the dielectric strength and reduce the electrode wear.

Overall, the use of nano-fluids in the machining process can offer several advantages, including improved heat transfer, enhanced lubricity, and reduced wear. However, more research is needed to fully explore the potential of nano-fluids in these applications and to address any potential safety concerns.

Metal Cutting Fluid

Metal cutting fluids are essential in the machining process to improve tool life and performance, reduce cutting forces, and improve surface finish. Nano-fluids have emerged as a promising alternative to conventional metal cutting fluids due to their unique properties, including improved heat transfer, enhanced lubricity, and reduced wear. Here are some ways in which nano-fluids can be used in metal cutting applications(Emami et al., 2020; Hegab et al., 2018):

- **Improving cutting performance:** Nano-fluids can be used as metal cutting fluids to improve the cutting performance by reducing the cutting forces and tool wear. The high thermal conductivity of nano-fluids can improve the heat transfer, which reduces the cutting temperature and the likelihood of thermal damage to the workpiece. The addition of nanoparticles can also improve the lubricity and reduce the friction between the tool and the workpiece, which reduces the cutting forces and the tool wear.

- **Enhancing tool life:** Nano-fluids can also enhance the tool life by reducing the wear and improving the cutting edge quality. The high lubricity of nano-fluids reduces the friction between the tool and the workpiece, which reduces the tool wear. The addition of nanoparticles can also improve the cutting edge quality by reducing the built-up edge formation and improving the chip control.
- **Improving surface finish:** Nano-fluids can be used to improve the surface finish of the machined parts. The high lubricity of nano-fluids reduces the friction between the tool and the workpiece, which reduces the surface roughness. The addition of nanoparticles can also improve the surface finish by reducing the tool wear and the chip deformation.
- **Reducing environmental impact:** Nano-fluids can offer a more environmentally friendly alternative to conventional metal cutting fluids. They can be formulated to have lower toxicity and better biodegradability than conventional cutting fluids.

Overall, the use of nano-fluids in metal cutting applications can offer several advantages, including improved cutting performance, enhanced tool life, improved surface finish, and reduced environmental impact. However, more research is needed to fully explore the potential of nano-fluids in these applications and to address any potential safety concerns.

Grinding Nano-Fluid

Nano-fluids have emerged as a promising alternative to conventional grinding fluids due to their unique properties, including improved heat transfer, enhanced lubricity, and reduced wear. Here are some ways in which nano-fluids can be used in grinding applications(Lee et al., 2010):

- **Improving surface finish:** Nano-fluids can be used to improve the surface finish of the ground parts. The high lubricity of nano-fluids reduces the friction between the grinding wheel and the workpiece, which reduces the surface roughness. The addition of nanoparticles can also improve the surface finish by reducing the wheel wear and the grinding forces.
- **Reducing grinding forces:** Nano-fluids can reduce the grinding forces by improving the lubrication and reducing the friction between the grinding wheel and the workpiece. This can lead to longer wheel life, reduced wheel wear, and improved dimensional accuracy.

- **Enhancing cooling:** Nano-fluids can also enhance the cooling of the grinding zone by improving the heat transfer. The high thermal conductivity of nano-fluids can help to dissipate the heat generated during grinding, which reduces the likelihood of thermal damage to the workpiece and the grinding wheel.
- **Reducing environmental impact:** Nano-fluids can offer a more environmentally friendly alternative to conventional grinding fluids. They can be formulated to have lower toxicity and better biodegradability than conventional fluids.

Overall, the use of nano-fluids in grinding applications can offer several advantages, including improved surface finish, reduced grinding forces, enhanced cooling, and reduced environmental impact. However, more research is needed to fully explore the potential of nano-fluids in these applications and to address any potential safety concerns.

Electrochemical Machining (ECM)

Electrochemical machining (ECM) is a non-traditional machining process that uses electrochemical reactions to remove material from a workpiece. The use of nano-fluids in ECM can offer several advantages, including improved surface finish, enhanced machining accuracy, and reduced tool wear. Electrochemical machining (ECM) is a non-traditional machining process that uses electrical energy to remove material from a workpiece. Here are some general steps that can be followed to use nano-fluids in ECM(Batyuk & Kizilova, 2019; Emami et al., 2020):

- **Select a suitable nano-fluid:** Choose a nano-fluid that is compatible with the electrolyte used in the ECM process and has the desired properties, such as thermal conductivity, viscosity, and stability.
- **Prepare the electrolyte solution:** Mix the nano-fluid with the electrolyte solution according to the recommended ratio. The mixing process can be done manually or using specialized equipment, depending on the specific ECM process and the type of nano-fluid being used.
- **Set up the ECM system:** Set up the ECM system according to the manufacturer's instructions, including the placement of the workpiece, the positioning of the electrodes, and the adjustment of the machining parameters.
- **Start the ECM process:** Apply the electrical energy to the electrodes to initiate the ECM process. Monitor the process to ensure that the desired material removal rate and surface finish are achieved.
- **Maintenance:** Regularly maintain the ECM system, including cleaning the electrodes and replacing the electrolyte solution as needed.

Some examples of nano-fluids used in ECM process are:

- **Graphene oxide nano-fluid:** Graphene oxide is a two-dimensional material with high thermal conductivity and excellent mechanical properties. Adding graphene oxide nano-fluid to the electrolyte solution can improve the heat transfer and reduce the tool wear in ECM process.
- **Carbon nanotube nano-fluid:** Carbon nanotubes are cylindrical structures with high aspect ratio and excellent mechanical properties. Adding carbon nanotube nano-fluid to the electrolyte solution can improve the surface finish and reduce the tool wear in ECM process.
- **Metal oxide nano-fluid:** Metal oxide nano-fluids, such as alumina and titania, have high thermal conductivity and can improve the heat transfer in ECM process. They can also enhance the surface finish and reduce the tool wear of the machined parts.

It is important to note that the specific procedure for using nano-fluids in ECM may vary depending on the specific ECM process and the type of nano-fluid being used. Therefore, it is recommended to consult the manufacturer's instructions and conduct thorough testing and optimization before using nano-fluids in ECM applications. Additionally, appropriate safety measures should be taken to ensure safe handling and disposal of the nano-fluids and the electrolyte solution.

Here are some ways in which nano-fluids can be used in ECM applications:

- **Improving surface finish:** Nano-fluids can be used as electrolytes in ECM to improve the surface finish of the machined parts. The addition of nanoparticles to the electrolyte can reduce the surface roughness by improving the dissolution rate of the material and reducing the formation of surface defects.
- **Enhancing machining accuracy:** Nano-fluids can enhance the machining accuracy by reducing the tool wear and improving the dimensional accuracy of the machined parts. The high lubricity of nano-fluids can reduce the friction between the tool and the workpiece, which reduces the tool wear. The addition of nanoparticles can also improve the accuracy by reducing the electrode wear and improving the current distribution.
- **Reducing tool wear:** Nano-fluids can reduce the tool wear in ECM by improving the lubrication and reducing the friction between the tool and the workpiece. The high lubricity of nano-fluids can also reduce the electrode wear and prolong the tool life.
- **Improving process efficiency:** Nano-fluids can improve the process efficiency of ECM by reducing the machining time and increasing the material removal rate. The high thermal conductivity of nano-fluids can help to dissipate the

heat generated during machining, which reduces the likelihood of thermal damage to the workpiece.

Overall, the use of nano-fluids in ECM can offer several advantages, including improved surface finish, enhanced machining accuracy, reduced tool wear, and improved process efficiency. However, more research is needed to fully explore the potential of nano-fluids in these applications and to address any potential safety concerns.

EDM

Electric discharge machining (EDM) is a non-traditional machining process that uses a series of electric discharges between the tool and the workpiece to remove material. The use of nano-fluids in EDM can offer several advantages, including improved machining accuracy, reduced tool wear, and improved surface finish. Here are some ways in which nano-fluids can be used in EDM applications(Boopathi, 2022a, 2022d; Boopathi et al., 2022; Trojovský et al., 2023):

- **Enhancing the discharge stability:** Nano-fluids can enhance the discharge stability in EDM by reducing the electrode wear and improving the current distribution. The addition of nanoparticles can also increase the dielectric strength of the fluid, which reduces the likelihood of arcing and improves the process stability.
- **Improving machining accuracy:** Nano-fluids can improve the machining accuracy in EDM by reducing the tool wear and improving the dimensional accuracy of the machined parts. The high lubricity of nano-fluids reduces the friction between the tool and the workpiece, which reduces the tool wear. The addition of nanoparticles can also improve the accuracy by reducing the electrode wear and improving the discharge stability.
- **Reducing tool wear:** Nano-fluids can reduce the tool wear in EDM by improving the lubrication and reducing the friction between the tool and the workpiece. The high lubricity of nano-fluids can also reduce the electrode wear and prolong the tool life.
- **Improving surface finish:** Nano-fluids can be used to improve the surface finish of the machined parts in EDM. The addition of nanoparticles to the dielectric fluid can reduce the surface roughness by improving the flushing action and reducing the formation of surface defects.

Overall, the use of nano-fluids in EDM can offer several advantages, including improved machining accuracy, reduced tool wear, improved surface finish, and enhanced discharge stability. However, more research is needed to fully explore the potential of nano-fluids in these applications and to address any potential safety concerns(Boopathi, 2022c, 2022b; Boopathi et al., 2023; Sampath & Myilsamy, 2021).

Powder-mixed electrical discharge machining (PMEDM) is an advanced machining process that combines the use of electric sparks and a conductive powder to remove material from a workpiece. The use of nano-fluids in PMEDM can offer several advantages, including improved machining efficiency, reduced tool wear, and improved surface quality. Here are some ways in which nano-fluids can be used in PMEDM applications:

- **Improving machining efficiency:** Nano-fluids can be used to improve the machining efficiency in PMEDM by enhancing the thermal properties of the dielectric fluid. The high thermal conductivity of nanoparticles can improve the heat dissipation during the machining process, which reduces the occurrence of thermal damage and improves the machining speed.
- **Reducing tool wear:** Nano-fluids can reduce the tool wear in PMEDM by reducing the sparking voltage and the discharge energy. The high lubricity of nanoparticles can reduce the sparking voltage and improve the discharge energy efficiency, which reduces the tool wear and extends the service life of the tool.
- **Improving surface quality:** Nano-fluids can be used to improve the surface quality in PMEDM by reducing the occurrence of micro-cracks and other surface defects. The addition of nanoparticles to the dielectric fluid can improve the spark erosion efficiency and reduce the occurrence of surface defects such as micro-cracks, pits, and notches.
- **Enhancing process control:** Nano-fluids can enhance the process control in PMEDM by providing better control over the sparking voltage, discharge energy, and other machining parameters. The high thermal conductivity and lubricity of nanoparticles can improve the process stability and consistency, which enhances the process control and reduces the risk of machining defects.

Overall, the use of nano-fluids in PMEDM can offer several advantages, including improved machining efficiency, reduced tool wear, improved surface quality, and enhanced process control. However, more research is needed to fully explore the potential of nano-fluids in these applications and to address any potential safety concerns.

FSW

Friction stir welding (FSW) is a solid-state welding process that uses a rotating tool to generate heat and plasticize the material, which is then joined together by mechanical deformation. The use of nano-fluids in FSW can offer several advantages, including improved tool life, reduced welding defects, and improved joint strength. Here are some ways in which nano-fluids can be used in FSW applications(Boopathi, 2022d):

- **Improving tool life:** Nano-fluids can be used to improve the tool life in FSW by reducing the tool wear and extending the service life of the tool. The addition of nanoparticles to the lubricant can reduce the friction between the tool and the workpiece, which reduces the tool wear.
- **Reducing welding defects:** Nano-fluids can reduce the occurrence of welding defects in FSW by improving the heat transfer and reducing the thermal gradient across the weld zone. The high thermal conductivity of nanoparticles can improve the heat transfer between the tool and the workpiece, which reduces the risk of weld defects such as porosity, voids, and cracks.
- **Improving joint strength:** Nano-fluids can be used to improve the joint strength in FSW by promoting the formation of a fine-grained microstructure and enhancing the metallurgical bonding at the joint interface. The addition of nanoparticles to the lubricant can refine the grain structure of the weld zone and promote the formation of a uniform and fine-grained microstructure.
- **Enhancing process control:** Nano-fluids can enhance the process control in FSW by providing better control over the welding temperature and the degree of plastic deformation. The high lubricity of nano-fluids can also provide better control over the welding speed and feed rate, which improves the process stability and consistency.

Overall, the use of nano-fluids in FSW can offer several advantages, including improved tool life, reduced welding defects, improved joint strength, and enhanced process control(Boopathi, 2022d; Boopathi et al., 2022; Trojovský et al., 2023). However, more research is needed to fully explore the potential of nano-fluids in these applications and to address any potential safety concerns.

Nano-Fluid in Thermal Barrier Coating Process

Thermal barrier coatings (TBCs) are used in high-temperature applications to protect the underlying materials from damage due to thermal stresses. The use of nano-fluids in the TBC process can offer several advantages, including improved thermal insulation, enhanced durability, and reduced cost. Here are some ways in which nano-fluids can be used in TBC applications(Mansourian-Tabaei et al., 2022; Yang et al., 2019):

- **Improving thermal insulation:** Nano-fluids can be used to improve the thermal insulation properties of the TBCs. The addition of nanoparticles to the coating material can increase the effective thermal conductivity of the coating and reduce the heat transfer from the underlying material to the environment.
- **Enhancing durability:** Nano-fluids can enhance the durability of TBCs by improving the adhesion strength between the coating and the substrate. The high surface area of nanoparticles can provide a greater contact area between the coating and the substrate, which improves the bonding strength.
- **Reducing cost:** Nano-fluids can reduce the cost of TBCs by improving the efficiency of the coating process. The high surface area of nanoparticles can improve the dispersion of the coating material, which reduces the material usage and increases the coating efficiency.
- **Improving process control:** Nano-fluids can be used to improve the process control in the TBC process. The addition of nanoparticles to the coating material can improve the rheological properties of the material and provide better control over the coating thickness and uniformity.

Overall, the use of nano-fluids in the TBC process can offer several advantages, including improved thermal insulation, enhanced durability, reduced cost, and improved process control. However, more research is needed to fully explore the potential of nano-fluids in these applications and to address any potential safety concerns.

Tissue Engineering

Nano-fluids can play an important role in tissue engineering by providing a favorable environment for the growth and differentiation of cells. Tissue engineering involves the use of living cells, scaffolds, and growth factors to create functional tissues that can be used to repair or replace damaged or diseased tissues. Here are some ways in which nano-fluids can be used in tissue engineering applications(Hamad et al., 2021):

- **Cell culture:** Nano-fluids can be used as a culture medium for growing and maintaining cells in vitro. The addition of nanoparticles to the culture medium can provide a favorable environment for the growth and differentiation of cells, which can improve the viability, proliferation, and functionality of the cells.

- **Scaffold fabrication:** Nano-fluids can be used in the fabrication of scaffolds, which are three-dimensional structures that support the growth and differentiation of cells. The addition of nanoparticles to the scaffold material can improve the mechanical properties, porosity, and surface properties of the scaffold, which can enhance the adhesion, proliferation, and differentiation of cells.
- **Drug delivery:** Nano-fluids can be used as carriers for the delivery of drugs and growth factors to specific target cells or tissues. The high surface area and reactivity of nanoparticles can facilitate the loading and release of drugs and growth factors, which can improve the efficacy and specificity of the therapy.
- **Imaging and diagnosis:** Nano-fluids can be used as contrast agents for imaging and diagnosis of tissues and organs. The addition of nanoparticles to the imaging agent can improve the contrast and sensitivity of the imaging technique, which can enhance the accuracy and resolution of the diagnosis.

Overall, the use of nano-fluids in tissue engineering can provide several advantages, including improved cell viability and functionality, enhanced scaffold properties, improved drug delivery, and improved imaging and diagnosis. However, more research is needed to fully explore the potential of nano-fluids in these applications and to address any potential safety concerns.

3D Printings

Nano-fluids can be used in additive manufacturing to improve the quality and efficiency of the manufacturing process. Additive manufacturing, also known as 3D printing, involves the layer-by-layer deposition of materials to create complex geometries. The procedure to use nano-fluids in 3D printing depends on the specific printing technology and the type of nano-fluid being used. Here are some general steps that can be followed to use nano-fluids in 3D printing(Lim et al., 2017):

- **Select a suitable nano-fluid:** Choose a nano-fluid that is compatible with the printing material and has the desired properties, such as thermal conductivity, viscosity, and stability.
- **Prepare the printing material:** Mix the nano-fluid with the printing material according to the recommended ratio. The mixing process can be done manually or using specialized equipment, depending on the type of printing technology.
- **Adjust the printing parameters:** Modify the printing parameters, such as printing temperature, nozzle size, and printing speed, to accommodate the

addition of the nano-fluid. The printing parameters may need to be optimized to achieve the desired printing quality and efficiency.

- **Perform printing:** Load the prepared printing material into the 3D printer and start the printing process. Monitor the printing process to ensure that the printed parts meet the desired specifications.
- **Post-processing:** After the printing is completed, perform any necessary post-processing steps, such as cleaning, curing, or annealing, to finalize the printed parts.

It is important to note that the specific procedure for using nano-fluids in 3D printing may vary depending on the specific printing technology and the type of nano-fluid being used. Therefore, it is recommended to consult the manufacturer's instructions and conduct thorough testing and optimization before using nano-fluids in 3D printing applications.

Here are some ways in which nano-fluids can be used in additive manufacturing:

- **Improving print quality:** Nano-fluids can be used as a printing medium to improve the quality of the printed parts. The addition of nanoparticles to the printing material can improve the mechanical properties, surface finish, and dimensional accuracy of the printed parts.
- **Reducing printing time:** Nano-fluids can reduce the printing time in additive manufacturing by improving the heat transfer properties of the printing process. The high thermal conductivity of nanoparticles can enhance the heat transfer between the printing material and the print bed, which reduces the cooling time and improves the printing speed.
- **Enhancing process control:** Nano-fluids can enhance the process control in additive manufacturing by providing better control over the printing parameters. The high thermal conductivity and stability of nanoparticles can improve the process stability and consistency, which enhances the process control and reduces the risk of printing defects.
- **Improving material properties:** Nano-fluids can improve the material properties in additive manufacturing by reinforcing the printing material with nanoparticles. The addition of nanoparticles can improve the strength, stiffness, and durability of the printed parts, which enhances the functional performance of the parts.

Overall, the use of nano-fluids in additive manufacturing can provide several advantages, including improved print quality, reduced printing time, enhanced process control, and improved material properties. However, more research is needed to fully explore the potential of nano-fluids in these applications and to address any potential safety concerns.

SUMMARY

- There are various types of nano-fluids, including metal-based, carbon-based, and polymer-based, each with unique properties suitable for different applications. To make nano-fluids, two-phase and one-phase methods are commonly used, which involve the dispersion of nanoparticles in a base fluid.
- In manufacturing, nano-fluids can be used as coolants and lubricants in machining processes, resulting in improved machining accuracy and reduced tool wear. In medical applications, nano-fluids can be used for drug delivery, medical imaging, and tissue engineering. They can improve the efficacy of drugs by enhancing their delivery to specific target cells or tissues. In 3D printing, nano-fluids can be used as printing inks to create intricate and precise structures.
- Nano-fluids have unique thermal properties, making them useful for improving the efficiency of heat exchangers and energy conversion systems, such as solar energy and fuel cells. They can also be used to enhance the properties of printed structures in 3D printing, such as mechanical strength and thermal conductivity.
- The influence of nano-fluids is significant in various applications, as they can enhance the performance and efficiency of systems, reduce tool wear, and improve the efficacy of drugs. However, there are still challenges associated with the production and application of nano-fluids, such as their stability, toxicity, and cost.
- In summary, nano-fluids have broad applications in micro-machining, medical, thermal, and 3D printing, among others. Different types of nano-fluids can be used depending on the application requirements, and there are various methods for making them. The influence of nano-fluids can significantly improve the properties and performance of systems, making them a crucial area of research and development.

REFERENCES

Batyuk, L. V., & Kizilova, N. M. (2019). Maghetohydrodynamic flows of micro/nano fluids through thin capillaries. *Bulletin of Taras Shevchenko National University of Kyiv. Series. Physics and Mathematics*, *3*(3), 32–36. doi:10.17721/1812-5409.2019/3.4

Boopathi, S. (2022a). An Extensive Review on Sustainable Developments of Dry and Near-Dry Electrical Discharge Machining Processes. *Journal of Manufacturing Science and Engineering*, *144*(5), 50801. doi:10.1115/1.4052527

Boopathi, S. (2022b). An investigation on gas emission concentration and relative emission rate of the near-dry wire-cut electrical discharge machining process. *Environmental Science and Pollution Research International*, *29*(57), 86237–86246. doi:10.100711356-021-17658-1 PMID:34837614

Boopathi, S. (2022c). Cryogenically treated and untreated stainless steel grade 317 in sustainable wire electrical discharge machining process: A comparative study. *Environmental Science and Pollution Research International*, 1–10. doi:10.100711356-022-22843-x PMID:36057706

Boopathi, S. (2022d). Experimental investigation and multi-objective optimization of cryogenic Friction-stir-welding of AA2014 and AZ31B alloys using MOORA technique. *Materials Today. Communications*, *33*, 104937. doi:10.1016/j.mtcomm.2022.104937

Boopathi, S., Balasubramani, V., & Sanjeev Kumar, R. (2023). Influences of various natural fibers on the mechanical and drilling characteristics of coir-fiber-based hybrid epoxy composites. *Engineering Research Express*, *5*(1), 15002. doi:10.1088/2631-8695/acb132

Boopathi, S., Haribalaji, V., Mageswari, M., & Asif, M. M. (2022). Influences of Boron Carbide Particles on the Wear Rate and Tensile Strength of Aa2014 Surface Composite Fabricated By Friction-Stir Processing. *Materiali in Tehnologije*, *56*(3), 263–270. doi:10.17222/mit.2022.409

Emami, S., Dibaei Bonab, M. H., Mohammadiun, M., Mohammadiun, H., & Sadi, M. (2020). Effect of nano-fluid types on the improvement of heat transfer in a micro-channel with regular hexagonal pattern. *Proceedings of the Institution of Mechanical Engineers. Part E, Journal of Process Mechanical Engineering*, *234*(6), 657–664. doi:10.1177/0954408920932349

Hamad, E. M., Khaffaf, A., Yasin, O., Abu El-Rub, Z., Al-Gharabli, S., Al-Kouz, W., & Chamkha, A. J. (2021). Review of Nanofluids and Their Biomedical Applications. *Journal of Nanofluids*, *10*(4), 463–477. doi:10.1166/jon.2021.1806

Hegab, H., Darras, B., & Kishawy, H. A. (2018). Sustainability Assessment of Machining with Nano-Cutting Fluids. *Procedia Manufacturing*, *26*, 245–254. doi:10.1016/j.promfg.2018.07.033

Lee, P. H., Nam, T. S., Li, C., & Lee, S. W. (2010). Environmentally-friendly nano-fluid minimum quantity lubrication (MQL) meso-scale grinding process using nano-diamond particles. *Proceedings - 2010 International Conference on Manufacturing Automation, ICMA 2010*, (pp. 44–49). IEEE. 10.1109/ICMA.2010.27

Lim, S. M., Shin, B. S., & Kim, K. (2017). Characterization of products using additive manufacturing with graphene/photopolymer - Resin nano-fluid. *Journal of Nanoscience and Nanotechnology*, *17*(8), 5492–5495. doi:10.1166/jnn.2017.14159

Mansourian-Tabaei, M., Asiaee, A., Hutton-Prager, B., & Nouranian, S. (2022). Thermal barrier coatings for cellulosic substrates: A statistically designed molecular dynamics study of the coating formulation effects on thermal conductivity. *Applied Surface Science*, *587*, 152879. doi:10.1016/j.apsusc.2022.152879

Mebarek-Oudina, F., & Chabani, I. (2022). Review on Nano-Fluids Applications and Heat Transfer Enhancement Techniques in Different Enclosures. *Journal of Nanofluids*, *11*(2), 155–168. doi:10.1166/jon.2022.1834

Moita, A., Moreira, A., & Pereira, J. (2021). Nanofluids for the next generation thermal management of electronics: A review. *Symmetry*, *13*(8), 1362. doi:10.3390ym13081362

Reddy, P. M., Kumar, T. G., & Reddy, C. V. R. G. (2020). *Nano-fluids and its Applications*.

Rubbi, F., Habib, K., Saidur, R., Aslfattahi, N., Yahya, S. M., & Das, L. (2020). Performance optimization of a hybrid PV/T solar system using Soybean oil/MXene nanofluids as A new class of heat transfer fluids. *Solar Energy*, *208*, 124–138. doi:10.1016/j.solener.2020.07.060

Sampath, B., & Myilsamy, S. (2021). Experimental investigation of a cryogenically cooled oxygen-mist near-dry wire-cut electrical discharge machining process. *Strojniski Vestnik. Jixie Gongcheng Xuebao*, *67*(6), 322–330. doi:10.5545v-jme.2021.7161

Sharma, A. K., Tiwari, A. K., & Dixit, A. R. (2016). Effects of Minimum Quantity Lubrication (MQL) in machining processes using conventional and nanofluid based cutting fluids: A comprehensive review. *Journal of Cleaner Production*, *127*, 1–18. doi:10.1016/j.jclepro.2016.03.146

Smaisim, G. F., Mohammed, D. B., Abdulhadi, A. M., Uktamov, K. F., Alsultany, F. H., Izzat, S. E., Ansari, M. J., Kzar, H. H., Al-Gazally, M. E., & Kianfar, E. (2022). Nanofluids: Properties and applications. *Journal of Sol-Gel Science and Technology*, *104*(1), 1–35. doi:10.100710971-022-05859-0

Trojovský, P., Dhasarathan, V., & Boopathi, S. (2023). Experimental investigations on cryogenic friction-stir welding of similar ZE42 magnesium alloys. *Alexandria Engineering Journal*, *66*(1), 1–14. doi:10.1016/j.aej.2022.12.007

Yang, L., Ji, W., Huang, J., & Xu, G. (2019). An updated review on the influential parameters on thermal conductivity of nano-fluids. *Journal of Molecular Liquids*, *296*, 111780. doi:10.1016/j.molliq.2019.111780

Yoo, D. H., Hong, K. S., & Yang, H. S. (2007). Study of thermal conductivity of nanofluids for the application of heat transfer fluids. *Thermochimica Acta*, *455*(1–2), 66–69. doi:10.1016/j.tca.2006.12.006

Yurtkuran, H. (2021). An Evaluation on Machinability Characteristics of Titanium and Nickel Based Superalloys Used in Aerospace Industry. *İmalat Teknolojileri ve Uygulamaları*, *2*(2), 10–29. doi:10.52795/mateca.940261

Chapter 9
Applications of Nano-Cutting Fluids in Advanced Machining Processes

Sampath Boopathi
Muthayammal Engineering College, India

Meka Umareddy
University of Technology and Applied Sciences, Oman

M. Elangovan
https://orcid.org/0000-0002-3579-7510
SNS College of Technology, India

ABSTRACT

Nano-cutting fluids, also known as nanofluids, have gained significant attention in recent years as a means of enhancing the efficiency and effectiveness of various cutting processes. They consist of a base fluid, such as oil or water, with suspended nanoparticles that can improve the lubrication and cooling properties of the fluid. In various cutting processes, such as lathe, electrochemical machining (ECM), electrical discharge machining (EDM), and ultrasonic machining (USM), the use of nanofluids has been shown to improve surface finish, reduce tool wear, and enhance material removal rates. Nanofluids have increased thermal conductivity and lubrication properties, as well as the ability to fill in micro-gaps and reduce friction. Further research is needed to optimize their use in industrial applications.

DOI: 10.4018/978-1-6684-9135-5.ch009

INTRODUCTION

Nano cutting fluid is a type of cutting fluid that contains nanoparticles, typically less than 100 nanometers in size, which are added to the fluid to enhance its lubrication and cooling properties during metal cutting operations. The nanoparticles can be made of various materials such as oxides, carbides, and diamond. The use of nano cutting fluid can improve machining performance, increase tool life, and reduce energy consumption and waste. Additionally, it can minimize the environmental impact of metalworking operations by reducing the amount of cutting fluid needed and the amount of waste generated. Nano-cutting fluid is a recent development in machining technology that has revolutionized traditional machining processes. Nano-cutting fluids, also known as nanofluids, are engineered fluids that contain suspended nanoparticles of materials such as oxides, carbides, and nitrides. These nanoparticles have unique physical and chemical properties that make them suitable for various machining applications(Khandekar et al., 2012).

Nano-cutting fluids have been introduced in various machining processes, including conventional machining, electrochemical machining (ECM), electrical discharge machining (EDM), ultrasonic machining, abrasive water jet machining (AWJM), and others. In this article, we will explore the introduction of nano-cutting fluids in these various machining processes. Conventional Machining: In conventional machining processes such as turning, drilling, and milling, nano-cutting fluids have been introduced to improve the machining performance. The use of nano-cutting fluids can improve tool life, reduce cutting forces, and improve surface finish. These fluids also act as a coolant, reducing the heat generated during machining and reducing the risk of thermal damage to the workpiece. Nano-cutting fluids have shown promising results in various machining processes, including conventional machining, electrochemical machining, ultrasonic machining, and abrasive jet machining. These fluids are specially designed to provide better cooling and lubrication during the machining process, which leads to improved tool life and surface finish.

In conventional machining processes such as turning, drilling, and milling, the use of nano-cutting fluids has shown significant improvements in tool life, surface finish, and dimensional accuracy. The addition of nanoparticles to the cutting fluid improves its properties, such as thermal conductivity, viscosity, and lubricity, which reduces the cutting force and temperature. This results in lower tool wear and improved surface finish. Some common types of nano-cutting fluids used in conventional machining processes are oil-in-water emulsions, water-based solutions, and vegetable oils.

Electrochemical machining (ECM) is a non-traditional machining process used for precision machining of complex shapes and hard-to-machine materials. The use of nano-cutting fluids in ECM has been reported to improve the material removal rate and surface finish. The addition of nanoparticles to the electrolyte enhances the ionic conductivity and reduces the overpotential, which increases the machining rate and improves the surface finish. Some common types of nano-cutting fluids used in ECM are oil-in-water emulsions, water-based solutions, and ionic liquids.

Ultrasonic machining (USM) is a non-traditional machining process that uses ultrasonic vibrations to remove material from the workpiece. The use of nano-cutting fluids in USM has been reported to improve the material removal rate and surface finish. The addition of nanoparticles to the cutting fluid improves its properties, such as thermal conductivity, viscosity, and lubricity, which enhances the efficiency of material removal and reduces tool wear. Some common types of nano-cutting fluids used in USM are water-based solutions, oil-in-water emulsions, and ionic liquids. Ultrasonic machining is another process where nano-cutting fluids have been used to improve machining efficiency. The addition of nano-cutting fluids in ultrasonic machining has been shown to reduce tool wear, improve surface quality, and increase material removal rate(Said et al., 2019).

Abrasive jet machining (AJM) is a non-traditional machining process that uses a high-velocity stream of abrasive particles to remove material from the workpiece. The use of nano-cutting fluids in AJM has been reported to improve the material removal rate and surface finish. The addition of nanoparticles to the cutting fluid improves the properties of the abrasive particles, such as hardness, toughness, and wear resistance, which enhances the efficiency of material removal and reduces tool wear. Some common types of nano-cutting fluids used in AJM are water-based solutions, oil-in-water emulsions, and ionic liquids. In recent years, the use of nano-cutting fluids has gained significant attention in various machining processes. These fluids have been found to provide better lubrication, cooling, and anti-wear properties as compared to traditional cutting fluids. Additionally, they have the ability to penetrate into micro-channels, reduce friction and wear, and improve surface quality.

In grinding, nano-cutting fluids have been used to improve surface roughness, reduce grinding temperature and wheel wear, and increase material removal rate. In ECM, the use of nano-cutting fluids has been shown to reduce tool wear, improve surface quality, and increase the machining rate. In EDM, the addition of nano-cutting fluids has been found to improve the machining efficiency, reduce tool wear, and increase surface quality.

There are various types of nano-cutting fluids that can be used in machining processes. Some common types include vegetable oil-based nano-cutting fluids, water-based nano-cutting fluids, and oil-based nano-cutting fluids. Vegetable oil-based nano-cutting fluids are environmentally friendly and have been found to provide better lubrication, cooling, and anti-wear properties as compared to traditional cutting fluids. Water-based nano-cutting fluids are also environmentally friendly and can reduce the environmental impact of machining processes. Oil-based nano-cutting

Figure 1. Classification of nano-cutting fluids

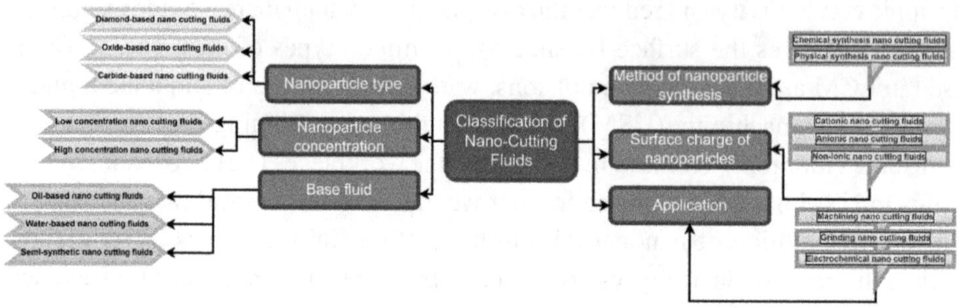

fluids have the advantage of providing better lubrication properties as compared to water-based nano-cutting fluids(Khandekar et al., 2012).

Thus, the use of nano-cutting fluids has provided significant benefits in various machining processes. The ability of these fluids to penetrate into micro-channels, reduce friction and wear, and improve surface quality has made them an attractive alternative to traditional cutting fluids. Additionally, the use of environmentally friendly nano-cutting fluids has reduced the environmental impact of machining processes. As the technology continues to evolve, it is likely that the use of nano-cutting fluids will become even more widespread in the manufacturing industry.

CLASSIFICATION OF NANO CUTTING FLUID

Nano cutting fluids can be classified based on various criteria, including the type of nanoparticles used, their concentration, the type of base fluid, and the intended application(Wu et al., 2021) (Figure 1). Here are some common classifications of nano cutting fluids:

Based on nanoparticle type:

- Oxide-based nano cutting fluids: contain nanoparticles of metal oxides such as alumina, silica, and titania.
- Carbide-based nano cutting fluids: contain nanoparticles of carbides such as silicon carbide, tungsten carbide, and boron carbide.
- Diamond-based nano cutting fluids: contain nanoparticles of diamond.

Based on nanoparticle concentration:

- Low concentration nano cutting fluids: contain nanoparticles at a concentration of less than 1%.
- High concentration nano cutting fluids: contain nanoparticles at a concentration of more than 1%.

Based on base fluid:

- Oil-based nano cutting fluids: use oil as a base fluid.
- Water-based nano cutting fluids: use water as a base fluid.
- Semi-synthetic nano cutting fluids: use a combination of oil and water as base fluids.

Based on application:

- General-purpose nano cutting fluids: suitable for a wide range of cutting operations.
- High-performance nano cutting fluids: designed for specific cutting operations such as high-speed machining, hard metal cutting, and dry machining.
- Environmentally friendly nano cutting fluids: formulated with biodegradable and non-toxic materials to minimize environmental impact.

Based on surface charge of nanoparticles:

- Cationic nano cutting fluids: contain positively charged nanoparticles.
- Anionic nano cutting fluids: contain negatively charged nanoparticles.
- Non-ionic nano cutting fluids: contain neutral nanoparticles.

Based on the method of nanoparticle synthesis:

- Chemical synthesis nano cutting fluids: nanoparticles are synthesized using chemical methods such as sol-gel and precipitation.
- Physical synthesis nano cutting fluids: nanoparticles are synthesized using physical methods such as laser ablation and sputtering.

Based on the intended use:

- Machining nano cutting fluids: used for metal cutting operations such as turning, milling, and drilling.
- Grinding nano cutting fluids: used for grinding operations such as surface grinding and cylindrical grinding.

- Electrochemical nano cutting fluids: used for electrochemical machining operations.

These are some of the common classifications of nano cutting fluids. The selection of a nano cutting fluid depends on the specific cutting operation and the material being machined.

NANO-CUTTING FLUIDS IN MACHINING PROCESS

Nano-cutting fluids are increasingly being used in conventional machining processes due to their ability to improve machining performance, increase tool life, and reduce environmental impact. Here are some common types of nano-cutting fluids and their applications in conventional machining processes:

- Alumina-based nano-cutting fluids: Alumina-based nano-cutting fluids are commonly used in turning and drilling operations of steel and aluminum alloys. They can reduce tool wear and improve surface finish due to their high lubrication and cooling properties.
- Silicon carbide-based nano-cutting fluids: Silicon carbide-based nano-cutting fluids are suitable for machining hard and brittle materials such as ceramics and composites. They can enhance the cutting performance by reducing tool wear and improving surface finish.
- Diamond-based nano-cutting fluids: Diamond-based nano-cutting fluids are used for high-speed machining operations on hard metals and alloys such as titanium and Inconel. They can reduce tool wear and improve machining efficiency due to their high thermal conductivity and lubrication properties.
- Carbon nanotube-based nano-cutting fluids: Carbon nanotube-based nano-cutting fluids are suitable for high-performance machining operations such as dry machining and micro-milling. They can reduce tool wear and improve machining accuracy due to their high strength and stiffness.
- Titanium dioxide-based nano-cutting fluids: Titanium dioxide-based nano-cutting fluids are used in grinding and polishing operations of ceramics, glasses, and semiconductors. They can enhance the surface finish and reduce the surface roughness by acting as a polishing agent.
- Hybrid nano-cutting fluids: Hybrid nano-cutting fluids are a combination of different types of nanoparticles and base fluids, and they are designed to improve the overall performance of the cutting fluid. For example, a hybrid nano-cutting fluid can contain both alumina and diamond nanoparticles,

which can improve both lubrication and thermal conductivity properties of the fluid.

- Magnetic nano-cutting fluids: Magnetic nano-cutting fluids are used in magnetic abrasive finishing and grinding operations. They contain magnetic nanoparticles that can be attracted to a magnetic field, which helps to keep the abrasive particles in contact with the workpiece surface.
- Graphene-based nano-cutting fluids: Graphene-based nano-cutting fluids are used for high-speed machining of aluminum and copper alloys. They can reduce tool wear and improve surface finish by providing a strong lubricating layer and high thermal conductivity.
- Polymer-based nano-cutting fluids: Polymer-based nano-cutting fluids are used in drilling and tapping operations of aluminum and titanium alloys. They can reduce tool wear and improve machining efficiency by providing high lubrication and cooling properties.

These are some additional types of nano-cutting fluids used in conventional machining processes. The choice of a nano-cutting fluid depends on the specific application and the material being machined, as well as the desired machining performance and environmental considerations.

NANO-CUTTING FLUIDS IN LATHE

To use nano-cutting fluid in lathe, you can follow these general steps(Arifuddin et al., 2022; Das et al., 2021; Khandekar et al., 2012):

- Choose the appropriate type of nano-cutting fluid based on the material being machined and the machining operation being performed. Consult the manufacturer's recommendations for guidance.
- Dilute the nano-cutting fluid with the recommended amount of water or oil, depending on the type of base fluid used. This will help to achieve the desired concentration of nanoparticles in the cutting fluid.
- Fill the lathe's coolant reservoir with the diluted nano-cutting fluid. Make sure that the coolant level is sufficient to cover the cutting tool and the workpiece.
- Turn on the lathe's coolant system and adjust the coolant flow rate and pressure to the desired level. This will help to ensure that the cutting tool and workpiece are properly lubricated and cooled during the machining operation.
- During the machining operation, monitor the coolant flow and adjust it as needed to maintain the desired coolant flow rate and pressure. Also, monitor

the cutting tool and workpiece for signs of overheating or wear, and adjust the coolant flow as needed to prevent these issues.

- After the machining operation is complete, turn off the lathe's coolant system and drain the coolant from the coolant reservoir. Dispose of the coolant in accordance with local environmental regulations.

By following these steps, you can effectively use nano-cutting fluid in a lathe to improve machining performance and extend the life of cutting tools.

Types of Nano-Cutting Fluid in Lathe

There are several types of nano-cutting fluids that can be used in lathe machining operations. Here are some examples:

- Alumina-based nano-cutting fluids: Alumina-based nano-cutting fluids are commonly used in turning and drilling operations of steel and aluminum alloys. They can reduce tool wear and improve surface finish due to their high lubrication and cooling properties.
- Silicon carbide-based nano-cutting fluids: Silicon carbide-based nano-cutting fluids are suitable for machining hard and brittle materials such as ceramics and composites. They can enhance the cutting performance by reducing tool wear and improving surface finish.
- Diamond-based nano-cutting fluids: Diamond-based nano-cutting fluids are used for high-speed machining operations on hard metals and alloys such as titanium and Inconel. They can reduce tool wear and improve machining efficiency due to their high thermal conductivity and lubrication properties.
- Polymer-based nano-cutting fluids: Polymer-based nano-cutting fluids are used in drilling and tapping operations of aluminum and titanium alloys. They can reduce tool wear and improve machining efficiency by providing high lubrication and cooling properties.
- Hybrid nano-cutting fluids: Hybrid nano-cutting fluids are a combination of different types of nanoparticles and base fluids, and they are designed to improve the overall performance of the cutting fluid. For example, a hybrid nano-cutting fluid can contain both alumina and diamond nanoparticles, which can improve both lubrication and thermal conductivity properties of the fluid.
- Carbon nanotube-based nano-cutting fluids: Carbon nanotube-based nano-cutting fluids are used for machining operations on hard and brittle materials such as ceramics and composites. They can reduce tool wear and improve

surface finish by providing a strong lubricating layer and high thermal conductivity.

- Graphene-based nano-cutting fluids: Graphene-based nano-cutting fluids are used for high-speed machining of aluminum and copper alloys. They can reduce tool wear and improve surface finish by providing a strong lubricating layer and high thermal conductivity.
- Magnetic nano-cutting fluids: Magnetic nano-cutting fluids are used in magnetic abrasive finishing and grinding operations. They contain magnetic nanoparticles that can be attracted to a magnetic field, which helps to keep the abrasive particles in contact with the workpiece surface.

These are some additional types of nano-cutting fluids that can be used in lathe machining operations. The choice of a nano-cutting fluid depends on the specific requirements of the machining operation and the material being machined. It's important to consult the manufacturer's recommendations to choose the appropriate type of nano-cutting fluid and ensure that it is used safely and effectively.

Effects of Nano-Cutting Fluid on the Lathe Performance

Nano-cutting fluids can have several effects on the performance of a lathe during machining operations. Some of these effects include:

- Improved tool life: Nano-cutting fluids can reduce tool wear and extend the life of cutting tools. This is due to their high lubrication and cooling properties, which help to reduce friction and heat during machining.
- Improved surface finish: Nano-cutting fluids can improve the surface finish of machined parts. This is due to their ability to reduce tool wear, which can result in less tool marks and surface defects.
- Increased machining efficiency: Nano-cutting fluids can improve the efficiency of machining operations by reducing tool wear and increasing cutting speeds. This can result in shorter machining times and increased productivity.
- Reduced environmental impact: Nano-cutting fluids can be designed to be environmentally friendly, by using biodegradable or non-toxic base fluids and reducing the amount of waste generated during machining.
- Health and safety benefits: Nano-cutting fluids can also provide health and safety benefits by reducing exposure to harmful chemicals and aerosols that can be generated during machining operations.
- Reduced power consumption: Nano-cutting fluids can help to reduce the power consumption of a lathe or other machining equipment by reducing

friction and heat generated during machining. This can result in energy savings and reduced operating costs.

- Enhanced chip removal: Nano-cutting fluids can improve the chip removal process by reducing the adhesion of chips to the cutting tool and workpiece. This can result in smoother chip flow, reduced chip clogging, and improved chip evacuation from the machining zone.
- Reduced machine downtime: By improving tool life and reducing the need for tool changes, nano-cutting fluids can help to reduce machine downtime and increase overall equipment effectiveness.
- Improved process control: Nano-cutting fluids can also help to improve process control by reducing variability in machining performance. This can result in improved consistency in part quality and reduced scrap rates.

However, it's important to note that the effectiveness of a nano-cutting fluid depends on several factors, including the type of fluid used, the machining operation being performed, and the material being machined. It's important to consult the manufacturer's recommendations and perform testing to ensure that a nano-cutting fluid is used safely and effectively in a lathe or any other machining equipment.

Overall, the use of nano-cutting fluids in lathe machining operations can provide several benefits, including improved tool life, surface finish, machining efficiency, environmental impact, health and safety, and reduced power consumption and machine downtime. However, it's important to carefully select and use these fluids in a safe and effective manner, and to monitor their performance regularly to ensure optimal results.

How to use Nano-cutting fluid in Grinding process

When using nano-cutting fluid in a grinding process, the following steps can be taken:

- Dilute the nano-cutting fluid with water to the recommended concentration.
- Apply the fluid to the grinding wheel or workpiece using a spray nozzle or other application method.
- Adjust the flow rate and pressure of the fluid to achieve the desired level of lubrication and cooling.
- Monitor the grinding process for signs of tool wear, surface finish, and other parameters.

The effects of nano-cutting fluids in a grinding process can include:

- Improved grinding performance: Nano-cutting fluids can improve the grinding performance by reducing tool wear and improving the surface finish of the workpiece.
- Reduced grinding forces: Nano-cutting fluids can reduce the grinding forces by providing a strong lubricating layer and reducing friction between the grinding wheel and workpiece.
- Improved surface quality: Nano-cutting fluids can improve the surface quality of the workpiece by reducing the formation of surface defects such as cracks, burns, and chatter marks.
- Improved chip evacuation: Nano-cutting fluids can improve the chip evacuation process by reducing the adhesion of chips to the grinding wheel and workpiece.
- Reduced energy consumption: Nano-cutting fluids can help to reduce the energy consumption of the grinding process by reducing the grinding forces and friction between the grinding wheel and workpiece.
- Improved process control: Nano-cutting fluids can also help to improve process control by reducing variability in grinding performance. This can result in improved consistency in part quality and reduced scrap rates.
- Reduced environmental impact: Nano-cutting fluids can be designed to be environmentally friendly, by using biodegradable or non-toxic base fluids and reducing the amount of waste generated during the grinding process.
- Health and safety benefits: Nano-cutting fluids can provide health and safety benefits by reducing exposure to harmful chemicals and aerosols that can be generated during grinding operations.

However, it's important to note that the effectiveness of a nano-cutting fluid depends on several factors, including the type of fluid used, the grinding operation being performed, and the material being ground. It's important to consult the manufacturer's recommendations and perform testing to ensure that a nano-cutting fluid is used safely and effectively in a grinding process. Additionally, proper safety measures such as wearing personal protective equipment should always be taken when handling any cutting fluid, including nano-cutting fluids.

NANO-CUTTING FLUIDS IN ECM PROCESS

Electrochemical machining (ECM) is a non-traditional machining process that uses an electrolytic solution to remove material from a workpiece(Khanafer et al., 2020; Rahman et al., 2020). When using nano-cutting fluids in an ECM process, the following steps can be taken:

- Dilute the nano-cutting fluid with water to the recommended concentration.
- Fill the ECM tank with the diluted nano-cutting fluid.
- Install the workpiece in the ECM machine and connect it to the appropriate electrodes.
- Adjust the process parameters such as current, voltage, and electrolyte flow rate according to the manufacturer's recommendations.
- Begin the ECM process and monitor the performance of the system.

The effects of nano-cutting fluids in an ECM process can include:

- Improved material removal rate: Nano-cutting fluids can improve the material removal rate by reducing the resistance of the electrolyte solution, which can result in faster and more efficient machining.
- Improved surface finish: Nano-cutting fluids can improve the surface finish of the workpiece by reducing the formation of surface defects such as burrs and roughness.
- Reduced tool wear: Nano-cutting fluids can reduce tool wear by providing a strong lubricating layer and reducing the formation of electrochemical deposits on the electrodes.
- Improved process control: Nano-cutting fluids can improve process control by reducing variability in machining performance, resulting in improved consistency in part quality and reduced scrap rates.
- Health and safety benefits: Nano-cutting fluids can provide health and safety benefits by reducing exposure to harmful chemicals and aerosols that can be generated during ECM operations. This can result in a safer working environment for operators and other personnel.
- Improved process efficiency: The use of nano-cutting fluids in an ECM process can lead to improved process efficiency, reduced operating costs, and increased production output.
- Reduced environmental impact: Nano-cutting fluids can be designed to be environmentally friendly, by using biodegradable or non-toxic base fluids and reducing the amount of waste generated during the ECM process.

Overall, the use of nano-cutting fluids in an ECM process can provide several benefits, including improved material removal rate, surface finish, tool wear, process control, health and safety, process efficiency, and reduced environmental impact. It's important to carefully select and use these fluids in a safe and effective manner, and to monitor their performance regularly to ensure optimal results.

Effects of Nano-Fluids on ECM Performance, Workpiece, Surface Finish

The use of nano-fluids in an electrochemical machining (ECM) process can have several effects on the performance of the process, the workpiece being machined, and the surface finish of the workpiece.

- Improved material removal rate: Nano-fluids can improve the material removal rate in an ECM process by reducing the resistance of the electrolyte solution and improving the mass transfer between the workpiece and the electrolyte.
- Improved surface finish: Nano-fluids can improve the surface finish of the workpiece by reducing the formation of surface defects such as burrs and roughness. This is achieved by reducing the formation of electrochemical deposits on the electrodes and the workpiece.
- Reduced tool wear: Nano-fluids can reduce tool wear in an ECM process by providing a strong lubricating layer and reducing the formation of electrochemical deposits on the electrodes.
- Improved process control: Nano-fluids can improve process control by reducing variability in machining performance, resulting in improved consistency in part quality and reduced scrap rates.
- Reduced heat generation: Nano-fluids can reduce heat generation in an ECM process by providing better cooling and lubrication, resulting in lower thermal stresses and reduced distortion of the workpiece.
- Improved workpiece surface integrity: Nano-fluids can improve the surface integrity of the workpiece by reducing the formation of micro-cracks and other defects that can occur during the machining process.

Furthermore, it is important to consider the potential risks associated with the use of nano-fluids in ECM processes. The small size of nanoparticles used in these fluids can potentially lead to adverse health effects, such as respiratory irritation, when inhaled or absorbed through the skin. It is recommended to use appropriate personal protective equipment and follow safe handling procedures when working with nano-fluids.

In summary, the use of nano-fluids in an ECM process can provide several benefits, including improved material removal rate, surface finish, tool wear, process control, reduced heat generation, and improved workpiece surface integrity. However, it is important to take appropriate safety precautions and to carefully consider the specific application before using nano-fluids in an ECM process.

NANO-CUTTING FLUID IN EDM PROCESS

The use of nano-cutting fluids in an electrical discharge machining (EDM) process can provide several benefits, including improved machining performance, reduced tool wear, and improved surface finish. Here are the steps to use nano-cutting fluids in an EDM process:

Select the appropriate nano-cutting fluid: There are several types of nano-cutting fluids available on the market. It is important to select the appropriate fluid based on the type of EDM process being used, the material being machined, and the desired outcome.

- Prepare the EDM machine: Before using the nano-cutting fluid, the EDM machine should be cleaned and prepared according to the manufacturer's recommendations.
- Mix the nano-cutting fluid with deionized water: Nano-cutting fluids are typically mixed with deionized water in a specific ratio. Follow the manufacturer's instructions for the correct mixing ratio.
- Fill the EDM machine's tank with the mixed solution: The mixed solution should be added to the EDM machine's tank, filling it to the recommended level.
- Start the EDM process: Once the tank is filled, the EDM process can be started as usual. The nano-cutting fluid will improve the machining performance, reduce tool wear, and improve the surface finish of the machined part.
- Monitor the process: It is important to monitor the EDM process and adjust the parameters as needed to ensure optimal performance. Regularly check the fluid level and replace the fluid as needed to maintain the correct ratio of nano-cutting fluid to deionized water.
- Dispose of the used fluid: Once the EDM process is complete, the used fluid should be disposed of properly according to local regulations. Some nano-cutting fluids may be recyclable or reusable, so it is important to check with the manufacturer for recommendations.

In summary, the use of nano-cutting fluids in an EDM process can provide several benefits, including improved machining performance, reduced tool wear, and improved surface finish. It is important to select the appropriate fluid, mix it with deionized water in the correct ratio, monitor the process, and dispose of the used fluid properly.

Impacts of Nano-Cutting Fluid in EDM Output Parameters

The use of nano-cutting fluids in electrical discharge machining (EDM) can have significant impacts on the output parameters of the process(Boopathi, 2022a, 2022d; Boopathi & Sivakumar, 2012a, 2012b, 2014; Dass james & Boopathi, 2016; Yupapin et al., 2023). Here are some of the impacts of using nano-cutting fluids in EDM:

- Improved machining performance: Nano-cutting fluids can improve the machining performance of the EDM process by increasing the material removal rate and reducing the tool wear. This is due to the unique properties of the nanoparticles in the fluid, such as their high surface area to volume ratio, which enhances their lubricating and cooling properties.
- Improved surface finish: Nano-cutting fluids can improve the surface finish of the machined part by reducing the number of microcracks and burrs that may form during the EDM process. The nanoparticles in the fluid help to reduce the heat generated during machining, which can minimize the formation of microcracks and burrs.
- Reduced electrode wear: Nano-cutting fluids can reduce the electrode wear during the EDM process, which can lead to longer electrode life and lower costs. This is due to the lubricating properties of the nanoparticles in the fluid, which help to reduce the friction between the electrode and the workpiece.
- Improved dimensional accuracy: Nano-cutting fluids can improve the dimensional accuracy of the machined part by reducing the thermal distortion caused by the EDM process. The nanoparticles in the fluid help to dissipate the heat generated during machining, which can minimize the thermal distortion of the workpiece.
- Reduced environmental impact: Nano-cutting fluids are typically more environmentally friendly than traditional cutting fluids. They have a lower environmental impact and are easier to dispose of.
- Reduced electric discharge duration: Nano-cutting fluids have been shown to reduce the electric discharge duration during the EDM process, which can lead to faster machining times and reduced energy consumption.
- Improved flushing capabilities: Nano-cutting fluids can improve the flushing capabilities of the EDM process by reducing the amount of debris and particles that can accumulate in the spark gap. This can help to improve the process stability and reduce the risk of damage to the workpiece.
- Reduced risk of thermal damage: Nano-cutting fluids can reduce the risk of thermal damage to the workpiece during the EDM process. The nanoparticles in the fluid help to dissipate the heat generated during machining, which can minimize the risk of thermal damage to the workpiece.

- Improved tool life: Nano-cutting fluids can improve the tool life during the EDM process by reducing the wear and tear on the tool. This can help to reduce costs associated with tool replacement and maintenance.

In conclusion, the use of nano-cutting fluids in EDM can have a significant impact on the output parameters of the process. These impacts include improved machining performance, improved surface finish, reduced electrode wear, improved dimensional accuracy, reduced environmental impact, reduced electric discharge duration, improved flushing capabilities, reduced risk of thermal damage, and improved tool life.

Various Process Parameters Applied in EDM Using Nano-Cutting Fluid

In electrical discharge machining (EDM), several process parameters are used to control the machining operation. When nano-cutting fluid is used in the EDM process, some of these process parameters may need to be adjusted to achieve the desired results(Boopathi, 2022c; Gowri et al., 2023; Kannan et al., 2022; Yupapin et al., 2023). Here are some of the process parameters that can be adjusted when using nano-cutting fluids in EDM:

Pulse current: The pulse current is the current that is applied to the workpiece during the EDM process. Adjusting the pulse current can help to control the material removal rate and the quality of the machined surface.

Pulse duration: The pulse duration is the length of time that the electrical discharge is applied to the workpiece. Adjusting the pulse duration can help to control the depth of cut and the material removal rate.

Pulse frequency: The pulse frequency is the number of pulses per unit time. Adjusting the pulse frequency can help to control the material removal rate and the quality of the machined surface.

Open circuit voltage: The open circuit voltage is the voltage that is applied to the electrode and the workpiece when the electrical discharge is not occurring. Adjusting the open circuit voltage can help to control the spark gap and the quality of the machined surface.

Dielectric fluid flow rate: The dielectric fluid flow rate is the rate at which the nano-cutting fluid is supplied to the spark gap. Adjusting the flow rate can help to control the cooling and flushing of the spark gap.

Dielectric fluid pressure: The dielectric fluid pressure is the pressure at which the nano-cutting fluid is supplied to the spark gap. Adjusting the pressure can help to control the cooling and flushing of the spark gap.

Gap distance: The gap distance is the distance between the electrode and the workpiece. Adjusting the gap distance can help to control the material removal rate and the quality of the machined surface.

Electrode material: The electrode material used in EDM can also impact the machining process when using nano-cutting fluid. The material properties can affect the rate of material removal and the surface finish of the machined part.

Particle concentration: The concentration of nanoparticles in the nano-cutting fluid can also be adjusted to optimize the EDM process. Higher concentrations may increase the cooling and flushing effects, but may also lead to more wear on the electrode.

Pulse-on time and pulse-off time: The pulse-on time is the time duration when electrical discharge is applied to the workpiece, and the pulse-off time is the time duration when no electrical discharge occurs. Adjusting these times can help to control the machining parameters and the quality of the machined surface.

Workpiece material: The workpiece material can also affect the EDM process when using nano-cutting fluid. Different materials may require different machining parameters, such as pulse current and frequency, to achieve the desired results.

By adjusting these process parameters, it is possible to optimize the use of nano-cutting fluid in the EDM process and achieve the desired results.

Overall, adjusting these process parameters when using nano-cutting fluid in EDM can help to optimize the machining process and achieve the desired results. It is important to carefully consider each parameter and how it impacts the machining operation to achieve the best performance.

Nano-Cutting Fluid (Vegetable Oil): EDM Process

Nano-cutting fluids made from vegetable oil can be used in the electrical discharge machining (EDM) process as an alternative to traditional dielectric fluids. Vegetable oils have several advantages over traditional dielectric fluids, including better biodegradability, lower toxicity, and higher flash point(Boopathi, 2021, 2022d, 2022b; Gunasekaran et al., 2022; Myilsamy & Sampath, 2021; Sampath & Myilsamy, 2021).

When vegetable oil-based nano-cutting fluids are used in EDM, they can improve the machining performance and reduce the environmental impact of the machining process. The nanoparticles in the cutting fluid can enhance the cooling and flushing effects, leading to better surface finish and reduced electrode wear.

However, using vegetable oil-based nano-cutting fluids in EDM requires careful consideration of the machining parameters and the choice of electrode material. Vegetable oil-based fluids have a higher viscosity and lower dielectric strength than traditional dielectric fluids, which can impact the machining performance.

Therefore, it is important to adjust the machining parameters, such as pulse current and frequency, to achieve optimal results.

The choice of electrode material can also impact the machining performance when using vegetable oil-based nano-cutting fluids. Some electrode materials may experience higher wear rates when used with vegetable oil-based fluids, which can reduce the machining accuracy and increase the cost of the process.

In summary, using vegetable oil-based nano-cutting fluids in EDM can offer several advantages over traditional dielectric fluids. However, careful consideration of the machining parameters and the choice of electrode material is necessary to achieve optimal results.

NANO-CUTTING FLUID USES IN ULTRASONIC MACHINING PROCESSES

Nano-cutting fluids can also be used in ultrasonic machining (USM) processes to improve machining performance. USM is a non-traditional machining process that uses high-frequency vibrations to remove material from a workpiece. The use of nano-cutting fluids in USM can enhance the cooling and lubrication effects during the machining process, leading to better surface finish and reduced tool wear(Abdelrazek et al., 2020; Das et al., 2021; Khan et al., 2020; Sharmin et al., 2020; Wu et al., 2021).

The following are some of the ways nano-cutting fluids can be used in USM processes:

- As a cooling and lubrication agent: The nanoparticles in the cutting fluid can help to reduce friction between the tool and the workpiece, leading to reduced tool wear and improved surface finish.
- To improve material removal rate: The use of nano-cutting fluids in USM can enhance the material removal rate by providing better cooling and lubrication, allowing for more efficient machining.
- To reduce tool wear: The use of nano-cutting fluids in USM can also help to reduce tool wear by providing a protective layer between the tool and the workpiece, reducing friction and heat generation.
- To improve surface finish: Nano-cutting fluids can help to improve the surface finish of the machined part by reducing tool wear and providing better lubrication during the machining process.
- To reduce environmental impact: Nano-cutting fluids can be an environmentally friendly alternative to traditional cutting fluids, as they are biodegradable and non-toxic.

Overall, the use of nano-cutting fluids in USM can provide several benefits, including improved machining performance, reduced tool wear, and improved surface finish. Careful consideration of the machining parameters and the choice of cutting fluid is necessary to achieve optimal results.

When using nano-cutting fluids in USM, the following factors should be considered:

- Choice of cutting fluid: The choice of cutting fluid can impact the machining performance in USM. It is important to select a cutting fluid that is compatible with the material being machined and can provide the desired cooling and lubrication effects.
- Concentration of cutting fluid: The concentration of cutting fluid in USM can impact the material removal rate and surface finish. Higher concentrations of cutting fluid can lead to improved cooling and lubrication effects, but may also result in reduced material removal rates.
- Machining parameters: The machining parameters in USM, such as vibration frequency, amplitude, and tool pressure, should be adjusted to optimize the performance of the cutting fluid.
- Type of workpiece: The type of workpiece being machined can also impact the performance of the cutting fluid. For example, some materials may require a higher concentration of cutting fluid or different machining parameters to achieve optimal results.

Overall, the use of nano-cutting fluids in USM can improve machining performance and reduce tool wear. However, careful consideration of the cutting fluid concentration, machining parameters, and workpiece material is necessary to achieve optimal results.

Nano-cutting fluids in Abrasive jet machining

Nano-cutting fluids can also be used in abrasive jet machining (AJM) to improve the machining performance and surface finish. AJM is a non-traditional machining process that uses a high-velocity stream of abrasive particles to remove material from a workpiece.

The use of nano-cutting fluids in AJM can provide the following benefits:

- Improved surface finish: The nanoparticles in the cutting fluid can help to reduce the surface roughness of the machined part by reducing the impact of abrasive particles and providing better cooling and lubrication.
- Reduced tool wear: The use of cutting fluids in AJM can help to reduce the wear of the nozzle and improve the lifespan of the abrasive particles.

- Better control of machining parameters: The use of cutting fluids can help to stabilize the abrasive particles and improve their penetration depth, resulting in better control of machining parameters such as depth of cut and material removal rate.
- Reduced environmental impact: Nano-cutting fluids are environmentally friendly and non-toxic, making them a good alternative to traditional cutting fluids.
- Type of cutting fluid: The choice of cutting fluid is critical in AJM, as it can impact the machining performance and surface finish. It is important to select a cutting fluid that is compatible with the abrasive particles and can provide the desired cooling and lubrication effects.
- Concentration of cutting fluid: The concentration of cutting fluid can impact the material removal rate and surface finish. Higher concentrations of cutting fluid can lead to improved cooling and lubrication effects, but may also result in reduced material removal rates.
- Machining parameters: The machining parameters in AJM, such as abrasive particle velocity, nozzle diameter, and abrasive flow rate, should be adjusted to optimize the performance of the cutting fluid.
- Type of workpiece: The type of workpiece being machined can also impact the performance of the cutting fluid. For example, some materials may require a higher concentration of cutting fluid or different machining parameters to achieve optimal results.

Overall, the use of nano-cutting fluids in AJM can provide several benefits, including improved surface finish, reduced tool wear, and better control of machining parameters. Careful consideration of the cutting fluid concentration, machining parameters, and workpiece material is necessary to achieve optimal results.

IMPACTS VERSUS TYPES OF NANO-CUTTING FLUIDS IN ABRASIVE-WATER JET MACHINING

Abrasive water jet machining (AWJM) is a non-traditional machining process that uses a high-pressure stream of water and abrasive particles to remove material from a workpiece. The use of nano-cutting fluids in AWJM can have different impacts depending on the type of cutting fluid used. Some of the common types of nano-cutting fluids used in AWJM are:

- Vegetable oil-based cutting fluids: These cutting fluids are biodegradable, non-toxic, and environmentally friendly. The use of vegetable oil-based cutting fluids in AWJM can help to improve the surface finish and reduce tool wear. They are also effective in reducing environmental pollution.

- Synthetic-based cutting fluids: These cutting fluids are usually composed of synthetic oil and additives that provide lubrication and cooling. The use of synthetic-based cutting fluids in AWJM can help to improve the machining performance and increase the material removal rate.
- Water-based cutting fluids: These cutting fluids are composed of water and additives that provide cooling and lubrication. The use of water-based cutting fluids in AWJM can help to improve the surface finish and reduce tool wear.

The impacts of nano-cutting fluids in AWJM can include:

- Improved surface finish: The use of nano-cutting fluids can help to reduce the surface roughness of the machined part by reducing the impact of abrasive particles and providing better cooling and lubrication.
- Reduced tool wear: The use of cutting fluids in AWJM can help to reduce the wear of the nozzle and improve the lifespan of the abrasive particles.
- Increased material removal rate: The use of cutting fluids can help to increase the material removal rate by improving the penetration of abrasive particles into the workpiece.
- Reduced environmental impact: Nano-cutting fluids are environmentally friendly and non-toxic, making them a good alternative to traditional cutting fluids.
- Cost-effective: Nano-cutting fluids can be cost-effective due to their longer lifespan and reduced tool wear.

In summary, the use of nano-cutting fluids in AWJM can provide several benefits, including improved surface finish, reduced tool wear, increased material removal rate, and reduced environmental impact. The choice of cutting fluid type should be based on the specific machining requirements and workpiece material.

SUMMARY

Nano-cutting fluids, or nanofluids, are a type of cutting fluid that contains suspended nanoparticles to improve lubrication and cooling properties during cutting processes. They are classified into two main types: oil-based and water-based, with different types of nanoparticles used for each.

The use of nanofluids in cutting processes, such as lathe, electrochemical machining (ECM), electrical discharge machining (EDM), and ultrasonic machining

(USM), has been shown to improve surface finish, reduce tool wear, and enhance material removal rates. The benefits of nanofluids can be attributed to their increased thermal conductivity, lubrication properties, and the ability of nanoparticles to fill in micro-gaps and reduce friction.

In lathe machining, nanofluids have been found to reduce cutting force and improve surface roughness. In ECM, nanofluids can improve machining accuracy and reduce electrolyte consumption. In EDM, nanofluids can reduce tool wear and improve machining speed. In USM, nanofluids can improve surface quality and efficiency.

The effects of nanofluids on cutting processes are still being researched, and more work is needed to fully understand their potential in industrial applications. However, it is clear that the use of nanofluids has the potential to improve machining performance and efficiency in a variety of cutting processes.

REFERENCES

Abdelrazek, A. H., Choudhury, I. A., Nukman, Y., & Kazi, S. N. (2020). Metal cutting lubricants and cutting tools: A review on the performance improvement and sustainability assessment. *International Journal of Advanced Manufacturing Technology*, *106*(9-10), 4221–4245. doi:10.100700170-019-04890-w

Arifuddin, A., Redhwan, A. A. M., Azmi, W. H., & Zawawi, N. N. M. (2022). Performance of Al2O3/TiO2 Hybrid Nano-Cutting Fluid in MQL Turning Operation via RSM Approach. *Lubricants (Basel, Switzerland)*, *10*(12), 366. doi:10.3390/lubricants10120366

Boopathi, S. (2021). Improving of Green Sand-Mould Quality using Taguchi Technique. *Journal of Engineering Research*. doi:10.36909/jer.14079

Boopathi, S. (2022a). An Extensive Review on Sustainable Developments of Dry and Near-Dry Electrical Discharge Machining Processes. *Journal of Manufacturing Science and Engineering*, *144*(5), 50801. doi:10.1115/1.4052527

Boopathi, S. (2022b). An investigation on gas emission concentration and relative emission rate of the near-dry wire-cut electrical discharge machining process. *Environmental Science and Pollution Research International*, *29*(57), 86237–86246. doi:10.100711356-021-17658-1 PMID:34837614

Boopathi, S. (2022c). Cryogenically treated and untreated stainless steel grade 317 in sustainable wire electrical discharge machining process: A comparative study. *Environmental Science and Pollution Research International*, 1–10. doi:10.100711356-022-22843-x PMID:36057706

Boopathi, S. (2022d). Performance Improvement of Eco-Friendly Near-Dry Wire-Cut Electrical Discharge Machining Process Using Coconut Oil-Mist Dielectric Fluid. *Journal of Advanced Manufacturing Systems*, 1–20. doi:10.1142/S0219686723500178

Boopathi, S., & Sivakumar, K. (2012a). Experimental Analysis of Eco-friendly Near-dry Wire Electrical Discharge Machining Process. *Archives des Sciences*, *65*(10), 334–346.

Boopathi, S., & Sivakumar, K. (2012b). Experimental comparative study of near-dry wire-cut electrical discharge machining (WEDM). *European Journal of Scientific Research*, *75*(4), 472–481.

Boopathi, S., & Sivakumar, K. (2014). Study of water assisted dry wire-cut electrical discharge machining. *Indian Journal of Engineering and Materials Sciences*, *21*(1), 75–82.

Das, A., Patel, S. K., Arakha, M., Dey, A., & Biswal, B. B. (2021). Processing of hardened steel by MQL technique using nano cutting fluids. *Materials and Manufacturing Processes*, *36*(3), 316–328. doi:10.1080/10426914.2020.1832688

Dass james, A., & Boopathi, S. (2016). Experimental Study of Eco-friendly Wire-Cut Electrical Discharge Machining Processes. *International Journal of Innovative Research in Science, Engineering and Technology*, *5*(1), 666–675.

Gowri, N. V., Dwivedi, J. N., Krishnaveni, K., Boopathi, S., Palaniappan, M., & Medikondu, N. R. (2023). Experimental investigation and multi-objective optimization of eco-friendly near-dry electrical discharge machining of shape memory alloy using Cu/SiC/Gr composite electrode. *Environmental Science and Pollution Research International*, 1–19. doi:10.100711356-023-26983-6 PMID:37126160

Gunasekaran, K., Boopathi, S., & Sureshkumar, M. (2022). Analysis of a Cryogenically Cooled Near-Dry Wedm Process Using Different Dielectrics. *Materiali in Tehnologije*, *56*(2), 179–186. doi:10.17222/mit.2022.397

Kannan, E., Trabelsi, Y., Boopathi, S., & Alagesan, S. (2022). Influences of cryogenically treated work material on near-dry wire-cut electrical discharge machining process. *Surface Topography : Metrology and Properties*, *10*(1), 15027. doi:10.1088/2051-672X/ac53e1

Khan, F., Soota, T., & ... (2020). Optimization of EDM process parameter for stainless steel D3. *Materials Today: Proceedings*, *25*, 635–638. doi:10.1016/j.matpr.2019.07.529

Khanafer, K., Eltaggaz, A., Deiab, I., Agarwal, H., & Abdul-Latif, A. (2020). Toward sustainable micro-drilling of Inconel 718 superalloy using MQL-Nanofluid. *International Journal of Advanced Manufacturing Technology*, *107*(7-8), 3459–3469. doi:10.100700170-020-05112-4

Khandekar, S., Sankar, M. R., Agnihotri, V., & Ramkumar, J. (2012). Nano-cutting fluid for enhancement of metal cutting performance. *Materials and Manufacturing Processes*, *27*(9), 963–967. doi:10.1080/10426914.2011.610078

Myilsamy, S., & Sampath, B. (2021). Experimental comparison of near-dry and cryogenically cooled near-dry machining in wire-cut electrical discharge machining processes. *Surface Topography : Metrology and Properties*, *9*(3), 35015. doi:10.1088/2051-672X/ac15e0

Rahman, M., Woon, K. S., & Neo, W. K. (2020). Tool-Based Micro/Nano Machining: Development of Innovative Machine and Machining Processes. *International Manufacturing Science and Engineering Conference, 84263*. IEEE. 10.1115/MSEC2020-8580

Said, Z., Gupta, M., Hegab, H., Arora, N., Khan, A. M., Jamil, M., & Bellos, E. (2019). A comprehensive review on minimum quantity lubrication (MQL) in machining processes using nano-cutting fluids. *International Journal of Advanced Manufacturing Technology*, *105*(5-6), 2057–2086. doi:10.100700170-019-04382-x

Sampath, B., & Myilsamy, S. (2021). Experimental investigation of a cryogenically cooled oxygen-mist near-dry wire-cut electrical discharge machining process. *Strojniski Vestnik. Jixie Gongcheng Xuebao*, *67*(6), 322–330. doi:10.5545v-jme.2021.7161

Sharmin, I., Gafur, M. A., & Dhar, N. R. (2020). Preparation and evaluation of a stable CNT-water based nano cutting fluid for machining hard-to-cut material. *SN Applied Sciences*, *2*(4), 1–18. doi:10.100742452-020-2416-x

Wu, X., Li, C., Zhou, Z., Nie, X., Chen, Y., Zhang, Y., Cao, H., Liu, B., Zhang, N., Said, Z., Debnath, S., Jamil, M., Ali, H. M., & Sharma, S. (2021). Circulating purification of cutting fluid: An overview. *International Journal of Advanced Manufacturing Technology*, *117*(9–10), 2565–2600. doi:10.100700170-021-07854-1 PMID:34465936

Yupapin, P., Trabelsi, Y., Nattappan, A., & Boopathi, S. (2023). Performance Improvement of Wire-Cut Electrical Discharge Machining Process Using Cryogenically Treated Super-Conductive State of Monel-K500 Alloy. *Iranian Journal of Science and Technology. Transaction of Mechanical Engineering*, *47*(1), 267–283. doi:10.100740997-022-00513-0

Chapter 10
Applications of Green Nanofluids in Thermal Systems:
An Innovative Study

A. Kalaiyarasan
https://orcid.org/0000-0002-1033-240X
Muthayammal Engineering College, India

S. Sundaram
https://orcid.org/0000-0002-9750-8139
Muthayammal Engineering College, India

K. Sankar
https://orcid.org/0000-0002-5189-2788
Muthayammal Engineering College, India

S. Priya
https://orcid.org/0009-0006-2358-1639
Muthayammal College of Engineering, India

ABSTRACT

Increased energy consumption is causing a constant rise in global temperature above pre-industrial levels, with harmful gas and radiation emissions causing catastrophic climatic conditions. To address this issue, it is essential to plan thermal solutions that are continuing and extremely efficient. Because of its exceptional qualities, scientists are doing research on diverse nanofluids to develop routine in thermal applications. has attracted substantial interest in recent years. However, because they are more reactive in solvents, it raises safety and health problems. Due to the presence of powerful chemicals, other undesirable consequences such as rusting of equipment and non-biodegradable by products develop. The current study provides a thorough overview of the many methods used to create green nanofluids utilizing a variety of natural extracts. Green nanoparticles are a good choice for improving the

DOI: 10.4018/978-1-6684-9135-5.ch010

efficiency of thermal systems due to their distinctive shape, optical characteristics, stability, large surface area, low toxicity, and improved thermo-physical properties.

INTRODUCTION

Due to their potential to improve heat transmission, nanofluids have garnered a lot of attention in recent years. As colloidal agents, aluminum alloy, iron, copper, gold, or their oxides are used. the the majority well-liked base fluid, such as wet, ethylene, toluene, glycol, or lubricate, to enhance the base fluid's thermo-physical qualities of heat transfer. The fluids containing Nanoparticles (1–100 nm) were referred to as "nanofluids" by Choi and Eastman (1995). Due to their uses, these nanofluids acquired pace and appeal, opening the door for a new class of heat transfer fluids. According to Choi (1995), adding Nanoparticles to ordinary liquids can boost their heat conductivity. A novel Lattice-Boltzmann technique (LBM) was used by Nt. Kamyar et al. (2012) to examine the intriguing behaviour of cooling fluids containing ultrafine Nanoparticles. It's interesting to note that the majority of these computer simulations support and concur with experimental findings. The nanofluids model, which describes how thermophoresis and Brownian motion affect nanofluids, was researched by Rana and Bhargava (2012). The effects of thermophoresis and Brownian motion on the boundary layer flow of a nanofluid were examined by Rana and Bhargava (2012). Many researches (Ahmada et al., 2014; Akyildiz et al., 2011; Goyal & Bhargava, 2014; Makinde & Aziz, 2011; Rashidi et al., 2014; Rashidi et al., 2015; Sadegh et al., 2015; Vajravelu et al., 2011; Van Gorder et al., 2010) have looked at the properties of nanofluids that improve thermal conductivity and heat transfer characteristics through various shapes. All of the preceding investigations examined while assuming constant thermo physical characteristics of the base fluid, the parameters of nanofluids flows. However, it is evident from the work of Lai and Kulacki (1990) that the fluid's physical characteristics, notably its viscosity and thermal conductivity, have an important role. may alter with temperature. Heat created by internal friction and the resulting rise in temperature modify the physical characteristics of lubricating fluids, and the fluid's qualities are no longer believed to remain constant.

The rise in temperature causes an increase in the transport phenomenon, which affects heat transfer at the wall. Heat transfer improvement has received a lot of attention in recent years because of its importance in engineering and industrial applications. Traditionally, warm remove fluid such as wet, ethylene glycol, and lubricate are utilized in most engineering equipment such as heat exchangers and electronic devices. However, because these base fluids have low thermal conductivity, heat transfer increase is limited. Thermal scientists enhance the aforementioned

Figure 1. Schematic diagram of nanofluids

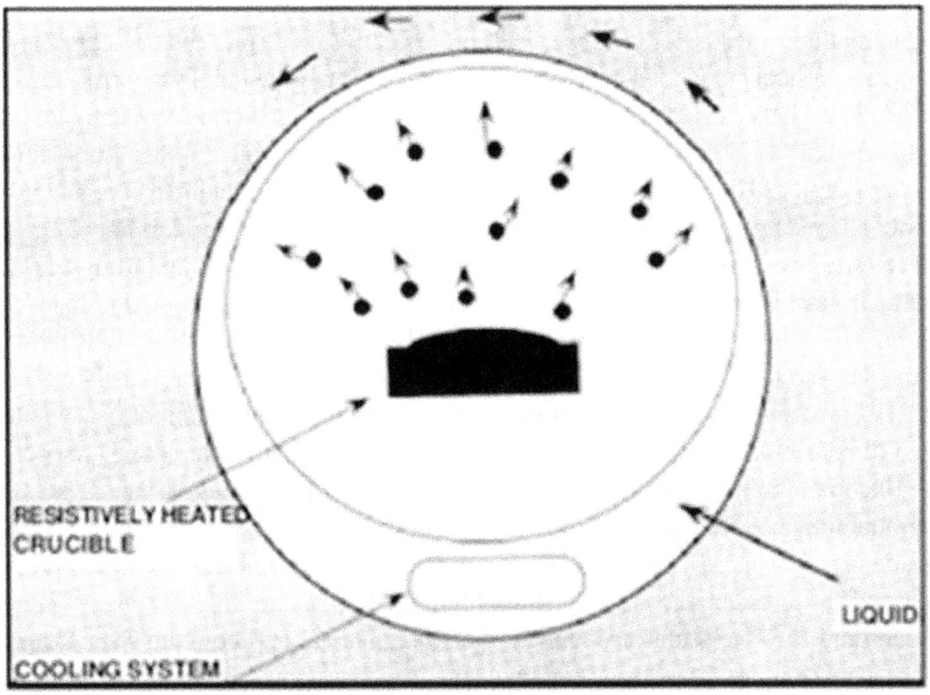

fluids by adding a single type of nanosized particle to make a combination dubbed 'nanofluids,' which was initially developed by Choi and Eastman (1995).

Since then, numerous researchers have studied the heat transmission of nanofluids for various features, either experimentally or statistically. Tiwari and Das (2007), for example, report on their research on nanofluids mathematical models. Since then, several scientists have investigated the heat transmission of nanofluids for various features, either experimentally or statistically. For instance, Tiwari and Das (2007) report on their work into the mathematical models of nanofluids. The purpose of the current work is to investigate the hybrid nanofluids flow and heat transfer across a sheet that is being stretched or contracted for uniform shear flow. The impact of transpiration is also taken into account. We are using the Tiwari and Das nanofluids equations model here (2007). This model examines the behavior of hybrid nanofluids while taking into account the solid volume fractions of copper (Cu) and alumina (Al_2O_3) Nanoparticles. They are then dispersed in water to create Al2O3-Cu/water hybrid nanofluids. For the goal of validation, the current numerical results are contrasted with those of previously published data. In this respect, the study on the improvement of a hybrid nanofluids heat transfer has been looked at as a novel idea in the problem of boundary layer flow.

Devi and Devi (2016) explore the subject of mixture Cu-Al_2O_3/water nanofluids flood crossways a holey extended piece by means of hydro magnetic effects by computational means, for instance. The authors of this paper presented a brand-new thermo physical model of a hybrid nanofluids. The experimental findings of Suresh et al. (2011) have been used to verify and validate this novel thermo physical model, which provides a good correlation with the pre-existing experimental data. According to Seth et al. (2019), viscous dissipation, Joule heating, thermal diffusion, and nth order chemical reaction were all taken into account in the numerical modeling of the Newtonian heating effect on unsteady MHD flow of Casson fluid through a flat vertical plate. Seth et al. (2018) studied the effects of Joule heating, viscous dissipation, thermo diffusion, and Newtonian heating on the unsteady MHD flow of a Casson fluid near a vertical oscillating plate through a non-Darcy porous material. Seth et al. (2019), investigated the simultaneous impacts of the Soret and Dufour effects on the two-dimensional MHD free convective flow of an electrically conducting, viscous, and incompressible visco-elastic fluid over a nonlinearly expanding surface. Systems for attaining global sustainability are constantly being improved by businesses and academics. The increased interest in system improvements spurred by enhancement approaches is essential for establishing highly effective and long-lasting systems. According to a particular account by the UN financial and communal board on sustainable progress objectives, employing improved systems might result in material savings of over 92 billion tonnes.

Low energy and material usage from highly efficient systems lowers the overall amount of trash produced. However, there is growing worry over a rise in global temperature as a result of these systems' radiation and gas emissions, which has become a very important socioeconomic issue (Choi & Eastman, 1995). Between 2030 and 2052, this rise is predicted to range from 1.5 to 1.7 °C over pre-industrial values, resulting in harsh climatic conditions (Buongiorno, 2006). In 2018 (Kamyar et al., 2012), the rise in worldwide energy consumption and associated CO2 emissions was around 2.3% and 1.7%, respectively. These increases are anticipated to continue, leading to a further increase in the earth's temperature. Therefore, it is crucial to create systems that are highly efficient, long-lasting, eco-friendly, and affordable so that they may serve as an alternative energy generating technique for reducing emissions. Additionally, in order to reduce the world temperature, these systems should aim to comply with the sustainability standards established by the Paris Climate Change Agreement and the joint Nations structure gathering on type of weather modify (Rana & Bhargava, 2012). The objective of the up to date review piece is to estimate the use of emerald nanofluids in thermal system while taking into account their benefits for both human health and the environment. It also identifies potential future research directions and technological advancements that could lead to safer actual applications.

Systems for transferring heat are used in many sectors. According to some reports, less effective thermal systems can result in a total heat loss of up to 50% of the energy intake (Rashidi et al., 2014). For heat transfer systems to achieve energy and financial savings, energy management is important. Furthermore, improving the performance of heat transfer systems requires a thorough understanding of the heat exchange process. Additionally, increasing efficiency for applications using non-renewable resources may lead to decreased emissions, which would slow down drastically changing climatic conditions (Choi & Eastman, 1995). The different types of analysis the parameter of thermal property and heat transfer property of the Nanoparticles, Nanofluids and Nano technology as shown in Figure 1.

Figure 2. Analysis of the parameter required to property of heat transfer application

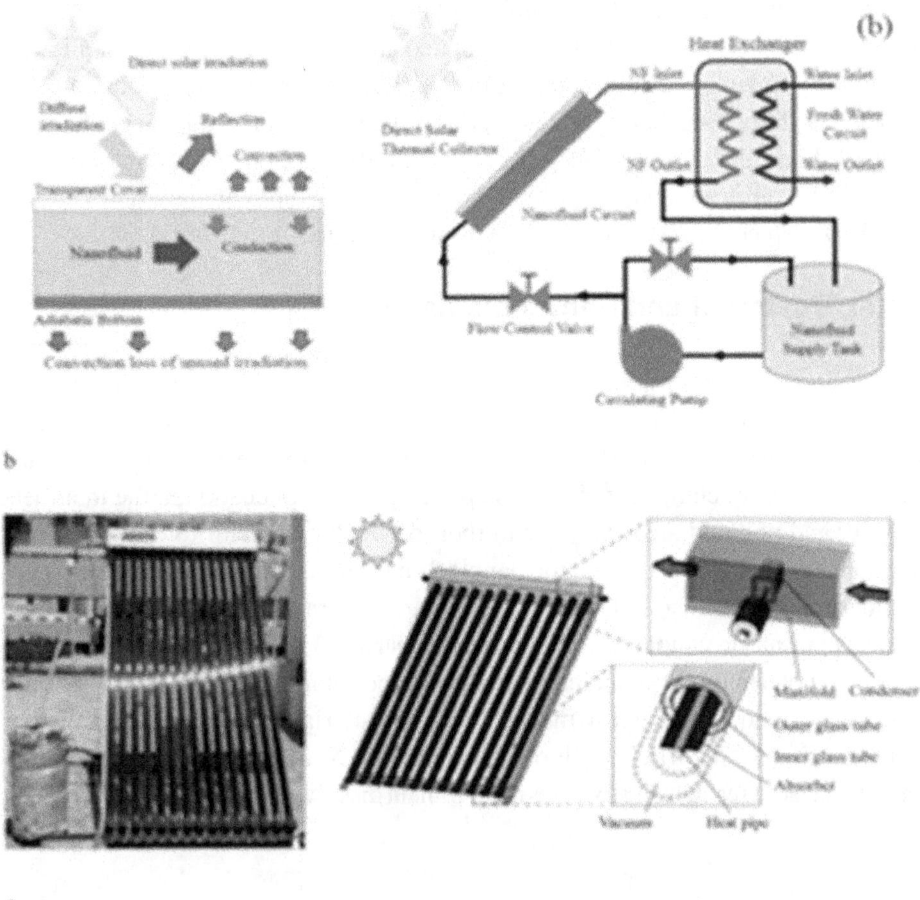

In the last 10 years, researchers have paid a lot of attention to the passive enhancement technique known as dispersion of Nanoparticles in various ordinary fluids including water, oil, glycols, etc. (Mehrali et al., 2016; Shirazi et al., 2015; Vidhu & Philip, 2015). Furthermore, When compared to other solid particles dispersed in conventional fluids, Nanoparticles with smaller sizes and higher thermo-physical properties added to a base fluid demonstrated better heat transfer capability, lubrication, enhanced cooling, and stability with some sedimentation; these conductive fluids are known as "Nanofluids." This technique is more promising and alluring than the active method due to the significant increase in thermal system efficiency. Numerous nanofluids (metals, metal oxides/nitrides/carbides, carbon, and hybrids) have been the subject of extensive investigation. conducted during the last decade to improve the overall performance of thermal systems. In thermal systems, nanofluids have greater energy absorption properties than ordinary working fluids, which increases the rate of heat transfer and results in increased efficiency (Khdher et al., 2016; Sadri, Hosseini, Kazi, Bagheri, Zubir, Ahmadi et al, 2017; Sadri, Hosseini, Kazi, Bagheri, Zubir, Solangi et al, 2017). Nanofluids' unique and enhanced properties make them a talented and possible operational liquid that can be used to get better presentation in a variety of applications, including heat exchangers (Luna-Sánchez et al., 2019; Sadri, Hosseini, Kazi, Bagheri, Ahmed et al, 2017; Sharmila et al., 2019; Sone et al., 2020), developed system (Okonkwo et al., 2018), lubricate and gab (Okonkwo et al., 2018), cool of electronics system (Ranjbarzadeh et al., 2019), cool of electronics system.

Water Treatment and Nanoremediation

Nanotechnology has the ability to create unique Nanomaterials for the treatment of harmful metal ions, organic and inorganic solutes, and microorganisms in surface water, groundwater, and wastewater, among other environmental materials. Many Nanomaterials are being researched and developed for application in the treatment of water and polluted areas because to their particular behavior towards stubborn pollutants. The current market for nanotech-based water treatment technologies includes reverse osmosis (RO), nanofiltration, and ultra filtration membranes. Nan fiber filters, carbon nanotubes, and other Nanoparticles are examples of developing goods. Nanotechnology is intended to remove pollutants that convectional water treatment techniques struggle to manage, such as bacteria, viruses, and heavy metals, more effectively. This efficiency is often attributed to Nanomaterials' extremely high specific surface area, which promotes contaminant dissolving, reactivity, and sorption.

Remediation of the Environment

The use of Nanoparticles for environmental cleanup is known as nanoremediation. The most common use of nanoremediation has been for groundwater treatment, with further studies in wastewater treatment. Nanoremediation has also been used to clean up soil and silt. More exploratory study is being conducted to investigate the use of Nanoparticles to remove harmful compounds from gases.

Some nanoremediation approaches, most notably the usage of Nano zerovalent iron for groundwater cleanup, have been used on full-scale cleanup sites. Nanoremediation is a new business; by 2009, nanoremediation methods have been reported at at least 44 cleanup sites worldwide, with the majority of them in the United States. A Nanoparticles agent must be brought into touch with the target pollutant under conditions that allow for a detoxifying or immobilizing response during nanoremediation. This procedure usually entails a pump-and-treat procedure or an in-situ application. Other approaches are still in the research stages. Scientists have been studying the possibilities of buckminsterfullerene in pollution management since it has the potential to influence certain chemical processes. Buckminster fullerene has been shown to be capable of producing reactive oxygen species protection and triggering lipid per oxidation. This substance might make hydrogen fuel more accessible to users.

Filtration of Water

Nan filtration is A relatively new membrane filtering technology used to soften (remove polyvalent cations) and remove disinfection by-product precursors such as natural organic matter and manufactured organic matter from low total dissolved solids water such as surface water and fresh groundwater. Nan filtration is also becoming increasingly popular in food processing applications such as dairy, where it is utilized for simultaneous concentration and partial (monovalent ion) demineralization.

Nan filtration is a membrane filtration technology that employs nanometer-sized cylindrical through-pores that travel at a 90° angle across the membrane. Nan filtration membranes have pore diameters ranging from 1 to 10 Angstrom, which are smaller than those used in microfiltration and ultra filtration but somewhat bigger than those used in reverse osmosis. Polymer thin films are commonly used to construct membranes. Polyethylene terephthalate and metals such as aluminum are examples of regularly used materials. During growth, pH, temperature, and time influence pore size, with pore densities varying from 1 to 106 pores per cm2. Membranes consisting of polyethylene terephthalate and other related materials are known as "track-etch" membranes, after the method by which the holes on the membranes are formed. The polymer thin film is "tracked" by blasting it with high-energy particles. This results in the holes, which are chemically formed into the membrane or "etched" into the membrane.

Nanotech to Disinfect Water

Nanotechnology provides an alternative solution to clean germs in water, a problem that has been getting worse due to the population explosion, growing need for clean water and the emergence of additional pollutants. One of the alternatives offered is antimicrobial nanotechnology stated that several Nanomaterials showed strong antimicrobial properties through diverse mechanisms, such as photo catalytic production of reactive oxygen species that damage cell components and viruses. There is also the case of the synthetically-fabricated nonmetallic particles that produce antimicrobial action called oligodynamic disinfection, which can inactivate microorganisms at low concentrations. Commercial purification systems based on titanium oxide photo catalysis also currently exist and studies show that this technology can achieve complete inactivation of fecal coli forms in 15 minutes once activated by sunlight. There are four classes of Nanomaterials that are employed for water treatment and these are dendrimers, zeolites, carbonaceous Nanomaterials, and metals containing Nanoparticles.The benefits of the reduction of the size of the metals (e.g. silver, copper, titanium, and cobalt) to the nanoscale such as contact efficiency, greater surface area, and better elution properties. Alumina membranes, for example, are formed by electrochemically generating a thin layer of aluminum oxide from aluminum metal in an acidic solution. Some nanotechnology-based water-treatment products are currently on the market, with more on the way. In a recent study, low-cost nanostructured separation membrane technologies were demonstrated to be successful in generating drinkable water.

Sensor Nanotechnology

Continuous exposure to heavy metal pollution and particle matter will result in health issues such as lung cancer, heart disease, and even motor neuron illnesses. However, precise and fast noncontact sensors capable of detecting contaminants at the atomic level can increase humanity's capacity to protect itself from these health concerns. Metal ions or radioactive materials are detected by these noncontact sensors with little effort. They can also be created in an automated manner so that they can be used at any time. Furthermore, because they are made with traditional microelectronic production equipment and electrochemical processes, these noncontact sensors are both energy and cost efficient.

Nano-based monitoring examples include:

1. Functionalized Nanoparticles capable of forming anionic oxidant bonds, allowing the detection of carcinogenic chemicals at extremely low concentrations.
2. Polymer nanospheres have been produced to detect organic contaminants in extremely low amounts.
3. "Peptide nanoelectrodes have been applied based on the notion of thermocouple. A peptide molecule is inserted in a 'Nano-distance separation gap to produce a molecular junction. When a certain metal ion is bonded to the gap, the electrical current results in a unique conductance value. As a result, the metal ion will be easily identified.
4. Composite electrodes, made of nanotubes and copper, have been developed to detect chemicals in low concentrations such as organ phosphorus insecticides, carbohydrates, and other wood harmful compounds.

It saves energy and gasoline by utilising less material and renewable inputs whenever possible. Furthermore, nanotechnological goods, processes, and applications are predicted to greatly contribute to environmental and climate protection by conserving raw materials, energy, and water, as well as lowering greenhouse gas emissions and hazardous waste. Green nanotechnology's key advantages include increased energy efficiency, reduced waste and greenhouse gas emissions, and reduced use of non-renewable raw resources. Green nanotechnology provides an excellent chance to prevent negative consequences from occurring. Green nanotechnology does not emerge from thin air; rather, it develops from the concepts of green chemistry and engineering. Green nanotechnology applications may include the use of nanomaterials in clean production processes that synthesise Nanoparticles using sunlight or by recycling industrial waste products into nanomaterials, in addition to obvious areas such as the development of solar cells, biofuels, and fuel cells. There is some "really" green nanotechnology, such as completely growing Nanoparticles in plants, but it will never reach the scale necessary for industrial nanomaterial manufacturing. Green nanotechnology, like other industrially created goods, requires a comprehensive process assessment in order to make a conclusive observation.

Plant Extracts Used to Create Nanoparticles

Gold and Silver Nanoparticles

Nanoparticles have gotten a lot of interest because of their tunable size, shape, and surface features. Because of their distinct features, gold Nanoparticles have been investigated for prospective uses in biosensors, hyperthermia treatment, antimicrobial medicines, genetic editing, and therapeutic delivery systems. Plants, which are biological factories, provide environmentally benign sources of Au Nanoparticles using green chemistry-based processes. The research of Nanoparticles syntheses also showed that depending on the pH of the reaction media, a range of forms, including rod-shaped, irregular, decahedral, icosahedra, and hexagonal, might be created. The major product of this investigation was spherical particles with sizes ranging from 20 to 80 nm, according to the findings.

Several publications have reported alloying Au and Ag to explore the characteristics of the resultant bimetallic Nanoparticles, in addition to manufacturing pure metal Nanoparticles by plants in this manner. This bimetallic Nanoparticles production involves a competitive reduction process between two aqueous solutions, each containing a distinct metallic ion precursor that reacts with a plant extract. Because gold has a higher reduction potential than silver in bimetallic Nanoparticles, it will develop first and form the core of the ensuing core-shell structure. As a result, the identical reduction of Ag ions results in Ag consolidating on the core to create the shell.

Nanoparticles of Copper and Copper Oxide

Plant extracts have been used to create a range of copper (Cu) and copper oxide (CuO) Nanoparticles. Cu Nanoparticles of magnolia leaf extract were biologically generated to produce stable Nanoparticles ranging in size from 40 to 100 nm. Furthermore, Cu Nanoparticles have demonstrated potential antibacterial action against Escherichia coli cells. Syzygium aromaticum (clove) extracts were employed in the production of Cu Nanoparticles with a spherical to granular shape and a mean particle size of 40 nm. Cu Nanoparticles were created by employing the stem latex of Euphorbia nivulia, often known as common milk hedge. Peptides and terpenoids found in latex help to stabilise these Nanoparticles. Furthermore, human adenocarcinomic alveolar basal epithelial cells have been shown to be toxic to these Nanoparticles.

Nanoparticles of Palladium and Platinum

Palladium Nanoparticles were created from C. zeylanicum (cinnamon) bark extract. The temperature, concentration, and reaction pH of the bark extract altered during synthesis, but morphology and particle size (12 to 19 nm) were unaffected. Palladium Nanoparticles ranging in size from 70 to 80 nm were also produced using Annona squamosa (custard apple) peel extract. Palladium Nanoparticles with diameters ranging from 12 to 55 nm and a cubic crystal symmetry in the core were synthesised using extracts of Camellia sinensis (tea) and Coffea arabica (coffee). Since then, the first platinum Nanoparticles of Diospyros kaki (persimmon) leaf extract with diameters ranging from 2 to 10 nm have been described. Recently, the biological manufacturing of platinum Nanoparticles with particle size and shape control has also been described.

Nanoparticles of Titanium Dioxide and Zinc Oxide

A variety of plant extracts were used to create these essential metal oxide Nanoparticles. Since it was demonstrated that TiO2 Nanoparticles could be efficiently generated from Annona squamosa peel, spherical particles ranging in size from 95 to 120 nm were discovered in Nyctanthes arbor-tristis leaf extracts. Eclipta prostrata leaf extracts may also yield particles ranging in size from 30 to 65 nm. Velayutham et al. created TiO2 Nanoparticles with irregular shapes and sizes ranging from 20 to 100 nm using Catharanthus roseus leaf extract. TiO2 suspensions were shown to be both larvicidal and adulticidal against Bovicola ovis (sheep louse) and Hippobosca maculate (hematophagous fly).

Nanoparticles of Indium Oxide (In2O3), Iron Oxide, Lead, and Selenium

A number of novel forms of metal oxide and metal Nanoparticles were organically produced using a range of plants. Aloe vera leaf extract was used to create indium oxide Nanoparticles (aloe barbadensis Miller). The precipitates were thermally treated at 400-600 °C after initial synthesis to produce Nanoparticles. The size of the produced spherical Nanoparticles varied from 5 to 50 nm and was temperature dependant. Iron Nanoparticles play a significant role in a variety of environmental cleanup approaches. As a result, several researches have concentrated on green chemistry in order to produce this iron (Fe) Nanoparticles. Sorghum bran aqueous extracts, for example, have been utilised to manufacture Fe Nanoparticles. Recently, spherical irons Nanoparticles with particle diameters of 100 nm were produced from Azadirachta indica (neem) leaf extract. Shah et al. previously generated iron Nanoparticles from plant extracts including Cymbopogon citratus (lemon grass tea), Datura innoxia, Tridax procumbens, Calotropis procera, Tinospora cordifolia, and Euphorbia milii. The smallest spherical Nanoparticles (13 nm) were produced from stem extract, while the largest size range (43-342 nm) was created from Cymbopogon citratus leaf extract. Lead (Pb) and selenium (Se) are two additional important Nanoparticles that have been biologically produced. Joglekar et al. used Jatropha curcas latex to create spherical Pb Nanoparticles with diameters ranging from 10 to 12.5 nm. Recently, spherical selenium (Se) Nanoparticles made from citrus reticulata peel have been synthesized.

Metal Nanoparticles Green Synthesis

Plants have long been recognized to have the ability to biologically reduce metallic ions and hyper-accumulate them. Plants have been deemed a more ecologically acceptable biological technique for production of metallic Nanoparticles, as well as effective for detoxifying applications, due to their outstanding qualities. Plant extracts contain a variety of bioactives, including alkaloids, proteins, phenolic acids, sugars, terpenoids, and polyphenols, all of which have been shown to play a vital role in first decreasing and then stabilising metallic ions.

AIM AND OBJECTIVE

The experimental work done using eco-friendly nanofluids in thermal applications is quite restricted and will expand in the future (with safety health, and environment in mind). This is the novelty of this article. The goal of this study is to explore the many viewpoints on environmentally friendly procedures used in the synthesis of nanofluids, as well as their stability, characterization, and thermo-physical characteristics for successful use in thermal systems. Also included are economic and environmental assessment tools for increased savings and the protection of human health and the environment.

In the current study, Section 2 discusses an overview of Nanomaterials and several approaches for synthesizing Nanoparticles, the one-step and two-step procedures for making nanofluids, assessment methodologies, and mechanisms for enhancing stability and thermo physical characteristics. Additionally, a general review of nanofluids production, mostly employing eco-friendly materials and methods, is included. Then, aspects influencing thermo-physical characteristics with enhancing approaches, characterization, stability assessment, and green Nano production processes are discussed. After that, Section 3 offers a thorough analysis of the crash of bottle green nanofluids (both investigational and mathematical) on the functionality of heat reassign system. For greater cost savings and environmental and human health protection, Section 4 highlights the economic and environmental assessment of employing nanofluids. This thorough analysis offers experts offering young researchers a fresh perspective on ecologically friendly nanofluids, especially for their practical use in thermal systems. Furthermore, assigning them a greater job for future study to choose and manufacture more environmentally acceptable Nanomaterials of diverse combinations with the primary focus of safer human health and the environment.

Figure 3. Thermal properties of nanofluids for current development

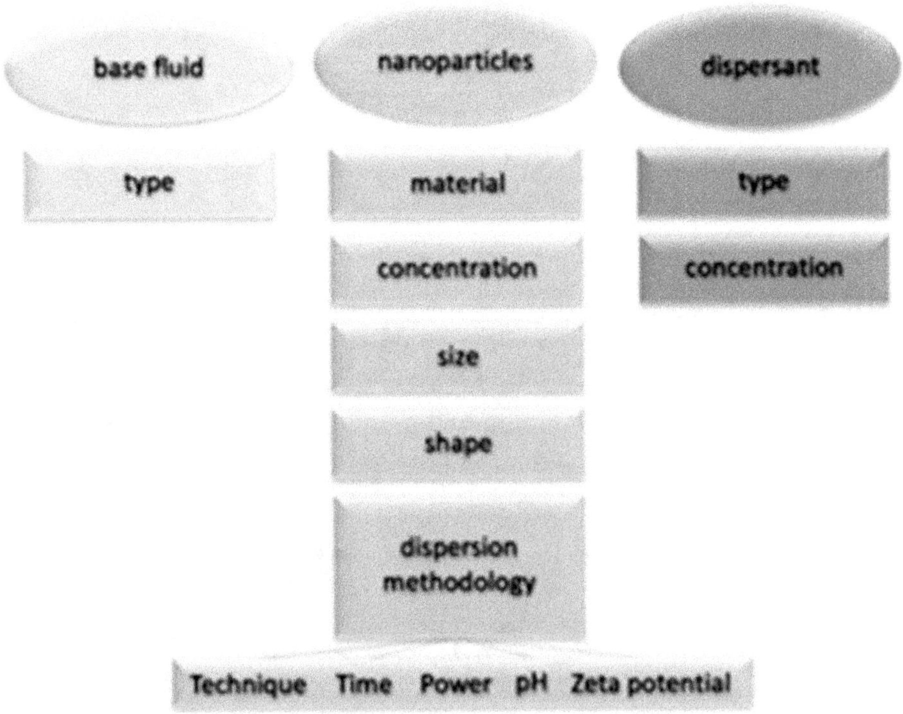

The amount of study on nanofluids and environmentally friendly nanofluids that has been published during the previous ten years (from 2011 to November 2020) in Science Direct.com has, however, shown that less work has been done utilizing environmentally friendly nanofluids in comparison.

NANOFLUIDS PREPARATION METHODS

The preparation of nanofluids is more than merely mixing the Nanoparticles with the base fluid. Under specific environmental circumstances, stability and correct mixing are necessary to get nanofluids with homogeneously distributed Nanoparticles. There are several ways for producing nanofluids, which may be further categorized into two categories:

Figure 4. Nanofluids preparation methods

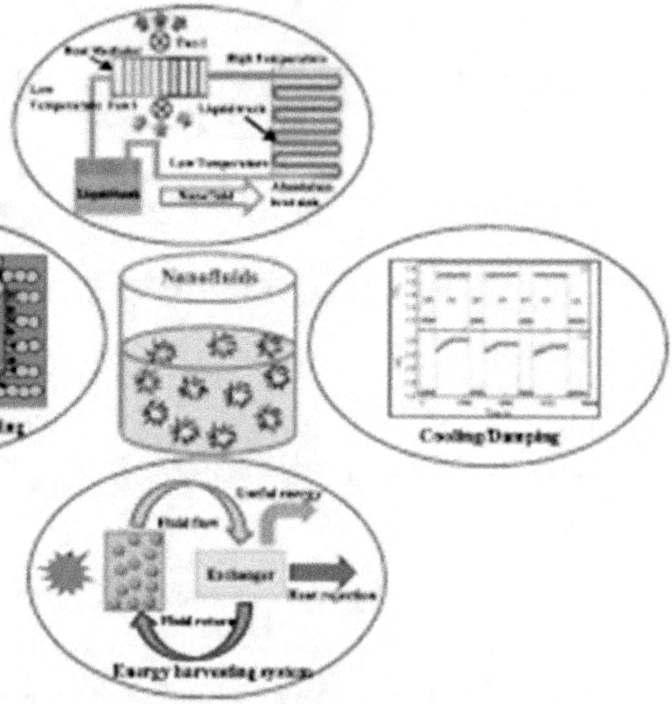

Single Stage Method

A single stage technique and system for creating nanofluids by evaporating a Nanoparticles source and depositing the evaporate into a base fluid. The base fluid, such as oil or ethylene glycol, is deposited in a revolving cylindrical drum equipped with an adjustable heater, boat evaporator, and heat exchanger cooler device. A thin liquid layer forms on the interior surface of the drum as it rotates. An insulated heater-boat-evaporator with a source of Nanomaterials put within it, the source of Nanomaterials being heated, which causes some of the evaporated material to evaporate and produce Nanoparticles, which are then absorbed by the liquid film to create Nanomaterials. produce nanofluids. The single stage procedure produces and disperses particles in the base fluids, which might be reduced to Nanomaterials aggregation. This approach improves the stability of nanofluids at the expense of the process's high cost (Stephen, 1998; Trisakti & Wongwises, 2007; Wong & De Leon, 2009; Yu & Huaqing, 2012).

Figure 5. Single stage method

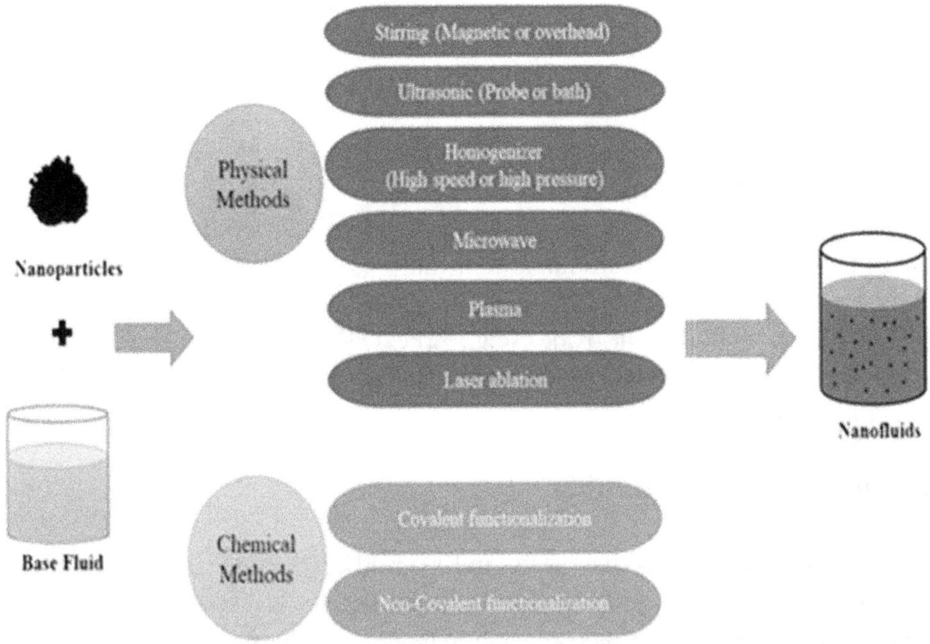

Double Stage Method

The most common approach for preparing nanofluids is a two-step procedure. This approach begins with the production of dry powders of Nanoparticles, nanofibres, nanotubes, or other Nanomaterials. In the second processing phase, Intensive magnetic force agitation, ultrasonic agitation, high-shear mixing, homogenizing,

Figure 6. Double stage method

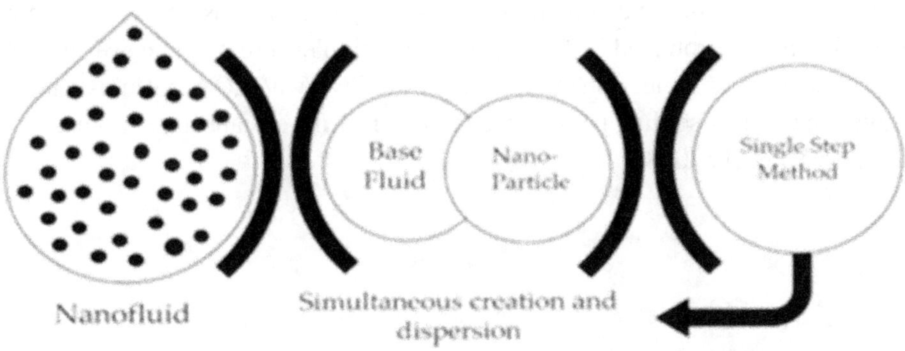

and ball milling will be used to spread the nanosized powder into a fluid. Because nanopowder synthesis techniques have been ramped up to commercial production levels in the past, the two-step process is the most cost-effective way to create nanofluids on a wide scale. Nanoparticles tend to agglomerate due to their high surface and surface activity.

Classification of Nanofluids

Nanofluids can be created from single element Nanoparticles (e.g., Cu,Fe, and Ag), single element oxides (e.g., Cu2O,CuO, TiO2, and Al2O3,) alloys (e.g., Cu-Zn and Fe-Ni,Ag-Cu), multielement oxides (e.g., ZrFe2O4, NiFe2O4, and CuZnFe4O4) metal carbides They are classified as single material nanofluids or hybrid nanofluids (Yu & Huaqing, 2012).

Single Material Nanofluids

Choi created this category of nanofluids in 1995, and it is regarded the most common type of nanofluids employed, in which a single type of Nanoparticles is used to make the suspension by various preparation processes. Many writers have claimed that nanofluids of this type perform better because they have substantially better thermo physical characteristics than their base fluid.

Hybrid Nanofluids

Hybrid nanofluids are a sort of advanced nanofluids that is composed of more than one type of Nanoparticles floating in a base fluid. Jana et al., in 2007, were the first to investigate this form of fluid experimentally in organize to advance the fluid thermal conductivity further than that of normal single material type nanofluids. Cu Nanoparticles, Au Nanoparticles and carbon nanotubes (CNTs) discrete in wet, as healthy as their hybrid (CNT-Au/H_2O) & CNT-Cu/H_2O along with subsist investigated in their study. The thermal conductivity of Cu/H_2O nanofluids be establish to be the furthermost surrounded by the experienced samples, greater than ever linearly with subdivision awareness. on the other hand, the CNT-Cu/H_2O nanofluids stability resulted in a longer settling period than the other forms of nanofluids (Otanicar et al., 2010; Yu et al., 2008).

APPLICATION OF GREEN NANOMATERIALS

The application of nanofluids will be the emphasis of this section. Nanofluids are employed in a variety of industries, including the automotive, medicinal, and energy sectors. Coolants, nanofluids as automobile brake fluid and distress absorbers, nanodrugs, melanoma remedy, astrophysical cells, planetary incorporation, power storage space, nanofluids in high temperature exchangers, breathing space and defense are several of the specific applications. The function of nanofluids in various applications is addressed in depth in the next section. Nanomaterials made from green processes have several uses in medicine, pharmaceuticals, in vitro testing, biomedicine, antimicrobials, optical imaging, molecular sensing, and labeling biological systems.

Nanofluids in Automobile Application

locomotive oil, automatic diffusion fluids, coolants, lubricants, and other unreal temperature heat transport fluids used in established motor vehicle thermal systems-radiators, engines, heating, ventilation, and air conditioning (HVAC)-are inherently poor heat transfer fluids. These could benefit from enhanced heat conductivity provided by nanofluids as a result of Nanoparticles addition (Chopkar et al., 2006).

Nanofluids in Biomedical Engineering Area

All Nanoparticles-containing suspensions or nanosuspensions were examined using the definition of nanofluids, and the utilization of such fluids in medicine administration and biomedical treatment is explored in this section (Ovsianikov et al., 2007).

Drug Liberation

The basic purpose of drug delivery system design is to deliver medicine in a regulated manner to the desired location. The early studies in drug delivery systems concentrated on drug release kinetics, pH and temperature sensitivity, polymers, nasal delivery, and oral drug administration. administration. The use of nanofluids in drug delivery systems is one of the subjects that many researchers have investigated in recent years to give certain benefits such as precise distribution to target cells, enhanced therapeutic qualities and safety, lower toxicity and biocompatibility.

Figure 7. Role of nanofluids in drug delivery

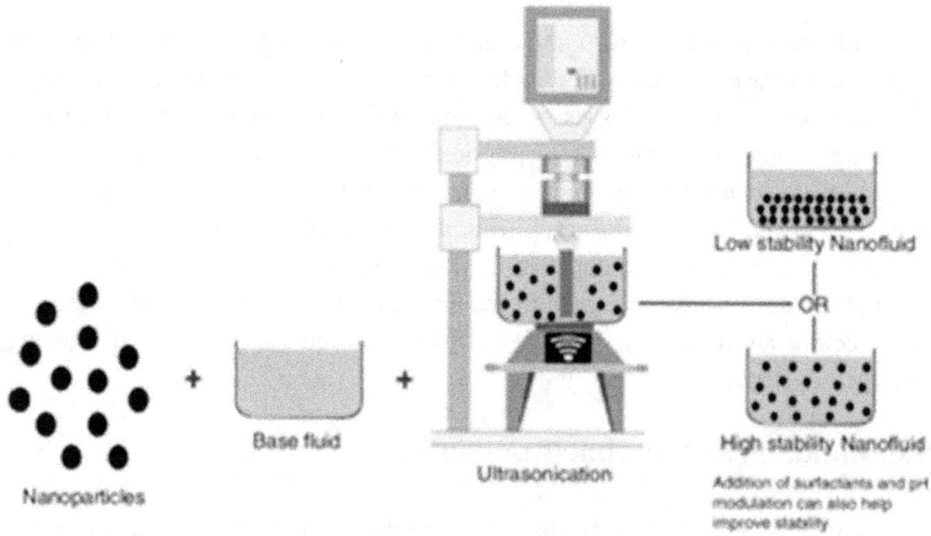

Cancer Therapeutics

A new project is using many features of particular nanofluids for use in cancer imaging and medication administration. This endeavor includes the use of iron-based Nanoparticles as medicine or radiation delivery vehicles in cancer patients. Magnetic nanofluids will be utilized to employ magnets to direct particles up the bloodstream to a tumour. It will enable doctors to provide high local doses of medications or radiation without causing damage to neighboring health tissue, which is a severe side effect of existing cancer therapy approaches (Khanam & Hasan, 2019; Labhasetwar & Leslie-Pelecky, 2007). Many cancer treatments and medications are already available, with many more being researched. Some therapies, such as surgery and radiation therapy, are "local," meaning they are intended to treat a specific tumour or region of the body. Because drug therapies (such as chemotherapy, immunotherapy, or targeted therapy) can influence the entire body, they are commonly referred to as "systemic" treatments. Learn about the most prevalent cancer treatments here.

Nano Cryosurgery

Cryosurgery is a method of destroying unwanted tissues by freezing them. Because of its significant therapeutic benefits, this treatment is gaining popularity. Although it is not yet considered a standard way of cancer therapy, cryosurgery is rapidly becoming an alternative to existing medicines (Ye et al., 2017). According

Figure 8. Role of nanofluids in cancer therapeutics

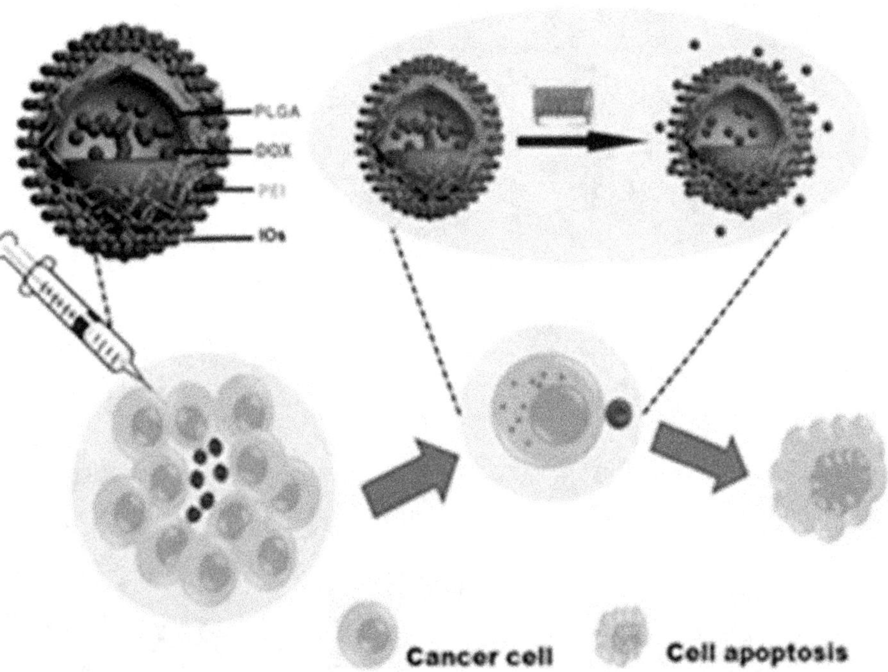

to theoretical interpretation and existing experimental findings, intentionally loading high thermal conductivity Nanoparticles In the absence of Nanoparticles, introducing them into target tissues can reduce the end temperature, increase the maximum freezing rate, and enhance the ice volume obtained. Furthermore, the use of nanofluids improved freezing may make traditional cryosurgery more versatile in a variety of ways, including the ability to manipulate the size, shape, image, and direction of ice ball creation.

Application of Nanofluids in Solar

Direct solar collectors based on nanofluids are solar thermal collectors in which nano particles in a liquid media scatter and absorb sun energy. They have recently garnered attention in distributing solar energy effectively.

Application of Nanofluids as Nuclear Reactor

Because nanofluids may be particularly successful in cooling over heated surfaces in emergency conditions, its use as a coolant in nuclear reactors is a extremely hopeful

Figure 9. Role of nanofluids in nano cryosurgery

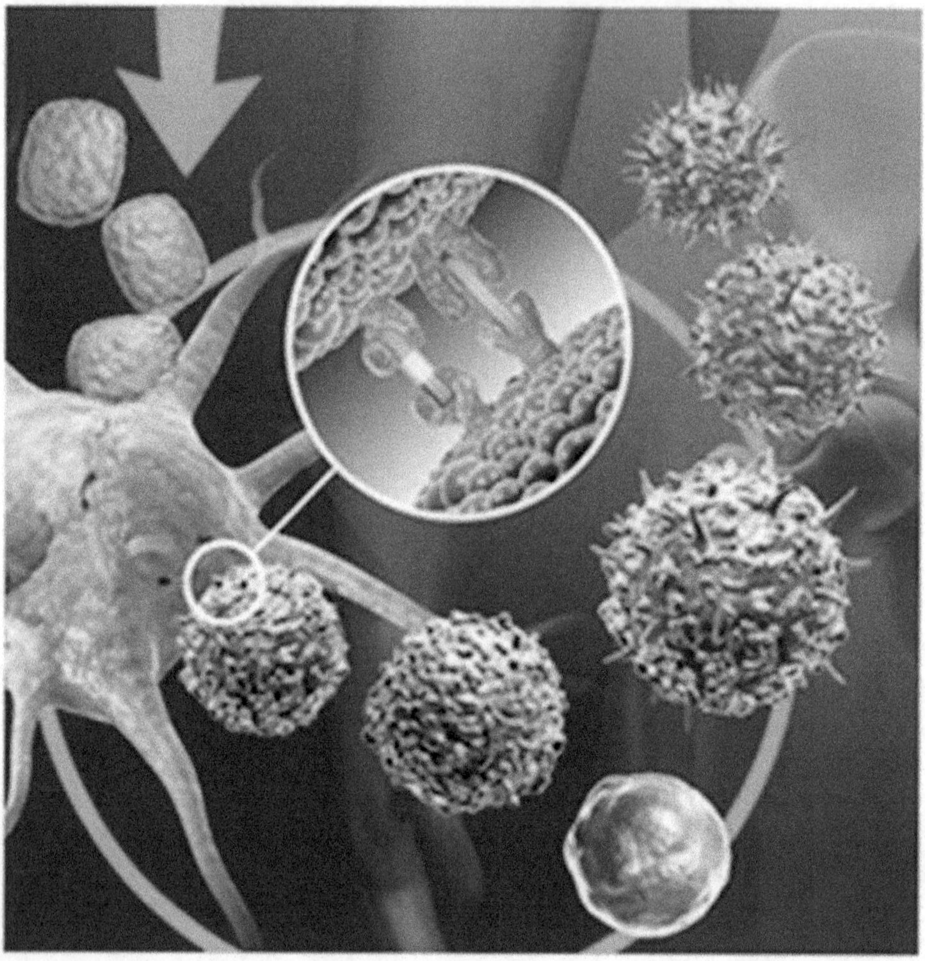

function. There are also worries about the loss of Nanoparticles through boiling vapour and the safety precautions for disposing of nanofluids.

OVERVIEW

Innovative chemical substances known as Nanomaterials, which also include particulates, are employed at the nanoscale (1 to 100 nm in diameter) range. When compared to bulk materials without nanoscale features, they exhibit improved chemical reactivity, conductivity, and distinctive overall qualities. Nanomaterials

Figure 10. Nanofluids in solar collector

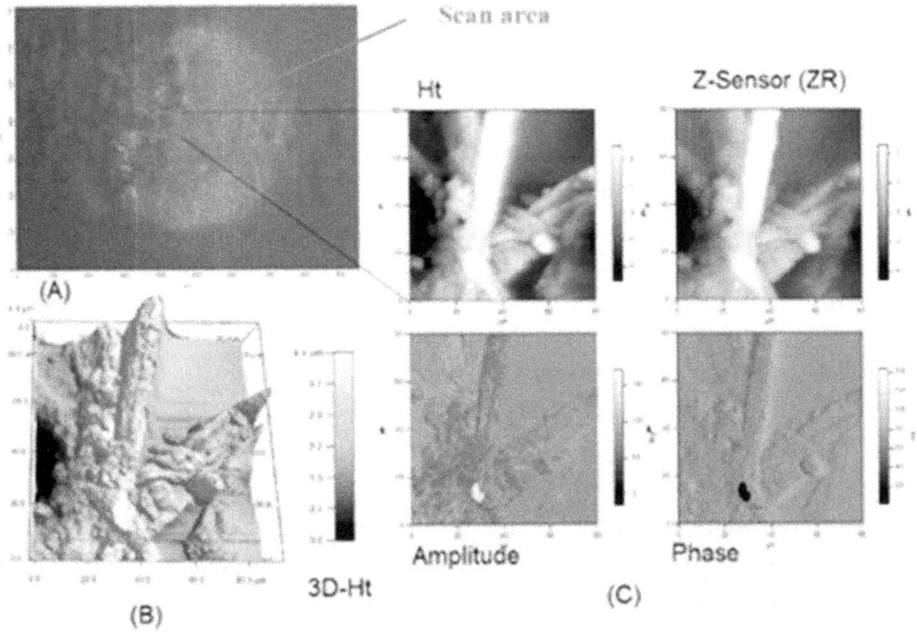

can be found as rods, wires, tubes, shells, cages, and pores, among other geometric shapes. They may be broadly divided into four categories: 0D, 1D, 2D, and 3D.

Thermal System Performance Analysis Utilizing Green Nanofluids

Researchers are becoming more and more interested in using environmentally friendly nanofluids in thermal applications to boost efficiency. In addition to being non-corrosive and cost-effective, using green nanofluids as alternative working fluids can provide benefits like improved "k" and "h" values in the system. Green nanofluids in thermal systems have showed improved thermal performance in recent experimental and computational investigated.

Analyses of the Economy and the Environment

A good thermal system should be affordable, highly efficient, and have a low environmental effect due to proper design, eco-friendly components, working fluids, and so on. Despite their high starting prices, a few thermal systems have demonstrated considerable energy savings with low repair costs. In terms of embodied energy, the total energy involved with the whole evolution of such systems is defined.

Challenges

Green nanofluids research for improving heat transfer system performance is growing at an exponential rate. However, there are always difficulties and obstacles associated with using green nanofluids in practical applications. As a result, new study possibilities for scientific writers and researchers all around the world have emerged.

Green Nanotechnology Risk Aspects

Risk Assessment

The same processing procedures that are used in the risk assessment of other types of materials/chemicals are employed in the risk assessment of Nanomaterials. These include hazard identification, hazard characterization, dose-response relationships, and exposure assessment for various situations. Unfortunately, the risk assessment procedure for Nanoparticles continues to be hampered by a lack of toxicological evidence for a wide range of Nanomaterials. Furthermore, in situations of long-term and low-dose exposure, the defining of important health impacts such as genotoxicity, lung toxicity, or carcinogenicity is approaching actual scenarios that demand attention. The measurement of exposure remains a critical criterion for the characterization of occupational Nanomaterials hazards.

Risk Management

The development of nanotechnology applications in the realm of green technology has sparked worries about the Nanomaterials influence on worker health and safety. This urgently necessitates scientific, technical, and governmental measures to address such labor concerns. This entails determining true dangers resulting from Nanomaterials exposure in the workplace, often known as "risk assessment": To plan the "risk management" control methods and, eventually, to explain the strategy. Overall, all of these risk-aspect processes are crucial and will be explored in the next part, which attempts to safeguard workers from damage while still providing all of the societal advantages of green nanotechnology.

Risk Assessment

The same processing procedures that are used in the risk assessment of other types of materials/chemicals are employed in the risk assessment of Nanomaterials. These include hazard identification, hazard characterization, dose-response relationships, and exposure assessment for various situations. Unfortunately, the risk assessment procedure for Nanoparticles continues to be hampered by a lack of toxicological evidence for a wide range of Nanomaterials. Furthermore, in situations of long-term and low-dose exposure, the defining of important health impacts such as genotoxicity, lung toxicity, or carcinogenicity is approaching actual scenarios that demand attention.

Future Scope

Nanofluids now have a wide range of uses, with many more planned in the future. The primary focus of research should most likely be on developing efficient energy transmission systems employing nanofluids. The thermal conductivity of the nanofluids has the greatest influence on the pace of heat transfer. The influence of particle shape and particle aggregation on Nanoparticles thermal conductivity should be carefully investigated in order to develop effective energy transfer systems. There are currently worries about the safety of nanofluids usage and manufacture, which are also subjects of future study. Nanofluids provide a revolutionary approach for obtaining energy from the earth's core. Nanofluids particles might potentially contribute in the development of smaller microchips. Smaller microchips may result in better computers. Nanofluids may potentially aid in targeted medication delivery, allowing the pharmaceutical sector to advance by leaps and bounds. Improving the thermal conductivity of nanofluids might potentially aid in the development of improved vehicle engines. Nanofluids technology can only be used if non-toxic or biodegradable Nanoparticles are developed (Lee & Mudawar, 2007).

The following major tasks will include the technical attempt to expand and describe the performance of nanofluids.

1) At ANL, Nanophase metal powders will be produced utilizing cutting-edge gas condensation preparation technology. Liquid Nanophase powders will be used examined for particle size and agglomeration behavior.
2) The technology for Nanoparticles suspension will be developed, and the stability, dispersion, and rheological/transport properties of these nanofluids will be investigated.
3) There will be practical investigations.
4) Create environmentally friendly and wear-resistant nanofluids.

5) Use of new efficient, low-GWP, and environmentally friendly fluids and refrigerants.
6) The use of dielectric liquid in micro channels allows working fluids to come into direct contact with electronics.
7) A better theoretical understanding of complicated nanofluids will have a greater influence.
8) Creation of computer-based models of events such as the physical and chemical interactions of Nanoparticles with base fluid molecules.
9) Conception of original new methodologies for characterize nanofluids in the laboratory and in natural world.
10) In adding together to the occupation potential listed above, nanofluids may be in employment in: A. astrophysical cool system B. photovoltaic system C. astrophysical combination cool system

Both scientific and economic advances in nanofluids continue, and this trend has the potential to accelerate the growth of nanofluids (Sizochenko et al., 2017; Sreelakshmy, Nair, Vidhya, Saranya, & Nair, 2014).

5.5.1 Future Difficulties

The make use of nanofluids in a extensive choice of application looks talented, however the field's development is hampered by the following factors:

1) a lack of consistency between results obtained in various laboratories
2) The suspensions are frequently mischaracterized.
3) A lack of theoretical knowledge of the mechanisms behind the observed property changes

(Saidur et al., 2011; Sreelakshmy, Nair, Vidhya, Saranya, & Nair, 2014).

CONCLUSION

The impression of present discovery in the grassland of nanofluids. The grounding, property, and uses of nanofluids have been methodically explored. Nanofluids are noteworthy since they might be utilizing in a variety of applications requiring heat transmission as well as other applications such as detergency. Colloids, which are also nanofluids, have long been employed in the biomedical area and will continue to do so. Future study on the manufacture and usage of nanofluids is required so that they may be used as expected. Nonetheless, several discoveries and improvements regarding the properties of nanofluids in the assessed applications have been recognized, and we are one step closer to building systems that are more efficient and smaller, thereby making the environment cleaner and healthier.

Many researchers are still working in this subject to improve the futuristic scope of nanofluids in industries and to better define the future of nanofluids.

REFERENCES

Ahmada, A., Asghara, S., & Alsaedi, A. (2014). Flow and heat transfer of a nanofluid over a hyperbolically stretching sheet. *Chinese Physics B*, *23*(7), 074401. doi:10.1088/1674-1056/23/7/074401

Akyildiz, F. T., Bellout, H., Vajravelu, K., & Van Gorder, R. A. (2011). Existence results for third order nonlinear boundary value problems arising in nano boundary layer fluid flows over stretching surfaces. *Nonlinear Analysis Real World Applications*, *12*(6), 2919–2930. doi:10.1016/j.nonrwa.2011.02.017

Buongiorno, J. (2006). *Convective transport in nanofluids*. Academic Press.

Choi, S. U., & Eastman, J. A. (1995). Enhancing thermal conductivity of fluids with nanoparticles (No. ANL/MSD/CP-84938; CONF-951135-29). Argonne National Lab.

Chopkar, M., Das, P. K., & Manna, I. (2006). Synthesis and characterization of nanofluid for advanced heat transfer applications. *Scripta Materialia*, *55*(6), 549–552. doi:10.1016/j.scriptamat.2006.05.030

Devi, S. P. A., & Devi, S. S. U. (2016). Numerical investigation of hydromagnetic hybrid Cu- Al2O3/water nanofluid flow over a permeable stretching sheet with suction. *International Journal of Nonlinear Sciences and Numerical Simulation*, *17*(5), 249–257. doi:10.1515/ijnsns-2016-0037

Ehsan, E.-B., Charjouei, M. M., Hamid, N., Weerapun, D., & Somchai, W. (2016). Experimental and numerical investigation of nanofluids heat transfer characteristics for application in solar heat exchangers. *International Journal of Heat and Mass Transfer*, *92*, 1041–1052. doi:10.1016/j.ijheatmasstransfer.2015.08.107

Goyal, M., & Bhargava, R. (2014). Numerical study of thermodiffusion effects on boundary layer flow. *Microfluidics and Nanofluidics*, *17*(3), 591–604. doi:10.100710404-013-1326-2

Kamyar, A., Saidur, R., & Hasanuzzaman, M. (2012). Application of computational fluid dynamics (CFD) for nanofluids. *International Journal of Heat and Mass Transfer*, *55*(15-16), 4104–4115. doi:10.1016/j.ijheatmasstransfer.2012.03.052

Khanam, P. N., & Hasan, A. (2019). Biosynthesis and characterization of graphene by using non-toxic reducing agent from Allium Cepa extract: Anti-bacterial properties. *International Journal of Biological Macromolecules*, *126*, 151–158. doi:10.1016/j.ijbiomac.2018.12.213 PMID:30584937

Khdher, A. M., Sidik, N. A. C., Hamzah, W. A. W., & Mamat, R. (2016). An experimental determination of thermal conductivity and electrical conductivity of bio glycol based Al2O3 nanofluids and development of new correlation. *International Communications in Heat and Mass Transfer*, *73*, 75–83. doi:10.1016/j.icheatmasstransfer.2016.02.006

Labhasetwar, V., & Leslie-Pelecky, D. L. (Eds.). (2007). *Biomedical applications of nanotechnology*. doi:10.1002/9780470152928

Lai, F. C., & Kulacki, F. A. (1990). The effect of variable viscosity on convective heat transfer along a vertical surface in a saturated porous medium. *International Journal of Heat and Mass Transfer*, *33*(5), 1028–1031. doi:10.1016/0017-9310(90)90084-8

Lee, J., & Mudawar, I. (2007). Assessment of the effectiveness of nanofluids for single-phase and two-phase heat transfer in micro-channels. *International Journal of Heat and Mass Transfer*, *50*(3-4), 452–463. doi:10.1016/j.ijheatmasstransfer.2006.08.001

Luna-Sánchez, J. L., Jiménez-Pérez, J. L., Carbajal-Valdez, R., Lopez-Gamboa, G., Pérez-González, M., & Correa-Pacheco, Z. N. (2019). Green synthesis of silver nanoparticles using Jalapeño Chili extract and thermal lens study of acrylic resin nanocomposites. *Thermochimica Acta*, *678*, 178314. doi:10.1016/j.tca.2019.178314

Mabood, F., Shateyi, S., Rashidi, M. M., Momoniat, E., & Freidoonimehr, N. (2016). MHD stagnation point flow heat and mass transfer of nanofluids in porous medium with radiation, viscous dissipation and chemical reaction. *Advanced Powder Technology*, *27*(2), 742–749. doi:10.1016/j.apt.2016.02.033

Makinde, O., & Aziz, A. (2011). Boundary layer flow of a nanofluid past a stretching sheet with a convective boundary condition. *International Journal of Thermal Sciences*, *50*(7), 1326–1332. doi:10.1016/j.ijthermalsci.2011.02.019

Mehrali, M., Sadeghinezhad, E., Akhiani, A. R., Latibari, S. T., Talebian, S., Dolatshahi-Pirouz, A., & Mehrali, M. (2016). An ecofriendly graphene-based nanofluid for heat transfer applications. *Journal of Cleaner Production*, *137*, 555–566. doi:10.1016/j.jclepro.2016.07.136

Okonkwo, E. C., Essien, E. A., Abid, M., Kavaz, D., & Ratlamwala, T. A. (2018). Thermal performance analysis of a parabolic trough collector using water-based green-synthesized nanofluids. *Solar Energy*, *170*, 658–670. doi:10.1016/j.solener.2018.06.012

Omiddezyani, S., Gharehkhani, S., Yousefi-Asli, V., Khazaee, I., Ashjaee, M., Nayebi, R., Shemirani, F., & Houshfar, E. (2021). Experimental investigation on thermo-physical properties and heat transfer characteristics of green synthesized highly stable CoFe2O4/rGO nanofluid. *Colloids and Surfaces. A, Physicochemical and Engineering Aspects*, *610*, 125923. doi:10.1016/j.colsurfa.2020.125923

Otanicar, T. P., Phelan, P., Prasher, R., Rosengarten, G., & Taylor, A. (2010). Nanofluid-based direct absorption solar collector. *Journal of Renewable and Sustainable Energy, 2*(3).

Ovsianikov, A., Chichkov, B., Mente, P., Monteiro-Riviere, N. A., Doraiswamy, A., & Narayan, R. J. (2007). Two photon polymerization of polymer–ceramic hybrid materials for transdermal drug delivery. *International Journal of Applied Ceramic Technology*, *4*(1), 22–29. doi:10.1111/j.1744-7402.2007.02115.x

Rana, P., & Bhargava, R. (2012). Flow and heat transfer of a nanofluid over a nonlinearly stretching sheet. *Communications in Nonlinear Science and Numerical Simulation*, *17*(1), 212–226. doi:10.1016/j.cnsns.2011.05.009

Ranjbarzadeh, R., Moradikazerouni, A., Bakhtiari, R., Asadi, A., & Afrand, M. (2019). An experimental study on stability and thermal conductivity of water/silica nanofluid: Eco-friendly production of nanoparticles. *Journal of Cleaner Production*, *206*, 1089–1100. doi:10.1016/j.jclepro.2018.09.205

Rashidi, M. M., Freidoonimehr, N., Hosseini, A., Anwar Bég, O., & Hung, T. K. (2014). Homotopy simulation of nanofluid dynamics from a non-linearly stretching isothermal permeable sheet with transpiration of nanofluids over a power law stretching sheet. *Meccanica*, *49*(2), 469–482. doi:10.100711012-013-9805-9

Rashidi, M.M., Freidoonimehr, N., Momoniat, E., & Rostami, B. (2015). Study of nonlinear MHD tribological squeeze film at generalized magnetic reynolds numbers using DTM. *PLoS One, 10*(8). http://dx.doi.org/.pone.0135004. e0135004 doi:10.1371/journal

Sadegh, K., Hossein, T., Arezoo, K., & Mehdi, R. M. (2015). Unsteady convective heat and mass transfer in pseudoplastic nanofluid over a stretching wall. *Advanced Powder Technology*, *26*(5), 1319–1326. doi:10.1016/j.apt.2015.07.006

Sadri, R., Hosseini, M., Kazi, S. N., Bagheri, S., Ahmed, S. M., Ahmadi, G., Zubir, N., Sayuti, M., & Dahari, M. (2017). Study of environmentally friendly and facile functionalization of graphene nanoplatelet and its application in convective heat transfer. *Energy Conversion and Management*, *150*, 26–36. doi:10.1016/j.enconman.2017.07.036

Sadri, R., Hosseini, M., Kazi, S. N., Bagheri, S., Zubir, N., Ahmadi, G., Dahari, M., & Zaharinie, T. (2017). A novel, eco-friendly technique for covalent functionalization of graphene nanoplatelets and the potential of their nanofluids for heat transfer applications. *Chemical Physics Letters*, *675*, 92–97. doi:10.1016/j.cplett.2017.02.077

Sadri, R., Hosseini, M., Kazi, S. N., Bagheri, S., Zubir, N., Solangi, K. H., Zaharinie, T., & Badarudin, A. (2017). A bio-based, facile approach for the preparation of covalently functionalized carbon nanotubes aqueous suspensions and their potential as heat transfer fluids. *Journal of Colloid and Interface Science*, *504*, 115–123. doi:10.1016/j.jcis.2017.03.051 PMID:28531649

Saidur, R., Leong, K. Y., & Mohammed, H. A. (2011). A review on applications and challenges of nanofluids. *Renewable & Sustainable Energy Reviews*, *15*(3), 1646–1668. doi:10.1016/j.rser.2010.11.035

Seth, G. S., Bhattacharyya, A., Kumar, R., & Mishra, M. K. (2019). Modelling and numerical simulation of hydromagnetic natural convection Casson fluid flow with n-th order chemical reaction and Newtonian heating in porous medium. *Journal of Porous Media*, *22*(9), 1141–1157. doi:10.1615/JPorMedia.2019025699

Seth, G. S., Bhattacharyya, A., & Mishra, M. K. (2019). Study of partial slip mechanism on free convection flow of viscoelastic fluid past a non-linearly stretching surface. Comput Thermal Sci. *International Journal (Toronto, Ont.)*, *11*(1–2), 105–117. doi:10.1615/ComputThermalScien.2018024728

Seth, G. S., Kumar, R., Tripathi, R., & Bhattacharyya, A. (2018). Double diffusive MHD Casson fluid flow in a non-Darcy porous medium with Newtonian heating and thermo-diffusion effects. *Int J Heat Tech*, *36*(4), 1517–1527. doi:10.18280/ijht.360446

Sharmila, G., Muthukumaran, C., Sangeetha, E., Saraswathi, H., Soundarya, S., & Kumar, N. M. (2019). Green fabrication, characterization of Pisonia alba leaf extract derived MgO nanoparticles and its biological applications. *Nano-Structures & Nano-Objects*, *20*, 100380. doi:10.1016/j.nanoso.2019.100380

Shirazi, S. F. S., Gharehkhani, S., Yarmand, H., Badarudin, A., Metselaar, H. S. C., & Kazi, S. N. (2015). Nitrogen doped activated carbon/graphene with high nitrogen level: Green synthesis and thermo-electrical properties of its nanofluid. *Materials Letters*, *152*, 192–195. doi:10.1016/j.matlet.2015.03.110

Sizochenko, N., Syzochenko, M., Gajewicz, A., Leszczynski, J., & Puzyn, T. (2017). Predicting physical properties of nanofluids by computational modeling. *The Journal of Physical Chemistry C*, *121*(3), 1910–1917. doi:10.1021/acs.jpcc.6b08850

Sone, B. T., Diallo, A., Fuku, X. G., Gurib-Fakim, A., & Maaza, M. (2020). Biosynthesized CuO nano-platelets: Physical properties & enhanced thermal conductivity nanofluidics. *Arabian Journal of Chemistry*, *13*(1), 160–170. doi:10.1016/j.arabjc.2017.03.004

Sreelakshmy, K. R., Nair, A. S., Vidhya, K., Saranya, T., & Nair, S. C. (2014). An overview of recent nanofluid research. *International Research Journal of Pharmacy*, *5*(4), 239-243.

Stephen, U. S. (1998). Nanofluid Technology. Current Status and Future Research, 1, 1-3.

Suresh, S., Venkitaraj, K. P., Selvakumar, P., & Chandrasekar, M. (2011). Synthesis of Al2O3-Cu/water hybrid nanofluids using two step method and its thermo physical properties. *Colloids and Surfaces. A, Physicochemical and Engineering Aspects*, *388*(1-3), 41–48. doi:10.1016/j.colsurfa.2011.08.005

Tiwari, R. K., & Das, M. K. (2007). Heat transfer augmentation in a twosided lid-driven differentially heated square cavity utilizing nanofluids. *International Journal of Heat and Mass Transfer*, *50*(9-10), 2002–2018. doi:10.1016/j.ijheatmasstransfer.2006.09.034

Trisakti, V., & Wongwises, S. (2007). Critical review of heat transfer characteristics of nanofluids. *Renewable & Sustainable Energy Reviews*, *11*(3), 512–523. doi:10.1016/j.rser.2005.01.010

Vajravelu, K., Prasad, K. V., Lee, J., Lee, C., Pop, I., & Van Gorder, R. A. (2011). Convective heat transfer in the flow of viscous Ag-water and Cu-water nanofluids over a stretching surface. *International Journal of Thermal Sciences*, *50*(5), 843–851. doi:10.1016/j.ijthermalsci.2011.01.008

Van Gorder, R. A., Sweet, E., & Vajravelu, K. (2010). Nano boundary layers over stretching surfaces. *Communications in Nonlinear Science and Numerical Simulation*, *15*(6), 1494–1500. doi:10.1016/j.cnsns.2009.06.004

Vidhu, V. K., & Philip, D. (2015). Phytosynthesis and applications of bioactive SnO2 nanoparticles. *Materials Characterization*, *101*, 97–105. doi:10.1016/j.matchar.2014.12.027

Wong, K. V., & De Leon, O. (2009). Applications of Nanofluids. *Current and Future*, *2*, 5.

Ye, W., Li, X., Luo, J., Wang, X., & Sun, R. (2017). Lignin as a green reductant and morphology directing agent in the fabrication of 3D graphene-based composites for high-performance supercapacitors. *Industrial Crops and Products*, *109*, 410–419. doi:10.1016/j.indcrop.2017.08.047

Yu, W., France, D. M., Routbort, J. L., & Choi, S. U. S. (2008). Review and comparison of nanofluid thermal conductivity and heat transfer enhancements. *Heat Transfer Engineering*, *29*(5), 432–460. doi:10.1080/01457630701850851

Yu, W., & Huaqing, X. A. (2012). Review on nano-fluids: Preparation, stability mechanisms, and applications. Journal of Nanomaterials.

Chapter 11
Applications of Nanofluids in Healthcare:
Future Perspectives

K. Radhika
https://orcid.org/0000-0003-4849-8680
Muthayammal Engineering College, India

N. Suthanthira Vanitha
https://orcid.org/0000-0002-9579-7651
Muthayammal Engineering College, India

D. Anitha
Muthayammal Engineering College, India

K. Bashkaran
Kongunadu College of Engineering and Technology, India

K. Shemalaa Ganthi
Muthayammal Engineering College, India

ABSTRACT

Advancement of nanotechnology leads to the creation of innovation in nanostructures which are characterized by their explorative utilization in a variety of applications in biosciences. Nanofluid technology is an interdisciplinary field, where nanoscience, nanotechnology, and thermal engineering have enhanced over the past decades. The diagnosis and therapeutic aspects of diseases are found to be improved very efficiently in drug delivery platform by using nanofluids. Nanofluids contain novel properties which are potentially employed in several applications like biotechnology, national security and defense, food and agriculture, information technology, aerospace, plastics and textile industries, energy and environment, cosmetics, and health and medicine. This chapter discusses the future perspectives and applications of nanofluids.

DOI: 10.4018/978-1-6684-9135-5.ch011

1. INTRODUCTION

High heat flux devices are produced as a result of microelectronics industry advancements, but heat dissipation is a crucial issue for their widespread use in industry and daily life since heat accumulation on electronic chips can cause these devices to malfunction or even entirely fail. The most crucial and difficult duty involved in these electrical chips and circuits is heat dissipation. However, due to their poor thermal conductivity, which causes poor heat dissipation and slower performance of electronics chips, the traditional coolants (water, oils, and ethylene glycols) have been shown to be ineffective. In order to increase the fluids' thermal conductivity and hence increase their capacity to dissipate heat, J.C. Maxwell proposed this in 1873. He suggested adding extremely small solid particles to the fluids (called base fluids). It is well known that micro and millimetre sized particles increase the thermal conductivity of base fluids, but experiments revealed additional issues that limit their use in the micro industry, including abrasive wear of pipelines, clogging of channels, sedimentation of particles, and pressure drop. Subsequently, nanosized particles were developed to avoid these issues and as nanoscale industry and nanotechnology advanced, they acquired popularity for use in research (Lim et al.,2013).

Nanoparticles, nanofibers, nanotubes, nanowires, nanorods, nanosheets, and droplets are examples of the new class of fluids called nanofluids that are created by distributing them in base fluids.In other words, condensed nanomaterials (or) nanometer-sized particles are suspended in nanoscale colloidal suspensions, or nanofluids. They consists of a two

phase system with a solid phase in one phase (liquid phase). In comparison to base fluids like oil or water, nanofluids are discovered to have improved thermal conductivity, thermal diffusivity, viscosity, and convective heat transfer coefficients. They can be utilized to improve heat transfer directly in a variety of industrial applications, including heat exchangers, transportation, electronics, biomedicine, and the food sector (Wong & De Leon, 2017). In compared to their solid counterparts, nanofluids physicochemical qualities alter significantly due to the growing interest in using them in nano biomedical technology. Nanofluids have recently been used in biological applications, including drug transport and antibacterial treatments(Lim et al.,2013).

The application of nanofluids has number of essential criteria that accurately describes the composition, size, crystallinity and shape. Other crucial characteristics are stability, non-agglomeration and biocompatibility (Stankic et al.,2016). The majority of the NPs partially scatter in water-based solutions. The composition, structure, size, and shape of the NPs play a significant role on the dispersion stability. Many techniques can be used to create a stable nanofluid, including the manufacture of nanomaterials, physical stirring, ultrasonic homogenization, surface engineering, and the use of surfactants and other chemical additives (Yu F et al.,2017). The main problem facing the researchers is stability in two phase systems, even with improved characteristics. The majority of research on nanofluids has focused on materials, surface science, and colloid theory. As a result, numerous characteristics of nano-based systems have been improved. The biological uses of nanofluids are far and call for more research.

2.TYPES OF NANOFLUIDS

The name nanofuid refers to fluids that contain distributed nanoscale particles. Nanofluids are classified into four different categories based on the type of dispersed particle: (1) metal (2) metal oxide (3) carbon (4) metal hybrid. Nanoparticles of single elements, such as Cu, Fe, and Ag, as well as single element oxides, such as CuO, Cu2O, Al2O3, and TiO2, as well as alloys, can be utilized to create nanofuids (Eg: Cu-Zn, Fe-Ni, and Ag-Cu), multielement oxides (Eg: CuZnFe4O4, NiFe2O4, and ZnFe2O4), metal carbides (SiC, B4C & ZrC), metal nitrides (SiN, TiN, and AlN), and carbon materials like graphite, carbon nanotubes, and diamond balanced in water, oil and other refrigents shown in figure 1.

2.1 . Nanofluids Based on Carbide, Metals, and Metal Oxides

The use of nanoparticles to enhance the thermophysical characteristics of thermo-fluids in 1995 (Florence,2012). Since then, extensive research has been done to create nanofluids with distinctive thermo-physical characteristics as thermal conductivity, thermal diffusivity, and viscosity (Otanicar et al., 2009 & Choi, 1995). According to studies, adding tiny concentrations of nanoparticles can enhance the thermo-physical characteristics of a variety of fluids. Pure metals (Au, Ag, Cu, Al, and Fe), metal oxides (Al2O3, CuO, Fe3O4, SiO2, TiO2, and ZnO), carbides (SiC, TiC), and a variety of carbon materials (diamond, graphite, single/multi wall carbon nanotubes) are among the types of nanoparticles that are studied. Many of these different kinds of nanoparticles have been used in liquids including water, water/ethylene glycol, ethylene glycol, and oils. Cu, Ag, Al2O3, SiO2, TiO2, graphite nanoparticles, and carbon nanotubes were some of the solid phases utilized in the corresponding nanofluids. Texatherm oil was employed as the fluid phase in all nanofluids.

2.2 Carbon Nanomaterial Based Nanofluids

Most prevalent element in the biosphere of the earth is carbon. Nature has created a wide variety of organic molecules using this element, primarily in combination with oxygen and hydrogen. Moreover, carbon atoms can join with one another in a variety of ways to create a range of carbon compounds known as allotropes of carbon. The particular material properties of each of these allotropes vary. Amorphous carbon, diamond, and graphite are examples of common carbon allotropes. The shapes and morphologies of carbon allotropes can also vary, including crystalline (such as diamond, which is three dimensional, 3D), graphite sheets (2D), and carbon nanotubes (1D) (Choi, 1995). Buckminster-fullerenes (or) Bucky balls, in the middle of the 1980s sparked a renewed interest in carbonaceous materials. Carbon nanomaterials are of special interest since they are black in color and perfect for sun absorption applications. They also make excellent additions for nanofluids because of their extremely high thermal conductivity.

Single and multiwall carbon nanotubes, nanoballs, crystalline fullerenes C70, C76, C84, and C60, as well as nanocones, nanohorns, and nanofilaments are examples of recent developments in carbon nanomaterials. Nanocapsules (Sano et al.,2003), graphite nanoparticles, grapheme and graphene oxide, black and carbon nanospheres (CNS) (Hong Wei Xian et al.,2020) are a few examples of the materials that are used.

2.3 Carbon Nanotube-Based Nanofluids and Others

Carbon-based, one dimensional cylinders with large aspect ratio are known as carbon nanotubes (CNTs).Singlewalled (SWCNTs) or multiwalled (MWCNTs) arrangements or both make up the cylindrical structure.

Due to its affordability and ease of availability in the market, water or distilled water is the most frequently utilised base fluid for the creation of nano fluid. To make base fluid, distilled water and ethylene glycol were combined in an equal amount. To create mono nanofluid, carboxyl-functionalized graphene nanoplatelets (COOH-GnP) powders were employed. Using a magnetic stirrer, titanium oxide and COOH-GnP powder were combined with the base fluid to create a hybrid nanofluid (Hong Wei Xian et al.,2020). For the purposes of the flat plate collector, alumina-water and alumina-iron/water hybrid nanofluids were employed (Eric et al.,2020). The impact of Aluminium oxide water nanofluid and effects of the time of ultrasonication on colloidal dispersion and thermophysical characteristics such density, thermal conductivity, viscosity and microstructure were examined.

Figure 1. Types of nanofluids

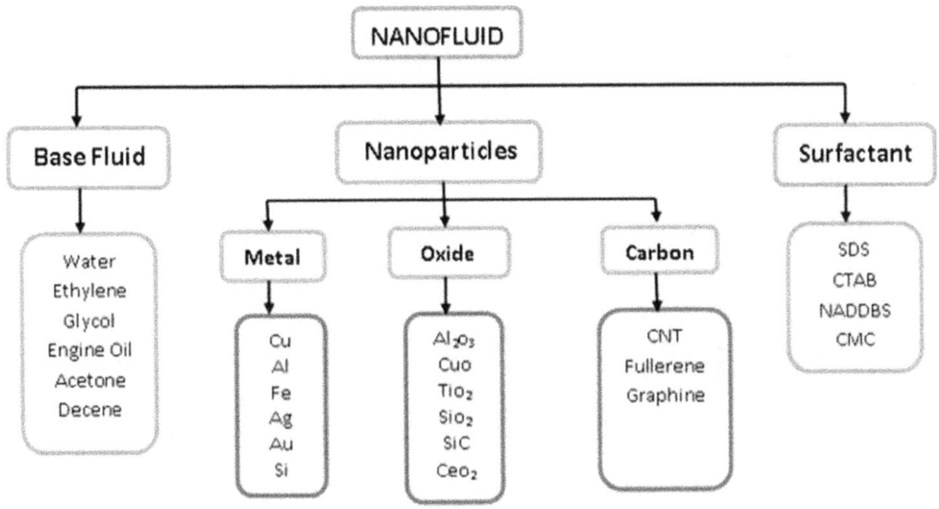

3. NANOFLUID PREPARATION METHODS

There are various methods for creating nanofluids, including first step, second step, and other methods. Surfactant may be added to them to prevent the sedimentation of nanoparticles during its operation. Stabilization and correct mixing are necessary under specific environmental conditions to produce nanofluids with homogeneously dispersed nanoparticles. The creation of nanofluids can be accomplished in a variety of ways are depicted in figure 2.

3.1 Single (or) One Step Method

The one-step approach combines the processes of NP synthesis and fluid dispersion into one step. A single-step method for creating NFs is shown in figure 3. By not requiring NPs to be stored, transported, dried, or dispersed, a one-step technique decreases agglomeration and increases fluid stability (Vallejo et al., 2019). A more stable NF with a constant NP distribution is produced when NFs are synthesized in one step. NFs are produced using

(i) Vacuum-SANSS (Submerged Arc NP Synthesis System)
(ii) Physical Vapour Deposition
(iii) Liquid Chemical Process
(iv) Chemical Solution Method

Figure 2. Nanofluid preparation methods

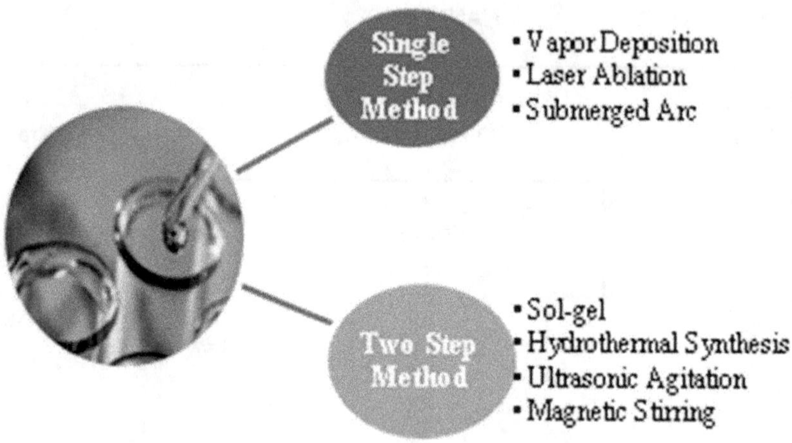

The base fluid, such as oil or ethylene glycol, is put into a revolving cylindrical drum with a heat exchanger cooler, an adjustable heater, a boat evaporator, and the like. A thin liquid layer forms on the interior surface of the drum as it rotates. A portion of the rotating thin liquid layer is in contact with an insulated heater-boat-evaporator that contains an evaporant material (nanomaterial source). As the evaporant material is heated, some of it evaporates and forms nanoparticles, which are then absorbed by the liquid film to form nanofluid. The one-step technique creates and

Figure 3. Single/One step method

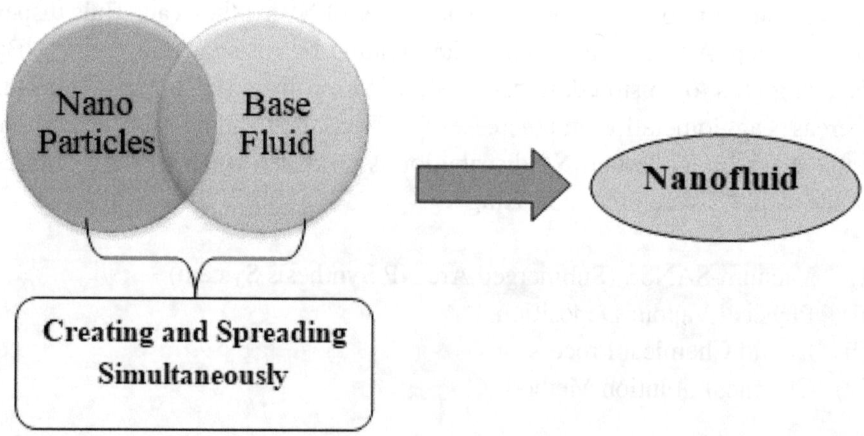

disperses particles in the basic fluids, which can be boiled down to an aggregation of nanomaterials. The high cost of the procedure is a limitation of this technology, which increases the stability of the nanofluids.

3.2. Two-Step Method

The most popular technique for creating nanofluids is a two-step process is shown in figure 4. Nanoparticles of various materials, such as nanofibers, nanotubes, or other nanomaterials, are initially created as nanoparticles between 10 and 100 nm in size via chemical or physical processes. Then, with the aid of vigorous magnetic force agitation, ultrasonic agitation, high-shear mixing, homogenising, and ball milling, the nanosized powder will be distributed in base fluids. Due to their large surface area and surface activity, nanoparticles have a propensity to agglomerate, which negatively affects the stability of nanofluid (Jailani et al.,2008,Yu W et al.,2012,Quddoos et al., 2014, Neto et al.,2010 & Micha et al 2012). The surfactant is added to the nanofluids to counteract that impact. This process is employed in mass production of nanofluids.

Particle agglomeration due to poor van der Waals strength during drying, storage, or transit is a drawback of this technology because the formulation of NPs and NFs is a separate procedure. To increase NPs stability surfactants are utilized. The surfactants capacity to be applied at high levels of temperature is still important, particularly for high levels of temperature applications.

Figure 4. Two step method

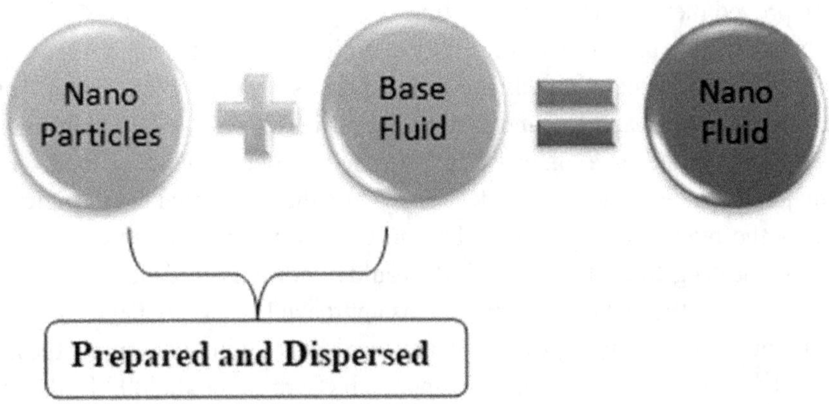

4. NANOFLUIDS IN MEDICAL APPLICATIONS

The way a disease is diagnosed and treated could be significantly changed by nanotechnology. Nanomaterials have been used for medical diagnosis for early disease detection, target-based clinical therapy, and regenerative medicine for the repair of damaged tissues due to their unmatched sensitivity and performance, improved durability and flexibility, and unique physicochemical properties. Many nano-devices and nano-biosensors have been developed to monitor biomolecules at very low concentrations, allowing for the early identification of disease. These may be a cutting-edge and potent tool for cancer detection. Traditional diagnostic methods are less accurate in distinguishing between benign and malignant stages of a tumour and unable to detect malignancies in their early stages. Novel nanoparticles have the ability to produce selective imaging of afflicted areas in comparison to existing approaches.

4.1 Drug Delivery

Drug delivery based NFs have been investigated recently to increase effectiveness and specificity. The qualities and attributes of NPs, including their small size, tailored surfaces, and multifunctionality, enable interactions with complicated biological functions in unexpected ways (Owens et al.,2006). The basic goal of a drug delivery system's design is to deliver the medicine to the desired spot in a regulated manner. The initial research into drug delivery methods focused on drug release kinetics, polymers, pH and temperature sensitivity, nasal delivery, and oral drug delivery (Park K,2014). One of the subjects that has attracted the attention of many researchers in recent years is the use of nanomaterials in drug delivery systems, which may offer advantages including more precise distribution to the target cells and improved therapeutic characteristics.

Magnetic NPs have drawn the most attention among the many types of NPs because they are effective medication transporters. This problem arises from the controllability, compactness, and surface properties of these NPs, which enable the carrier to be guided by a magnetic field to the appropriate place. Blood serves as the major fluid in magnetic drug delivery, while the magnetic nanoparticles (NPs) serve as the medication's carriers. Due to the strong and concentrated magnetic gradient, the drug-loaded magnetic NPs would be administered close to the tumour; the tumour may then absorb the drug. This approach lessens the negative effects of chemotherapy. Using stronger anticancer drug dosages without harming the healthy side tissues is another benefit of this approach (Barcena et al.,2009,Kothandapani et al.,2015,Pankhurst et al.,2003).

Using magnetic nanoparticles (NPs), which involve inserting a magnet adjacent to the tumour and administering the medication into the blood arteries that are closest to the tumour, considered as the most effective method of cancer treatment.. The waves that are produced in the cone-shaped asymmetric channel walls may have an impact on the dynamics of these NPs through peristaltic motion. Treatment of malignant tissues can be made much more effective by analysing the nanofluid flow that is affected by this kind of movement (Kothandapani et al.,2015 & Frazier et al.,2015). For magnetic drug delivery in clinical applications, superparamagnetic iron oxide nanoparticles (SPIOs) are the most used magnetic nanoparticle.

Due to their low toxicity and lack of immunogenicity, carbon nanotubes can be utilised to deliver biomolecules and medications to specific cells or organs. Anticancer medication delivery methods may be able to utilize graphene-based NFs. 1 A drug coated NPs catheter was used to study the rheological characteristics of blood flow through a stenotic tapering artery .Due to high mass and HT at the extremes, the results demonstrated that drug delivery through catheter is beneficial in treating arterial obstruction with fewer adverse effects. To prevent arterial obstructions, this model can be employed in biomedical and drug delivery applications (Gunasekaran et al.,2022).

4.2 Cancer Treatment

Cancer can be treated with hyperthermia. “Hyper” and “thermia,” two Greek words that denote growth and heat, respectively. By reducing the disease-causing agent’s ability to multiply, the body raises its temperature to ward against illness (Saleh et al., 2017). For the purpose of treating cancer, an organ or tissue is heated to between 41 and 46 °C for anywhere between 20 and 60 minutes. This method is known as hyperthermia. With the help of magnetic NPs, a uniform dispersion of heat was made achievable. Recent studies have shown that MNFs enter cells effectively and have stable temperature increase properties .Blood perfusion rate and the volume of NF delivered to the target area are crucial for affecting the temperature distribution of tissue when employing MNPs hyperthermia for cancer therapy. The uses of Fe3O4 and -Fe2O3 for hyperthermia are largely being studied due to their biocompatibility, low toxicity, strong chemical stability, mild saturation magnetization, high magnetocrystalline anisotropy, high conductivity, and high specific absorption rate.Since an excessive amount of iron in live tissue might be hazardous, tissue iron uptake needs to be researched.

Specific Absorption Rate (SAR) which measures energy distribution in terms of W/m3 or W/g, has been employed (Wiedeman,1963). This value represents the ability of magnetic NPs to destroy malignant cells as determined by several temperature sensors inserted into the agarosegel. Moreover, a method known as multiple injections has been devised to consistently raise tumour temperature in tumours with uneven shapes. The cubic shape of NPs has more SAR than the other forms, according to research (spherical, rod, start, polyhedron, and cone) (Ebaid & Aly, 2013). The temperature, the frequency of the magnetic field changes, the intensity of the magnetic field, and the size and shape of the NPs are some of the variables that affect the SAR readings. The size of magnetic NPs is one of the key elements in determining their effectiveness. A few benefits of NPs' small size include their ultra-paramagnetic (ultra-magnetic) states, greater heat dispersion, prevention of agglomeration, and an increase in SAR. Notwithstanding these benefits, their performance might be compromised if the NP size was less than usual (2–5 nm).

Magnetic Fluid Hyperthermia The use of magnetic NPs in the presence of an external magnetic field to selectively apply heat to the targeted tissue is known as magnetic fluid hyperthermia. Using magnetic NPs,(Salloum et al.,2008) examined the rate of temperature rise in animal tissues during hyperthermia in their 2008

Figure 5. Controlled drug delivery system

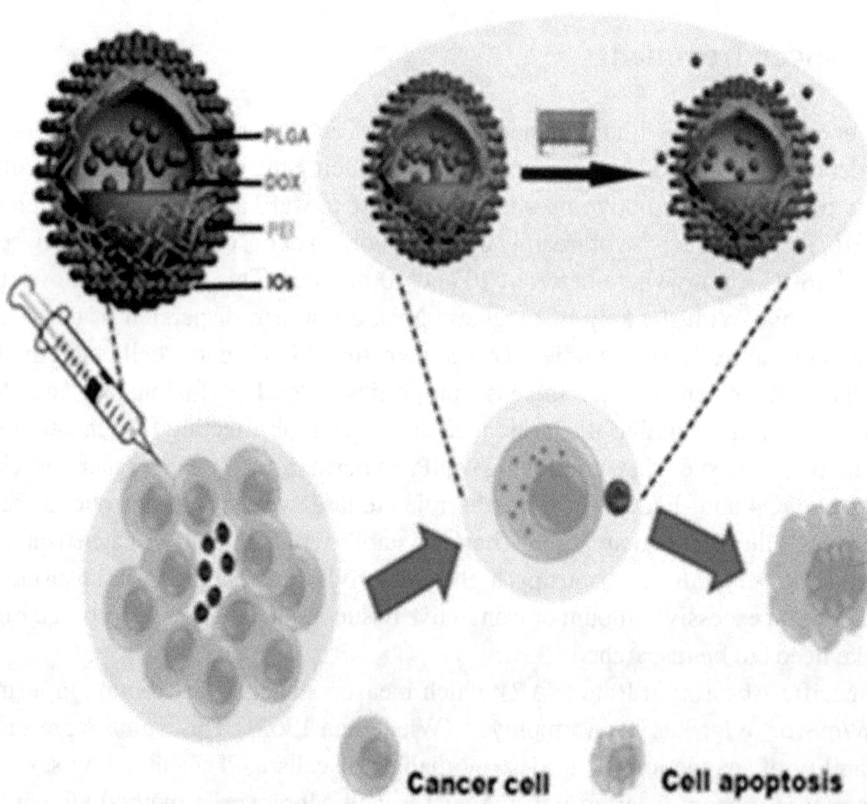

work on experimental lab mice. The rate of blood perfusion and the quantity of nanofluid supplied to the tissue were two crucial elements in this study's discussion of magnetic hyperthermia (Salloum et al.,2008). Under the influence of a magnetic field, carbon nanotubes (CNTs) may be employed to cure cancer cells. In the linked research studies, NPs are injected into the blood arteries close to the tumour, and a magnet is positioned close to the tumour (Ebaid & Aly, 2013, Fang et al., 2015). Doxorubicin-loaded magnetic PLGA(poly(lactic-co-glycolic acid)), microspheres have been created as a controlled drug delivery device for inhibiting tumour development in breast cancer (Mojgan Sheikhpour et al.,2020) shown in figure 5.

4.3 Nano-Cryosurgery

One of the best methods for completely treating and controlling tumour cells is cryosurgery. This technique makes use of liquid nitrogen or solid carbon dioxide. A reduction in the amount of liquid water results in the production of ice crystals, which increases the risk of the target cell's membrane breaking (dehydration) (Singh & Bhargava,2014).However, this approach has several drawbacks, such as improper freezing, which might harm the good side tissues. To increase the freezing capacity, the tumour tissue is loaded with highly-conductive NPs (Hou et al.,2018).

4.4 Anemia Treatment

The stability of magnetic iron oxide NPs in the fluid interface is one of the prerequisites for employing them in medical applications. (Tran & Webster,2010) produced a nanofluid with anti-anemic capabilities that contained sustained magnetic NPs. PVP (polyvinyl pyrrolidine) and mexidol were used to stabilize these NPs (2-ethyl-6-methyl-3-hydroxy pyridine succinate). Following substantial blood loss, they evaluated the information on 109 male albino mice used as experimental animals. According to the study's findings, healthy animals' production of red blood cells (RBCs), haemoglobin, and hematocrit was boosted by the nanofluid. A nanofluid aids in recovering blood loss in cases of severe anaemia caused by blood loss and boosts the quantity of reticulocytes. The erythropoiesis of intact animals is stimulated by this nanofluid, and anaemia induced by blood loss promotes the erythron's regenerative response.

4.5 Image Diagnosis

The use of nanomaterials in magnetic resonance imaging is one of the most fascinating diagnostic technology. MRI contrast and sensitivity are improved by using magnetic NPs . One of the finest options for improving contrast in an MRI is super-paramagnetic nanoparticles (SPMNPs). Iron oxides, gold, and gadolinium NPs are a few possible ingredients for these substances (Shokrollahi,2013). The core of the SPMNPs, which could also contain a Fe/Mn composite, is made of iron oxide, and it is surrounded by a polymeric matrix that provides colloidal stability. The US Food and Drug Administration (FDA) and the European Medicines Agency (EMEA). have approved many injectable magnetic contrast agents.Contrast materials based on gadolinium are routinely utilised in MRI (Su Hy et al.,2015). Its applicability is constrained by gadolinium toxicity issues, particularly in patients with renal insufficiency. From the application point, the MRI contrast contrast agents can be divided into three groups: extracellular fluids, intravascular contrast agent and target-specific agents. Oral administration of various MRI contrast agents is an option, and some compositions have been created for intravenous injection (Shokrollahi,2013).

The contrast media tolerance, safety are significant and important variables. A good contrast agent must be completely removed from the body without any side effects or negative reactions, as well as being physiologically and chemically inert. Typically, the targeting agents adhere to the target organ's tissue-specific antibodies. For liver imaging, the majority of target-specific contrast agents have been used (Su Hy et al.,2015).Targeting contrast compounds based on liposomes and superparamagnetic iron oxide are readily available in the market. In this regard, contrast has been produced in MRI imaging using Fe3O4O-dextran NPs (Fe3O4 with an oxidised dextran coating). These NPs have crucial qualities such high stability, minimal cytotoxicity, and strong biocompatibility and bioconjugation. Cancer cells that were treated to Fe3O4O-dextran NPs (with antibody titers) at varied doses were seen in MRI images by Huang et al. The image of cancer cells became darker as iron quantities raised from 0.75 mM to 1.15 mM, demonstrating that the use of NPs improved MRI contrast.

Due to their ability to generate robust CT attenuation, cellular absorption, and targeting abilities, nanosized CT contrast agents have been employed in an wide range of applications. For instance, red blood cells have been employed to scan blood flow by absorbing gold nanoparticles. Liposomal iodine has been utilized to assess tumour vasculature because of its lengthy circulation duration and significant CT enhancement. Prostate cancer has been imaged using aptamer and prostate-specific membrane antigen embedded in gold nanoparticles. The buildup of zirconium dioxide nanoparticles has also been used to scan tumours and track therapeutic absorption.

4.6 Antibacterial Function

Antibacterial compounds are important to the industry, which includes food packaging, the textile and medical sector, as well as water disinfection. Zinc oxide (ZnO), one of the several forms of metal oxide NPs, has the greatest capacity to inhibit bacterial growth in a wide variety of bacteria (Usman et al.,2013). Among other nanofluids, ZnO nanofluids that are safe for human cells have mostly been explored for their antibacterial properties. Many elements of ZnO toxicity were identified based on morphological studies on the cell walls of bacteria exposed with ZnO NPs by SEM and FESEM. In addition, TiO2 and SiO2 NPs has antibacterial properties lower than that of ZnO. These three Nanoparticles are photosensitive. Also, due to their uses in the pharmaceutical industry, MgO and Mg(OH)2 nanomaterials have received a lot of attention throughout the years (Dong et al.,2010).According to reports, E. coli, B. phytofirmans, and a few other bacteria are significantly inhibited by Mg(OH)2 nanomaterials. According to Dong et al. the antibacterial mechanism of Mg(OH)2 NPs would be extremely distinct from that of metal-based compounds. The buildup of OH- ions in bacterial cells causes the pH level to rise to 10, which causes the bacteria to die.

Cu and CuO NPs have antibacterial properties against a variety of microorganisms (Usman et al.,2013 & Mahapatra et al.,2008).The size of NPs and the nanoparticle's production temperature affect its antibacterial activity. More antibacterial activity is produced with smaller CuO NPs. Due to their smaller size than the pore size in bacteria, nanoparticles have the unusual ability to pass through cell membranes without being blocked.

Silver (Ag) NPs having strong antimicrobial activity against Gram-positive and Gram-negative bacteria, including multiple-resistant species like Staphylococcus aureus, were synthesized using the single-step technique. A broad range of toxicity mechanisms exist for silver nanoparticles, the main one being the disruption of mitochondrial function caused by changes in the permeability of the mitochondrial membrane. In addition, silver nanoparticles trigger inflammatory reactions because they produce reactive oxygen species (ROS) (Akter et al.,2018).

Gold nanoparticles have a wide range of uses and are known to be biocompatible and significantly less cytotoxic. Inhibiting ATP generation by changing membrane potential and inhibiting tRNA binding to the ribosome are the two main mechanisms through which gold nanoparticles' antibacterial activity is commonly based. ROS generation has no effect on the antibacterial activity of gold nanoparticles.

4.7 Cell Preservation (Cryopreservation)

The quartz micro-capillary system for novel ultra-fast regeneration designed would be a viable and effective method for the cryopreservation of murine embryonic stem cells at vitrification at a low concentration of intracellular cryoprotectants. Traditional vitrification techniques frequently involve harmful cryoprotectants in extremely high concentrations (> than 4 M). The amount of cryoprotectant employed in this investigation was 2M (He X et al.,2008), the same amount that is often used for the conventional slow-freezing techniques. Hence, this approach combined the advantages of both fast chilling and traditional slow freezing. Murine embryonic stem cells were vitrified using this technique, with a viability rate comparable to the unfrozen control cells. In addition to the aforementioned biomedical applications, NFs have extensive uses in the areas of electronics, energy, and mechanical engineering due to their superior qualities over those of ordinary fluids.

5. CHALLENGES AND FUTURE PERSPECTIVES

5.1 Challenges

Colloidal theory states that as the particle size goes below a critical radius, sedimentation in suspensions occurs due to Brownian forces that balance gravity when the particle size is above the critical radius. Nanoparticles with a smaller size may be more advantageous for a variety of uses. On the other hand, because of its high surface, agglomerates develop in their spaces. Making a stable nanofluid with the ideal particle diameter and concentration is therefore a considerable challenge. Hence,it is necessary to maximize experimental studies of nanofluids in terms of stability, preparation method, temperature, crystalline size and shape.

The interaction of NPs with the environment may create a harm to health and the environment due to their relatively small size and low concentration . The ecological system could also be in trouble. The ecology is impacted by the physical and chemical characteristics of NPs. NPs on human skin can be hazardous since they are simple to penetrate into the lymphatic and blood cell systems, where they can cause significant adverse effects on the body.

The toxicity of NPs and their effects on healthy tissues have been studied by several researchers. From these studies, they have discovered that Al2O3, Fe3O4, and Au NPs are biologically appropriate and may find extensive use in medical applications. 29,30,37,38 Additionally, it has been demonstrated that among all used NP kinds, Au NPs had the highest freezing effectiveness and the largest possible ice-ball size. In addition to speeding up the freezing process, the presence of NPs may also be able to regulate the size and expansion of the ice crystals that develop during the freezing process. This characteristic results in the removal of unwanted malignant tissues and producing the least amount of damage to the surrounding healthy tissues.Different types and concentrations of NP solutions could have varied effects on the freezing process and result in different freezing efficiencies. Thus, choosing the right NPs and optimizing the NP suspension dose might be a crucial decisions.

Metallic nanoparticles have some drawbacks, such as stability problems, biocompatibility concerns, and improper elimination from the body. Problems with solubility are also significant for metal and metallic oxide NPs. A connection exists between the reported toxicity and the release of metal ions from dissolved NPs. Significant relationships exist between NPs' biological properties, their nature and structural traits. Hence, it is important to take into account any modifications to them that could lessen their toxicity and increase their biocompatibility, such as surface modification, artificially controlled size and structure.

5.2 Future Perspective

- Nanofluids with low viscosity and strong conductivity have a wide range of potential uses. One of the potential solutions is to improve the interaction among nanomaterials and the base fluids by altering the characteristics between two phases
- Establish more accurate correlations that can forecast changes in pH values induced by temperature, particle concentration, base fluid type, and other factors that affect the stability and thermophysical characteristics of nanofluids
- Nanofluids in targeted MRI is the most fascinating imaging and diagnosis technique, has grown substantially. In the MRI, contrast agents play a crucial function, and some nanofluids, such those based on Fe_3O_4, have been taken into consideration . In order to achieve the optimal contrasting behaviour of various agents in different diagnosis, it may be helpful to experiment with other nanofluids.
- Increasing the number of in vivo and in vitro tests conducted on nanofluid agents
- Evaluating various nanofluid types for use in bacterial disinfection, drug delivery, and therapy
- In diagnostic part, it compare various contrast agents
- Future predictions indicate that nanofluid technology will play an essential role in medical field to create non-toxic or biodegradable nanoparticles in order to produce stable and effective nanofluids.

6. CONCLUSION

Nanofluids plays an significant role in all fields of science and technology. Many intriguing characteristics of nanofluids have been reported throughout the last few decades. This chapter provides an overview of preparation methods of nanofluids, classification and potential applications in biomedical field. Also,chapter outlines the challenges and future perspectives of nanofluids.

REFERENCES

Akter, M., Sikder, M.T., & Rahman, M.M. (2018). A systematic review on silver nanoparticles-induced cytotoxicity: physicochemical properties and perspectives. *J Adv Res*, *9*, 1–16. doi:. jare.2017.10.008 doi:10.1016/j

Bárcena, C., Sra, A. K., & Gao, J. (2009). Applications of Magnetic Nanoparticles in Biomedicine. In *Nanoscale Magnetic Materials and Applications* (pp. 591–626). Springer. doi:10.1007/978-0-387-85600-1_20

Choi, S. U. S. (1995). *Enhancing Thermal Conductivity of Fluids with Nanoparticles, Developments and Applications of Non-Newtonian Flows*. ASME.

Dong, C., Cairney, J., Sun, Q., Maddan, O. L., He, G., & Deng, Y. (2010). Investigation of Mg(OH)2 nanoparticles as an antibacterial agent. *Journal of Nanoparticle Research*, *12*(6), 2101–2109. doi:10.100711051-009-9769-9

Ebaid, A., & Aly, E. H. (2013). Exact analytical solution of the peristaltic nanofluids flow in an asymmetric channel with flexible walls and slip condition: Application to the cancer treatment. *Computational and Mathematical Methods in Medicine*, *2013*, 1–8. doi:10.1155/2013/825376 PMID:24151526

Fang, K., Song, L., Gu, Z., Yang, F., Zhang, Y., & Gu, N. (2015). Magnetic field activated drug release system based on magnetic PLGA microspheres for chemo-thermal therapy. *Colloids and Surfaces. B, Biointerfaces*, *136*, 712–720. doi:10.1016/j.colsurfb.2015.10.014 PMID:26513754

Florence, A. T. (2012). Targeting Nanoparticles: The constraints of physical laws and physical barriers. *Journal of Controlled Release*, *164*(2), 115–124. doi:10.1016/j.jconrel.2012.03.022 PMID:22484196

Frazier, N., & Ghandehari, H. (2015). Hyperthermia approaches for enhanced delivery of nanomedicines to solid tumors. *Biotechnology and Bioengineering*, *112*(10), 1967–1983. doi:10.1002/bit.25653 PMID:25995079

Gunasekaran, K., Suthanthira Vanitha, N., Radhika, K., & Suresh, P. (2022). Nanomedicine in healthcare: Impact and challenges for future generation. Taylor & Francis. doi:10.1201/9781003239888-12

Harish, S., Ishikawa, K., Einarsson, E., Aikawa, S., Chiashi, S., Shiomi, J., & Maruyama, S. (2012). Enhanced thermal conductivity of ethylene glycol with single-walled carbon nanotube inclusions. *International Journal of Heat and Mass Transfer*, *55*(13-14), 3885–3890. doi:10.1016/j.ijheatmasstransfer.2012.03.001

He, X., Park, E. Y. H., Fowler, A., Yarmush, M. L., & Toner, M. (2008). Vitrification by ultra-fast cooling at a low concentration of cryoprotectants in a quartz micro-capillary: A study using murine embryonic stem cells. *Cryobiology*, *56*(3), 223–232. doi:10.1016/j.cryobiol.2008.03.005 PMID:18462712

Hong, W. X., Che Sidik, N., & Saidur, R. (2020). Impact of different surfactants and ultrasonication time on the stability and thermophysical properties of hybrid nanofluids. *ICHMT, 110*, 104389.

Hou, Y., Sun, Z., & Rao, W. (2018). Nanoparticle-mediated cryosurgery for tumor therapy. *Nanomedicine, 14*(2), 493–506. doi:. nano.2017.11.018 doi:10.1016/j

Jailani, S., Franks, G. V., & Healy, T. W. (2008). The potential of nanoparticle suspensions: Effect of electrolyte concentration, particle size and volume fraction. *Journal of the American Ceramic Society*, *91*(4), 1141–1147. doi:10.1111/j.1551-2916.2008.02277.x

Kothandapani, M., & Prakash, J. (2015). The peristaltic transport of Carreau nanofluids under effect of a magnetic field in a tapered asymmetric channel: Application of the cancer therapy. *Journal of Mechanics in Medicine and Biology*, *15*(03), 1550030. doi:10.1142/S021951941550030X

Lim, E.-K., Jang, E., Lee, K., Haam, S., & Huh, Y.-M. (2013). Delivery of cancer therapeutics using nanotechnology. *Pharmaceutics*, *5*(2), 294–317. doi:10.3390/pharmaceutics5020294 PMID:24300452

Mahapatra, O., Bhagat, M., Gopalakrishnan, C., & Arunachalam, K. D. (2008). Ultrafine dispersed CuO nanoparticles and their antibacterial activity. *Journal of Experimental Nanoscience*, *3*(3), 185–193. doi:10.1080/17458080802395460

Micha,. D., Marcin, L., Grzegorz, D., & Andrzej, G.(2012). Preparation of metal oxide-water nanofluids by two-step method. *Prosimy Cytowa Jako: Inżynieria I Aparatura Chemiczna*, *51*, 213-215.

Neto, E. T., & Filho, E. P. B. (2010). Preparation methods of nano-fluids to obtain stable dispersions. *In: 13th Brazilian Congress of Thermal Sciences and Engineering*. IEEE.

Okonkwo, E. C., Wole-Osho, I., Kavaz, D., Abid, M., & Al-Ansari, T. (2020). Thermodynamic evaluation and optimization of a flat plate collector operating with alumina and iron mono and hybrid nanofluids. *Sustainable Energy Technologies and Assessments*, *37*, 100636. doi:10.1016/j.seta.2020.100636

Otanicar, T. P., Phelan, P. E., & Golden, J. S. (2009). Optical properties of liquids for direct absorption solar thermal energy systems. *Solar Energy*, *2009*(83), 969–977. doi:10.1016/j.solener.2008.12.009

Owens, D. E. III, & Peppas, N. A. (2006). Opsonization, biodistribution, and pharmacokinetics of polymeric nanoparticles. *International Journal of Pharmaceutics*, *307*(1), 93–102. doi:10.1016/j.ijpharm.2005.10.010 PMID:16303268

Pankhurst, Q. A., Connolly, J., Jones, S. K., & Dobson, J. (2003). Applications of magnetic nanoparticles in biomedicine. *Journal of Physics. D, Applied Physics*, *36*(13), R167–R181. doi:10.1088/0022-3727/36/13/201

Park, K. (2014). Controlled drug delivery systems: Past forward and future back. *Journal of Controlled Release*, *190*, 3–8. doi:10.1016/j.jconrel.2014.03.054 PMID:24794901

Quddoos, A., Anand, A., Mishra, G. K., & Nag, P. (2014). Nano-fluids: Introduction, preparation, stability analysis and stability enhancement techniques. *IJPAS*, *1*, 31–36.

Saleh, H., Alali, E., & Ebaid, A. (2017). Medical applications for the flow of carbon-nanotubes suspended nanofluids in the presence of convective condition using Laplace transform. *J Assoc Arab Univ Basic Appl Sci.*, *24*(1), 206–212. doi:10.1016/j.jaubas.2016.12.001

Salloum, M., Ma, R., & Zhu, L. (2008). An in-vivo experimental study of temperature elevations in animal tissue during magnetic nanoparticle hyperthermia. *International Journal of Hyperthermia*, *24*(7), 589–601. doi:10.1080/02656730802203377 PMID:18979310

Sano, N., Akazawa, H., Kikuchi, T., & Kanki, T. (2003). Separated synthesis of iron-included carbon nanocapsules and nanotubes by pyrolysis of ferrocene in pure hydrogen. *Carbon*, *41*. 2159-2162. doi: 00215-X. doi:10.1016/S0008-6223(03)

Sheikhpour, M., Arabi, M., Kasaeian, A., Rokn Rabei, A., & Taherian, Z. (2020). Role of Nanofluids in Drug Delivery and Biomedical Technology: Methods and Applications, *Dove Press journal. Nanotechnology, Science and Applications*, *13*, 47–59. doi:10.2147/NSA.S260374 PMID:32801669

Shokrollahi, H. (2013). Contrast agents for MRI. *Materials Science and Engineering C*, *33*(8), 4485–4497. doi:10.1016/j.msec.2013.07.012 PMID:24094150

Singh, S., & Bhargava, R. (2014). Simulation of phase transition during cryosurgical treatment of a tumor tissue loaded with nanoparticles using meshfree approach. *Journal of Heat Transfer*, *136*(12), 121101. doi:10.1115/1.4028730

Stankic, S., Suman, S., Haque, F., & Vidic, J. (2016). Pure and multi metal oxide nanoparticles: Synthesis, antibacterial and cytotoxic properties. *Journal of Nanobiotechnology*, *14*(1), 1–20. doi:10.118612951-016-0225-6 PMID:27776555

Su, H. Y., Chang-Qiang, W., & Dan-Yang, L. (2015). Self-assembled superparamagnetic nanoparticles as MRI contrast agents - A review. *Chinese Physics B*, *24*(12), 127506. doi:10.1088/1674-1056/24/12/127506

Tran, N., & Webster, T. J. (2010). Magnetic nanoparticles: Biomedical applications and challenges. *Journal of Materials Chemistry*, *20*(40), 8760–8767. doi:10.1039/c0jm00994f

Usman, M. S., El Zowalaty, M. E., & Shameli, K. (2013). Synthesis, characterization, and antimicrobial properties of copper nanoparticles. *International Journal of Nanomedicine*, *8*, 4467–4479. doi:10.2147/IJN.S50837 PMID:24293998

Vallejo, J., Sani, E., Żyła, G., & Lugo, L. (2019). Tailored silver/graphene nanoplatelet hybrid nanofluids for solar applications. *Journal of Molecular Liquids*, *296*, 112007. doi:10.1016/j.molliq.2019.112007

Wiedeman, M.P. (1963). Dimensions of blood vessels from distributing artery to collecting vein. *Circ Res.*, *12*(4), 375–378. doi:. RES.12.4.375 doi:10.1161/01

Wong, K. V., & De Leon, O. (2017). Applications of Nanofluids: Current and Future, in Nanotechnology and Energy. Jenny Stanford Publishing.

Yu F, Chen Y, & Liang X. (2017),Dispersion stability of thermal nanofluids. *Prog Nat Sci., 27*(5), 531–542. doi:. pnsc.2017.08.010 doi:10.1016/j

Yu, W., & Huaqing, X. A. (2012). Review on nano-fluids: Preparation, stability mechanisms, and applications. *Journal of Nanomaterials*, *2012*(435873), 17p.

APPENDIX: ABBREVIATIONS

NPs: Nanoparticles
NFs: Nanofluids
ZNO: Zinc Oxide
ROS: Reactive Oxygen Species
SEM: Scanning Electron Microscopy
FESEM: Field Emission Scanning Electron Microscopy
MRI: Magnetic Resonance Imaging
SPIOs: Superparamagnetic Iron Oxide Nanoparticles
CNS: Carbon Nanospheres
CNTs: Carbon Nanotubes
SWCNTs: Single Walled Carbon Nanotubes
MWCNTs: Multi Walled Carbon Nanotubes
MNF: Micro Nano Fluids
MNPs: Micro Nano Particles
CT-Computed Tomography
RBC: Red Blood Cells
EMEA: European Medicines Agency
FDA: Food and Drug Administration
tRNA: Transfer Ribonucleic Acid
SPMNPs: Super Paramagnetic Nanoparticles
PLGA: Poly(lactic-co-glycolic acid)
SAR: Specific Absorption Rate

Chapter 12
Advancements and Future Challenges in Nanoparticle and Nanofluid Research

Sagnika Pradhan
Jharkhand Rai University, India

Khandekar Hussan Reza
Maulana Abul Kalam Azad University of Technology, India

ABSTRACT

Nanoparticles and nanofluids are dynamic area of research due to their broad applications in various emerging field such as medicine, energy, and environmental science. Recent advancements in this field include the development of new synthesis techniques for creating nanoparticles with specific properties, important uses of nanoparticles in drug delivery systems in medical field, and the exploration of using nanofluids as coolants in thermal management systems. However, there are also challenges that need to be addressed in this field. One of the main challenges is understanding the behavior and properties of nanoparticles in different environments, as they can be affected by factors such as size, shape, and surface chemistry. Additionally, there are concerns about the possible health and environmental hazards associated with the use of nanoparticles, and more research is needed to understand and mitigate these risks.

DOI: 10.4018/978-1-6684-9135-5.ch012

INTRODUCTION

Nanoparticles and nanofluids are a rapidly growing field of research, with latent applications in a broad range of industries including medicine, energy and solar energy, computer cooling system, heat exchangers, and environmental science (Wong & De Leon, 2010; Khanafer & Vafai, 2018). The unique properties of nanoparticles, such as their small size and high surface-to-volu me ratio, make them useful for a variety of applications including drug delivery, imaging, and sensing. Similarly, nanofluids, which are fluids containing nanoparticles, have potential uses in cooling systems, lubrication, and heat transfer (Rafati et al., 2012; Huminic & Huminic, 2012; Devendiran & Amirtham, 2016). Recent advancements in nanoparticle and nanofluid research have led to new and improved synthesis techniques, allowing for the creation of nanoparticles with specific properties and sizes (Huminic & Huminic, 2012). Additionally, researchers have been exploring the use of nanoparticles in drug delivery systems, where they have the potential to improve the effectiveness of drugs while reducing side effects. Furthermore, the use of nanofluids as coolants in thermal management systems has also been gaining attention as they have higher thermal conductivity compared to the traditional fluids. Despite these advancements, there are still many trials that need to be addressed in demand to fully realize the potential of nanoparticles and nanofluids. One of the main challenges is understanding the behavior and properties of nanoparticles in different environments, as they can be affected by characteristics such as size, shape, and surface chemistry. Additionally, there are concerns about the potential health and environmental hazards associated with the use of nanoparticles and nanofluids, and more research is needed to understand and mitigate these risks. Nanoparticles are particles with at least one dimension less than 100 nanometers (Pradhan & Mishra, 2015).

They have exclusive properties due to their small size, such as increased surface area and reactivity. Nanoparticles can be made of a variability of materials, including metals, ceramics, and polymers. Nanoparticles and nanofluids have potential applications in a wide range of fields, including:

- Medicine: Nanoparticles can be used for targeted drug delivery, as well as imaging and diagnostic applications. For example, gold nanoparticles have been used in cancer treatment to deliver drugs directly to tumor cells.
- Materials science: Nanoparticles can be used to create new and improved materials with unique properties such as increased strength, conductivity, and optical properties.
- Energy: Nanofluids can be used in solar thermal systems to improve heat transfer, and in heat exchangers to increase efficiency. Nanoparticles can also be used in batteries and fuel cells to improve performance.

- Environmental remediation: Nanoparticles can be used to remove pollutants from water and air, and to clean up contaminated soil.
- Cosmetics and personal care: Nanoparticles can be used in sunscreens and other personal care products to improve effectiveness and stability.
- Food industry: Nanoparticles can be used as food additives to advance preservation and packaging of food items.

These are just a few examples of the potential applications of nanoparticles and nanofluids, and research is ongoing to explore new ways to utilize their unique properties.

Nanofluids are liquids that contain nanoparticles suspended in them. The nanoparticles can be made of metals, oxides, or other materials. Due to the unique properties of nanoparticles, nanofluids can have enhanced thermal conductivity and heat transfer properties associated to the base fluid alone. They have important applications in areas such as electronics cooling, solar thermal systems, and heat exchangers.

The research focus and objectives for nanomaterials and nanofluids vary depending on the specific application, but some common themes include:

- Synthesis and characterization: Developing new methods for synthesizing nanoparticles and nanofluids, and developing techniques for characterizing their properties (Wong & De Leon, 2010).
- Property enhancement: Exploring ways to enhance the properties of nanomaterials and nanofluids, such as thermal conductivity, mechanical strength, and optical properties.
- Modeling and simulation: Developing models and simulations to predict the behavior of nanomaterials and nanofluids, and to understand the underlying mechanisms that govern their properties (Ali et al., 2018).
- Environmental and health safety: Investigating the potential environmental and health impacts of nanomaterials and nanofluids, and developing ways to minimize any negative effects.
- Applications: Developing new applications for nanomaterials and nanofluids, such as in energy, medicine, and environmental remediation.
- Scale-up: Developing strategies to produce and handle large quantities of nanoparticles and nanofluids with reproducible quality, in order to make their use in industry viable.

Overall, research on nanomaterials and nanofluids aims to fully exploit their unique properties and potential benefits, while minimizing any potential risks (Pradhan & Mishra, 2015;).

In summary, recent advancements in nanoparticle and nanofluid research have opened up exciting possibilities for their use in various applications, but there are still challenges that need to be addressed in order to fully realize their potential. Continued research and development in this field will be needed to address these challenges and to develop new and innovative applications for nanoparticles and nanofluids.

ADVANCEMENTS IN NANOPARTICLE SYNTHESIS

Overview of Current Synthesis Technique

Nanoparticle synthesis refers to the process of creating particles that are between 1 and 100 nanometers in size These particles can be made up of from a widespread range of materials, including metals, ceramics, and polymers. There are several different methods for synthesizing nanoparticles, including physical methods (such as laser ablation and high-energy ball milling), chemical methods (such as chemical vapor deposition and sol-gel synthesis), and biological methods (such as bacterial synthesis and bio mineralization). The choice of method will depend on the desired properties of the nanoparticles and the materials from which they are made (Pradhan & Mishra, 2015; Attar & Kamble, 2022). Table 1 Recent Examples that shows different methods for synthesizing nanoparticles (Wong & De Leon, 2010; Khanafer & Vafai, 2018).

Advancements in Size and Shape Control

Advancements in size and shape control of nanoparticles have been made in current years over the development of new synthesis methods and the optimization of existing ones.

One important advancement has been the development of bottom-up synthesis methods, such as self-assembly and directed assembly. These methods allow for the precise control of particle size and shape by controlling the conditions under which the particles are synthesized, such as temperature, pH, and reactant concentrations(Table 2).

Another area of advancement has been in the use of templates and masks in the synthesis of nanoparticles. These techniques involve the use of a pre-existing structure to guide the growth of the nanoparticles, allowing for precise control of their size and shape. Furthermore, the use of innovative characterization techniques such as transmission electron microscopy (TEM) and scanning electron microscopy (SEM) have also enabled superior control over the size and shape of nanoparticles

Table 1. General methods of synthesis of nanoparticles

Method	Description	Advantages	Disadvantages
Chemical Reduction	Precursors are chemically reduced to form nanoparticles	Simple, low cost, high yield	Hazardous chemicals, difficult to control particle size
Sol-Gel	Precursors are mixed in a liquid and then gelled to form nanoparticles	Easy to control particle size, high purity	Long processing time, expensive precursors
Physical Vapor Deposition (PVD)	Material is vaporized and deposited on a substrate to form nanoparticles	High purity, good control over particle size	Expensive equipment, low yield
Electrochemical Deposition	Material is deposited on a substrate using an electric current	Good control over particle size and shape	Limited to certain materials, low yield
Co-Precipitation	Precipitation of the metal ions is triggered by changing pH or temperature	Simple, low cost, high yield	Difficult to control particle size and shape
Green Synthesis	Nanoparticles are synthesized using natural sources such as plants and bacteria	Environmentally friendly, cost-effective	Limited control over particle size and shape

Figure 1. (a): SEM image of Crystalline transformation of alumina
(Aluminium Oxide,a New Generation of Indispensable Hard Materials | TRUNNANO, n.d.-b)

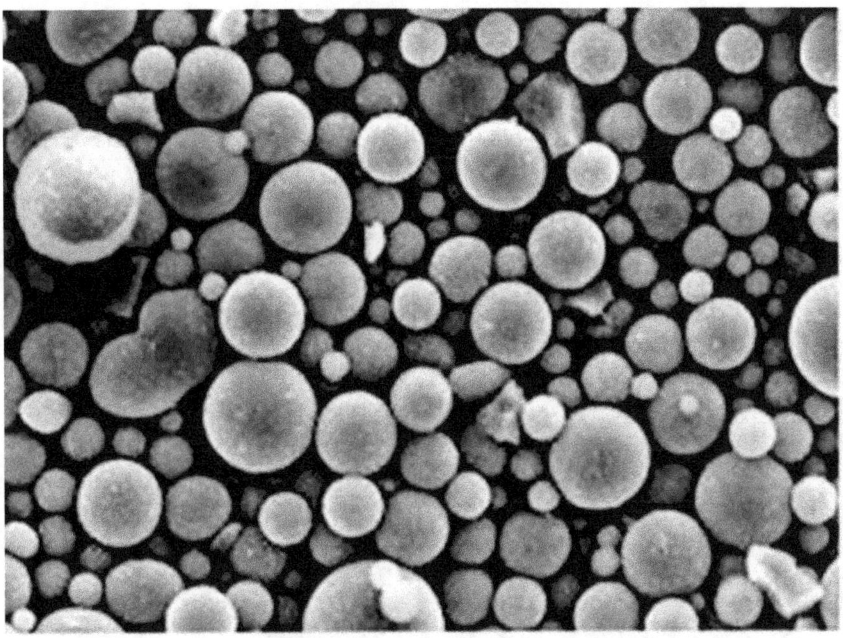

Figure 2. (b): TEM image of colloidal gold, 2015

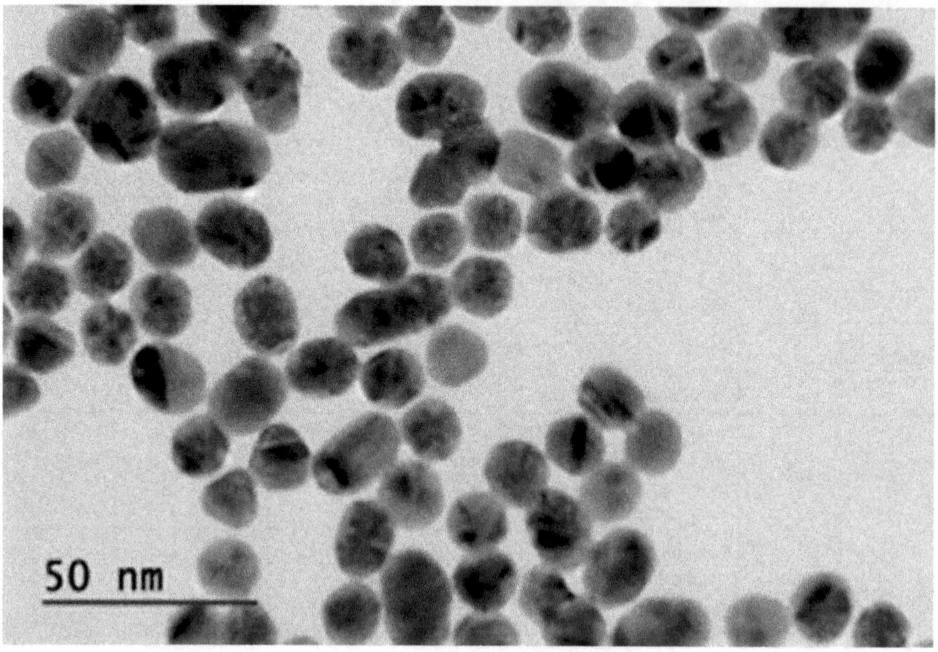

Table 2. Summarizing some of the methods used to synthesize nanoparticles with control over their shape and size

Shape	Size	Synthesis Method
Spheres	< 10 nm	Colloidal synthesis using a reducing agent and stabilizer
	10-100 nm	Colloidal synthesis using a surfactant or template
	> 100 nm	Emulsion-based methods or spray drying
Rods	Length: 20-200 nm; diameter: 5-50 nm	Seed-mediated growth or electrochemical synthesis
Wires	Length: 1-100 μm; diameter: 10-100 nm	Electrochemical synthesis
Plates	Edge length: 50-500 nm; thickness: < 10 nm	Solvothermal synthesis
Cubes	Edge length: 10-100 nm	Seed-mediated growth
Octahedra	Edge length: 10-50 nm	Seed-mediated growth
Tetrapods	Arm length: 100-500 nm	Seed-mediated growth

by allowing for more precise measurements of these properties as shown in figures 1 and 2.

Advancements in Surface and Structural Modification

There are also some recent advancements in the use of biomolecules and enzymes to control the shape of nanoparticles. Biomolecules like DNA, proteins, and enzymes are used as templates for the growth of nanoparticles, which can be used to control the shape and size of the resulting nanoparticles(Table 3).

Overall, ability to control the size and shape of nanoparticles is important for many applications, including in materials science, catalysis, and biomedicine, and ongoing research in this area promises to continue to lead to new and improved methods for synthesizing nanoparticles with precise control over their size and shape (Pradhan & Mishra, 2015; Pradhan et al., 2017).

Advancements in surface modification of nanoparticles and nanofluids have been made in recent years concluded the development of new methods and the optimization of prevailing ones (Begum et al., 2022).

One important advancement has been the development of chemical modification methods, such as coating and encapsulation. These methods involve the use of chemical agents to modify the surface of the nanoparticles, which can be used to improve their stability, biocompatibility, and targeting capabilities (Singh et al., 2019).

Another area of advancement has been the use of physical methods to modify the surface of nanoparticles, such as plasma treatment and high-energy ball milling. These methods can be used to change the surface properties of the nanoparticles, such as their charge and wettability.

Furthermore, the use of biomolecules and enzymes have also enabled greater control over the surface modification of nanoparticles. Biomolecules such as proteins, antibodies, and DNA can be used to functionalize the surface of nanoparticles, which can be used to improve their targeting capabilities and biocompatibility.

In adding, the use of nanofluids has also been an area of current advancement. Nanofluids are suspensions of nanoparticles in a liquid, and surface modification of these nanoparticles can be used to improvise the stability, thermal conductivity and other properties of the nanofluids (Ali et al., 2018).

Overall, surface modification of nanoparticles and nanofluids is important for many applications, including in materials science, catalysis, and biomedicine, and ongoing research in this area promises to continue to lead to new and improved methods for modifying the surfaces of nanoparticles and nanofluids to improve their properties and performance.

Table 3. Recent advancement in the surface and structural modifications of nanoparticles

Type of Modification	Advancement	Application
Ligand Exchange	Ligand exchange strategies for more stable and biocompatible nanoparticles, e.g. "click chemistry"	Biomedical applications, e.g., drug delivery
Surface Functionalization	Controlled surface functionalization to impart specific properties, e.g., stimuli-responsive surfaces	Sensors, catalysis, and biomedical applications
Core-Shell Nanoparticles	Synthesis of core-shell nanoparticles with controlled composition and thickness of the shell	Biomedical applications, e.g, drug delivery and imaging
Doping	Controlled doping of nanoparticles with different elements to modify their physical and chemical properties	Energy conversion, catalysis, and biomedical applications
Structural Design	Structural design of nanoparticles with specific shapes, e.g. hollow or porous nanoparticles	Catalysis, energy conversion, and biomedical applications
Surface Coatings	Development of surface coatings to enhance stability and biocompatibility, e.g. polymers, silica, and metal-organic frameworks	Biomedical applications, e.g. drug delivery and imaging

(Attar & Kamble, 2022)

APPLICATIONS OF NANOPARTICLES IN VARIOUS FIELD

Nanoparticles as Drug Delivery System

Nanoparticles are being researched as a potential drug delivery system due to their small size and ability to target specific cells or tissues in the body. They can be made from a variety of materials such as lipids, polymers, metals and non-metals can be designed to release drugs over a specific time period (Singh et al., 2019). Additionally, nanoparticles can protect drugs from degradation and enhance their bioavailability. These characteristics make nanoparticles a promising option for treating cancer, inflammation, and other diseases. However, more research is required to fully understand the safety and efficacy of nanoparticles as a drug delivery system. Nanoparticles are being researched as a potential drug delivery system due to their small size and ability to target specific cells or tissues in the body (Rajpoot, 2019). They can be made from a variability of materials such as lipids, polymers, and metals, and can be designed to release drugs over a specific time period (Rajpoot, 2019). Additionally, nanoparticles can protect drugs from degradation and enhance their bioavailability. These characteristics make nanoparticles a promising option for treating cancer, inflammation, and other diseases. However, more research is desired to fully understand the safety and efficacy of nanoparticles as a drug delivery system (Singh et al., 2019).

Figure 3. SEM photomicrography of liposome
(Ghatnur et al., 2012)

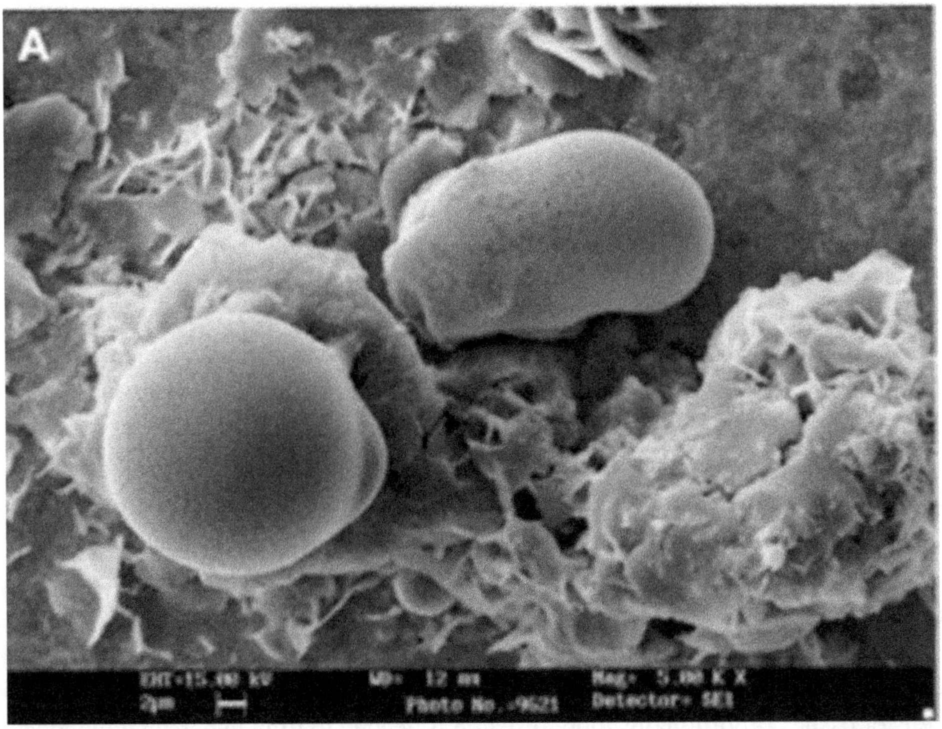

There are several types of nanoparticle-based drug delivery systems, including liposomes, polymeric nanoparticles, dendrimers, and inorganic nanoparticles. Liposomes are spherical structures composed of a lipid bilayer that can encapsulate drugs and target specific cells as shown in figure 3.

Polymeric nanoparticles are composed of polymers and can be designed to release drugs over a prolonged period of time. Dendrimers are highly branched and highly monodisperse polymeric nanoparticles that can be used for targeted drug delivery (Begines et al., 2020). Inorganic nanoparticles, such as gold nanoparticles and iron oxide nanoparticles, can be used for imaging and therapeutic applications. The targeting ability of nanoparticles can be achieved through various methods such as surface modification with targeting moieties, pH sensitivity, temperature sensitivity, and by exploiting the enhanced permeability and retention (EPR) effect (Ahamed et al., 2019).

Despite the potential of nanoparticles as drug carriers, there are still challenges to overcome, such as toxicity and the ability to effectively target specific cells or tissues. Additionally, more research is needed to fully understand the long-term effects of nanoparticles in the body. Nanofluids are suspensions of nanoparticles in a liquid medium, and they have been proposed as a potential drug delivery system. Due to their small size, nanofluids can penetrate into small spaces and crevices, making them useful for targeting specific cells or tissues.

Nanofluids have been proposed as a potential drug delivery system for cancer treatment (Awasthi et al., 2021). The nanoparticles can be engineered to target cancer cells specifically, and the high surface area of the particles allows for a higher drug loading capacity. Additionally, the small size of the nanoparticles allows them to bypass the body's natural defense mechanisms, which can help to increase the efficacy of the drugs (Zahmakıran & Özkar, 2011).

Nanofluids have also been proposed for use in wound healing. The nanoparticles can be engineered to release drugs over a prolonged period of time, which can help to speed up the healing process and reduce the risk of infection (Devendiran & Amirtham, 2016).

However, more research is needed to fully understand the safety and efficacy of nanofluids as a drug delivery system. The toxicity of the nanoparticles and the potential for long-term effects in the body are areas of concern that need to be addressed. Additionally, the stability and shelf life of nanofluids also need to be studied (Ali et al., 2018).

Despite the potential of nanoparticles as a drug delivery system, there are still challenges to overcome, such as toxicity and the ability to effectively target specific cells or tissues. Additionally, more research is needed to fully understand the long-term effects of nanoparticles in the body.

Design of Imaging and Sensing of Nanomaterial

Nanoparticles have a wide range of applications in imaging and sensing, due to their unique optical, magnetic, and electrical properties. They can be used to enhance the sensitivity and specificity of imaging techniques, and to create new types of sensors (Singh et al., 2019).

One application of nanoparticles in imaging is in medical imaging. Gold nanoparticles, for example, have been used as contrast agents for in vivo imaging, such as in computed tomography (CT) and magnetic resonance imaging (MRI). They can also be used to enhance the imaging of tumors and other diseased tissue. Iron oxide nanoparticles have been used as contrast agents for MRI, and have also been used in imaging of cells and tissues as shown in figure 4.

Nanoparticles have also been used in biosensors and chemical sensors. For example, gold nanoparticles have been used in sensors for the detection of small molecules such as glucose, DNA, and proteins. The unique optical properties of gold nanoparticles, such as their plasmon resonance, can be used to enhance the

Figure 4. Iron nanoparticles as a contrasting agent
(Li, 2013)

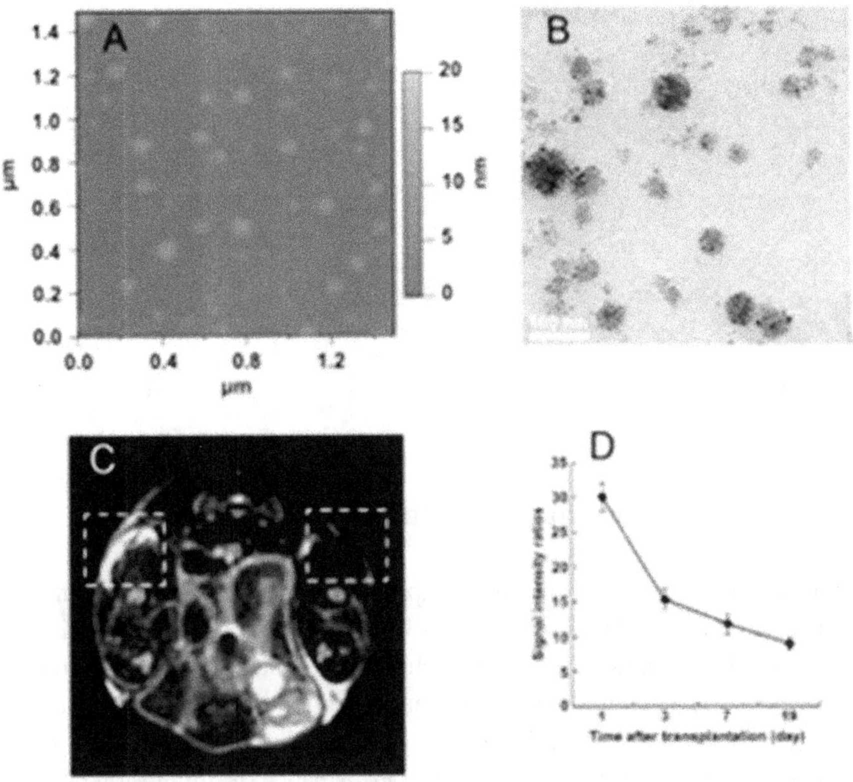

sensitivity of the sensors. Additionally, the small size of the nanoparticles allows them to penetrate into small spaces and crevices, which can increase the specificity of the sensors (Myung et al., 2011).

Nanoparticles have been also used to create new types of sensors, such as chemical and biological sensors, by functionalizing their surface with specific receptors. In addition, nanoparticles have been used to create highly sensitive strain and pressure sensors, and to create sensors that can detect changes in temperature and pH (Myung et al., 2011).

Despite the potential of nanoparticles in imaging and sensing, more research is needed to fully understand their safety and potential long-term effects in the body. Additionally, the stability and shelf life of nanoparticles-based sensors also need to be studied.

Figure 5. Silicon nanoparticles have been used to improve the performance of lithium-ion batteries
(锂离子电池硅纳米粒子/碳复合材料, *2019*)

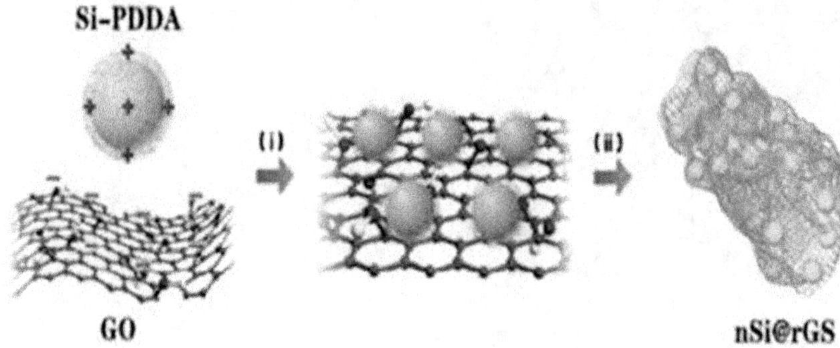

Energy and Environmental Applications

Nanoparticles have a wide range of potential applications in energy and environmental fields due to their unique properties such as high surface area, high reactivity, and high thermal stability. In the energy field, nanoparticles have been used to improve the efficiency of solar cells and batteries. For example, titanium dioxide nanoparticles have been used to improve the efficiency of dye-sensitized solar cells, and silicon nanoparticles have been used to improve the performance of lithium-ion batteries.

In environmental applications, nanoparticles have been used to remove pollutants from water and air. For example, iron oxide nanoparticles have been used to remove heavy metals from water, and titanium dioxide nanoparticles have been used to remove pollutants such as volatile organic compounds (VOCs) and nitrogen oxides (NOx) from air. Nanoparticles have also been used in catalytic applications to convert pollutants into less harmful compounds (Pradhan et al., 2016). For example, gold nanoparticles have been used as catalysts to convert carbon monoxide into carbon dioxide, and platinum nanoparticles have been used as catalysts to convert nitrogen oxides into nitrogen and oxygen. Nanoparticles have also been used in the oil and gas industry to improve the recovery of oil and gas. For example, silica nanoparticles have been used as proppants to keep fractures in shale rock open during hydraulic fracturing, and aluminum oxide nanoparticles have been used as catalysts to improve the efficiency of catalytic cracking. (Begum et al., 2022)

Despite the potential of nanoparticles in energy and environmental applications, more research is needed to fully understand the safety and potential long-term effects of the nanoparticles on the environment. Additionally, the stability and shelf-life of nanoparticles-based catalysts need to be studied.

Nanoparticles as Catalytic Application

Nanoparticles have a wide range of catalytic applications due to their unique physical and chemical properties. Some common catalytic applications of nanoparticles are:

Environmental catalysis: Nanoparticles can be used in environmental catalysis for the removal of pollutants from air and water. For example, metal nanoparticles such as palladium, platinum, and gold are used to catalyze the reduction of pollutants like nitric oxide, carbon monoxide, and volatile organic compounds.

Energy conversion: Nanoparticles are used in energy conversion processes, such as fuel cells and solar cells (Khanafer & Vafai, 2018). For example, platinum nanoparticles are used as a catalyst in the electrodes of fuel cells to facilitate the conversion of hydrogen and oxygen into water.

Chemical synthesis: Nanoparticles can be used as catalysts in chemical synthesis processes. For example, palladium nanoparticles are commonly used in organic synthesis reactions, such as the Suzuki coupling reaction (Attar & Kamble, 2022).

Biocatalysis: Nanoparticles can be used in biocatalysis, which is the use of biological systems to catalyze chemical reactions. For example, gold nanoparticles are used as a catalyst in the oxidation of alcohols to aldehydes and ketones (Pradhan & Mishra, 2018).

Nano catalysis: Nanoparticles can be used as a catalyst in the synthesis of other nanoparticles. For example, gold nanoparticles can be used as a catalyst in the synthesis of silver nanoparticles (Ahamed et al., 2019).

Overall, nanoparticles have immense potential for catalytic applications, and ongoing research is likely to uncover even more uses for these tiny particles in the future.

Nanoparticles as Modern Electronic Application

Nanoparticles have several potential applications in modern electronics, due to their unique properties such as high surface area, high conductivity, and quantum confinement effects. Some of the electronic applications of nanoparticles are:

Display technology: Nanoparticles such as quantum dots are being explored for their potential use in display technology. Quantum dots emit light of a specific wavelength, making them ideal for use in LED displays and lighting (A. Ali et al., 2021).

Optoelectronics: Nanoparticles are being used in optoelectronic devices, such as photodetectors, solar cells, and light-emitting diodes (LEDs). For example,

Figure 6. Gold nanoparticle (AuNP)-labeled biosensor
(Wang & Alocilja, 2015)

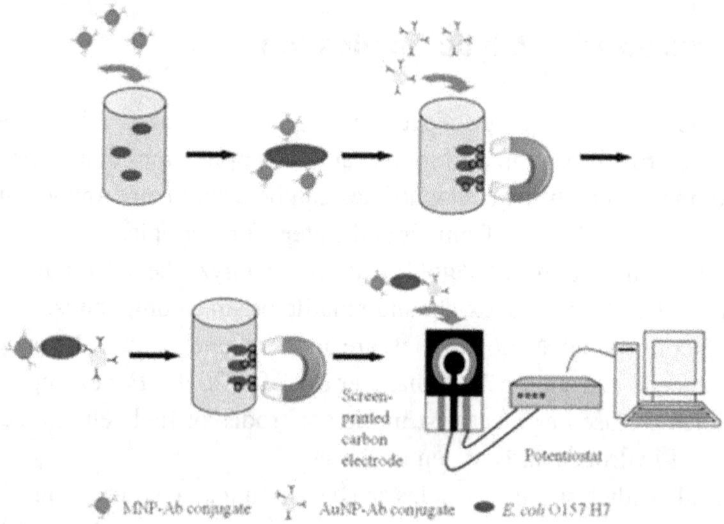

nanoparticles made of materials such as silicon, titanium dioxide, and zinc oxide have been used to enhance the efficiency of solar cells.

Data storage: Nanoparticles are being explored for use in data storage, particularly in high-density storage media such as hard drives. Iron oxide nanoparticles, for example, have been shown to have excellent magnetic properties, making them useful for data storage applications (A. Ali et al., 2021).

Sensors: Nanoparticles are being used in the development of sensors for various applications, such as environmental monitoring, food safety, and medical diagnostics. For example, gold nanoparticles can be used as biosensors to detect specific molecules or pathogens.

Printed electronics: Nanoparticle inks can be used in the production of flexible, printed electronics. Silver nanoparticles, for example, can be used as conductive inks to print electronic circuits onto flexible substrates.

Some Recent modern applications of nanoparticles are given in Table 4. Overall, the unique properties of nanoparticles make them promising candidates for a wide range of electronic applications, and ongoing research is likely to uncover even more uses for these tiny particles in the future.

Table 4. Here are common data for the most versatile applications of nanoparticles

Field/Industry	Modern Applications of Nanoparticles
Biomedical	Gene therapy, personalized medicine, targeted drug delivery, immunotherapy, biosensors, regenerative medicine, etc.
Energy	Photovoltaic cells, hydrogen production and storage, supercapacitors, thermoelectric materials, batteries, etc.
Electronics	Wearable devices, flexible displays, high-performance computing, quantum computing, etc.
Environmental	Carbon capture and storage, water desalination, pollution control, oil spill cleanup, etc.
Construction	High-strength concrete, smart concrete, self-healing concrete, transparent insulation, etc.
Textiles	Smart fabrics, adaptive textiles, self-cleaning fabrics, color-changing textiles, etc.
Cosmetics	Nanocosmetics, Nano dispersions, Nanocapsules, sunscreens, etc.
Food	Smart packaging, food preservation, food safety, food quality, etc.
Automotive	Lightweight materials, catalytic converters, fuel additives, sensors, etc.
Agriculture	Nanopesticides, nanofertilizers, nanosensors, etc.

CHALLENGES OF NANOMATERIAL

Challenges of Nanotechnology and Nanomaterial for Energy and Health

The challenges of nanotechnology and nanomaterials in energy and health include (Ali et al., 2018).

- Health and safety risks: There are concerns about the potential toxic effects of nanomaterials on human health and the environment
- Regulatory challenges: There is a lack of regulations and standards for the development and use of nanomaterials, which can impede the commercialization of new technologies.
- Environmental impact: The potential environmental impact of nanomaterials, such as their potential to accumulate in the food chain and affect wildlife, is not yet fully understood.
- Cost-effectiveness: The cost of producing nanomaterials and incorporating them into new technologies is often high, which can limit their widespread adoption.

- Technical challenges: There are technical challenges associated with the large-scale production of nanomaterials and their integration into new products and technologies.
- Ethical issues: There are ethical concerns associated with the use of nanotechnology, such as its potential to increase social and economic inequalities, and its impact on privacy and civil liberties.

Challenges for Assessing Toxicity of Nanomaterial

The challenges for assessing the toxicity of nanomaterials include:

- Complexity: Nanomaterials have unique physical and chemical properties that make them challenging to study and assess for toxicity (Zahmakıran & Özkar, 2011).
- Lack of standardization: There is a lack of standard methods and protocols for testing the toxicity of nanomaterials, which can lead to inconsistent results.
- Limited in vitro methods: In vitro methods for assessing the toxicity of nanomaterials are limited and may not accurately reflect real-life exposure scenarios.
- Dosage and exposure: The dose and exposure pathways of nanomaterials in the human body are not well understood, making it difficult to assess their toxicity.
- Interactions with biological systems: Nanomaterials have the potential to interact with biological systems in complex and unpredictable ways, making it challenging to assess their toxicity.
- Variability: The toxicity of nanomaterials can vary based on their size, shape, composition, and surface properties, which makes it difficult to generalize results.
- Data gaps: There are significant gaps in our understanding of the toxicity of nanomaterials, particularly for long-term exposure and chronic effects.

Challenges on Nano Architectonics

The recent challenges in nano architectonics, the design and organization of nanoscale building blocks, include: (Ali et al., 2018)

- Synthesis and scalability: The synthesis of complex and well-defined nanostructures is still a major challenge, particularly in terms of scalability and cost-effectiveness.

- Assembly and integration: The integration of nanoscale building blocks into functional systems remains a challenge, as it requires precise control over the assembly process and compatibility with the target application.
- Material compatibility: The compatibility of nanoscale building blocks with various materials and substrates is still a challenge, as it affects their performance and stability.
- Characterization: Characterizing nanoscale structures, especially in terms of their composition, structure, and properties, is a major challenge due to their small size and complexity.
- Environmental and health impacts: The potential environmental and health impacts of nanoscale building blocks, such as toxicity and biocompatibility, must be carefully considered and addressed.
- Performance and durability: The performance and durability of nanoscale systems, especially under challenging conditions, must be improved to meet the requirements of various applications.
- Interdisciplinary collaboration: Nano architectonics requires interdisciplinary collaboration and a cross-functional approach, as it involves advances in various fields, including materials science, chemistry, physics, and engineering.

Promises and Challenges of Nanomaterial for Lithium-Based Rechargeable Batteries

Nanomaterials have attracted significant attention in the field of lithium-based rechargeable batteries due to their unique properties and potential for improving the performance of these batteries. Here are some promises and challenges of nanomaterials for lithium-based rechargeable batteries:

Promises

High energy density: Nanomaterials have a large surface area to volume ratio, which means that they can store more lithium ions, leading to a higher energy density.

Improved cycling stability: Nanomaterials can provide better cycling stability by reducing the stress on the active materials during charging and discharging cycles (Khanafer & Vafai, 2018; Rafati et al., 2012).

Faster charging and discharging: Nanomaterials can improve the rate performance of lithium-based batteries by allowing for faster charging and discharging rates due to their high surface area and shorter diffusion distances. Enhanced safety: Nanomaterials can improve the safety of lithium-based batteries by reducing the risk of short circuits and thermal runaway reactions.

Challenges

Cost: The cost of producing nanomaterials is often higher than traditional materials, which can limit their widespread use in lithium-based batteries.

Scalability: The synthesis of nanomaterials is often done in small quantities, and scaling up production can be challenging (Bhattad et al., 2018).

Stability: Some nanomaterials can be unstable and prone to degradation over time, leading to a reduction in battery performance.

Toxicity: Some nanomaterials can be toxic to humans and the environment, and their use must be carefully evaluated to ensure that they do not pose a risk to public health or the environment (Ali et al., 2018).

In conclusion, nanomaterials hold great promise for improving the performance of lithium-based rechargeable batteries, but their widespread use in commercial applications requires addressing the challenges related to cost, scalability, stability, and toxicity

ADVANCEMENTS IN NANOFLUID RESEARCH

Overview of Current Nanofluid Synthesis Techniques

There are several techniques used to synthesize nanofluids, each with its own advantages and disadvantages. Physical mixing: This is the simplest method of synthesizing nanofluids, where nanoparticles are simply mixed with a base fluid using a mechanical stirrer. This method is easy to implement, but the resulting nanofluid may have poor stability, with the nanoparticles quickly settling out of the liquid(Devendiran & Amirtham, 2016; Angayarkanni & Philip, 2015).

Ultrasonication: In this method, the nanoparticles are suspended in a base fluid and ultrasonically agitated to create a homogenous mixture. This method can produce stable nanofluids with a high concentration of nanoparticles, but it can also cause damage to the nanoparticles and the base fluid.

Co-precipitation: In this method, the nanoparticles are created in situ by the precipitation of precursor solutions. This method can produce a high concentration of nanoparticles in the nanofluid, but it can be a complex and time-consuming process (Pradhan et al., 2016).

Micro emulsion: This method involves the formation of a micro emulsion, which is a stable mixture of oil, water, and surfactant. The nanoparticles are then introduced into the micro emulsion, resulting in a stable nanofluid. This method can be used to produce nanofluids with a high concentration of nanoparticles, but it requires the use of surfactants, which can be toxic. Self-assembly: This method uses the natural tendency of nanoparticles to self-assemble in a liquid medium. This method can produce stable nanofluids with a high concentration of nanoparticles, but it is a complex process that requires the use of specific precursors and conditions (Pradhan & Mishra, 2018).

Electrostatic stabilization: This method uses the electrostatic repulsion between the nanoparticles to keep them suspended in the base fluid. The nanoparticles are first coated with a material that carries an opposite charge to the nanoparticle's charge, this coating will repel the particles from each other and keep them suspended. This method can produce stable nanofluids with a high concentration of nanoparticles, but it requires the use of specific precursors and conditions (Rafati et al., 2012; Huminic & Huminic, 2012; Harilal et al., 2017).

Overall, each method has its own advantages and disadvantages, and the choice of the method depends on the specific application and the desired properties of the nanofluid.

Advancements in Thermal Conductivity Enhancement

Thermal conductivity enhancement of fluids using nanoparticles, also known as nanofluids, has been an active area of research in recent years. The main goal of this research is to increase the thermal conductivity of liquids, which can be used to improve the efficiency of heat transfer in various applications such as cooling systems, solar thermal systems, and electronics cooling (Rafati et al., 2012).

One of the main advancements in thermal conductivity enhancement is the development of new methods for synthesizing stable nanofluids with high concentrations of nanoparticles. This has been achieved through the use of methods such as ultra-sonication, co-precipitation, micro emulsion, and self-assembly. These methods have been shown to produce stable nanofluids with a high concentration of nanoparticles, which can lead to a significant increase in thermal conductivity (Angayarkanni & Philip, 2015).

Another advancement is the use of advanced characterization techniques to study the properties of nanofluids. These techniques include transmission electron microscopy (TEM), scanning electron microscopy (SEM), dynamic light scattering (DLS), and small-angle X-ray scattering (SAXS). These techniques have been used to study the size, shape, and distribution of nanoparticles in nanofluids, which can help to understand how these properties affect the thermal conductivity of the nanofluids (Awasthi et al., 2021).

Additionally, researchers have been developing new types of nanoparticles that can enhance thermal conductivity, such as carbon nanotubes, grapheme, and boron nitride. These nanoparticles have been shown to have a higher thermal conductivity than traditional nanoparticles, such as metal oxide nanoparticles, which can lead to even greater thermal conductivity enhancement when used in nanofluids (Zahmakıran & Özkar, 2011).

Finally, the use of computational simulations and theoretical models have been used to understand the thermal conductivity enhancement mechanisms of nanofluids and to predict the thermal conductivity of nanofluids with different properties.

Overall, there have been many advancements in thermal conductivity enhancement using nanofluids. However, more research is needed to fully understand the mechanisms of thermal conductivity enhancement and to optimize the properties of nanofluids for specific applications.

Advancements in Viscosity and Stability

Viscosity and stability are two important properties of nanofluids that are closely related to their performance in various applications, such as heat transfer and fluid flow. There have been several advancements in the field of viscosity and stability of nanofluids in recent years.

One advancement is the use of new methods for synthesizing stable nanofluids with high concentrations of nanoparticles. These methods include ultrasonication, co-precipitation, microemulsion, and self-assembly, which have been shown to produce stable nanofluids with a high concentration of nanoparticles (Agarwal et al., 2016).

Another advancement is the use of advanced characterization techniques to study the properties of nanofluids. These techniques include transmission electron microscopy (TEM), scanning electron microscopy (SEM), dynamic light scattering (DLS), and small-angle X-ray scattering (SAXS). These techniques have been used to study the size, shape, and distribution of nanoparticles in nanofluids, which can help to understand how these properties affect the viscosity and stability of the nanofluids (Awasthi et al., 2021).

Additionally, researchers have been developing new types of nanoparticles that can improve the viscosity and stability of nanofluids, such as polymeric nanoparticles and dendrimers. These nanoparticles have been shown to have a lower tendency to aggregate and to have a higher stability in the base fluid than traditional nanoparticles (Agarwal et al., 2016).

The use of surfactants and polymers as stabilizing agents has also been studied. These agents can adsorb on the surface of nanoparticles and prevent them from aggregating and settling, which leads to an increase in the stability of the nanofluid (Pradhan & Mishra, 2018).

Finally, the use of computational simulations and theoretical models have been used to understand the viscosity and stability of nanofluids and to predict the behavior of nanofluids with different properties (Rafati et al., 2012).

Overall, there have been many advancements in viscosity and stability of nanofluids. However, more research is needed to fully understand the mechanisms of viscosity and stability enhancement and to optimize the properties of nanofluids for specific applications

Challenges and Risks in Nanoparticle and Nanofluid Research Nanoparticles and nanofluids have many potential applications, but there are also challenges and risks associated with their research and development (Ali et al., 2018).

One major challenge is the lack of understanding of the toxicity and long-term effects of nanoparticles on human health and the environment. Some studies have suggested that certain types of nanoparticles may be toxic to cells and organs, and that they may accumulate in the body over time. More research is needed to fully understand the potential risks of exposure to nanoparticles (Baig et al., 2021).

Another challenge is the lack of standardization in the synthesis and characterization of nanoparticles and nanofluids. This can make it difficult to compare results across different studies and to replicate findings. Additionally, the lack of standardization in the characterization of nanoparticles and nanofluids can make it difficult to accurately measure their properties, such as size and shape (Baig et al., 2021).

Another risk is the potential release of nanoparticles into the environment, which can have unknown effects on ecosystems and wildlife. Additionally, the release of nanoparticles into the environment can also pose a risk to human health (Ali et al., 2018).

Finally, the lack of regulation in the use of nanoparticles and nanofluids can also be a concern. There are currently no specific regulations in place for the use of nanoparticles in consumer products, and there is a lack of guidelines for their safe use in industrial applications. Overall, while nanoparticles and nanofluids have many potential applications, it is important to continue research in order to better understand their properties, behavior, and potential risks (Ali et al., 2018)

Environmental and Health Risks Associated With Nanoparticles

Nanoparticles have the potential to cause environmental and health risks due to their small size and unique properties. Some of the main risks include: (Begum et al., 2022)

Toxicity: Some studies have suggested that certain types of nanoparticles may be toxic to cells and organs, and that they may accumulate in the body over time. This could lead to health problems such as inflammation, DNA damage, and cancer.

Environmental release: The release of nanoparticles into the environment can have unknown effects on ecosystems and wildlife. Additionally, the release of nanoparticles into the environment can also pose a risk to human health. Bioaccumulation: Nanoparticles have the potential to bio accumulate in living organisms, which can cause long-term health effects. They can also be transported through the food chain, potentially affecting a wide range of organisms (Ali et al., 2018).

Persistence: Certain types of nanoparticles, such as carbon-based nanoparticles, may persist in the environment for a long time and can be transported over large distances through air and water (Agarwal et al., 2016).

Interference with natural processes: Nanoparticles can also interfere with natural processes such as photosynthesis, nutrient cycling, and soil microorganism activity.

Accidents or spills: The handling and transportation of nanoparticles present a risk of accidents or spills, which can release large quantities of nanoparticles into the environment.

To minimize these risks, it is important to conduct more research to understand the behavior and properties of nanoparticles in the environment and in living organisms. Additionally, regulations and guidelines for the safe handling, use, and disposal of nanoparticles are needed to protect the environment and human health (Zahmakıran & Özkar, 2011).

Challenges in Understanding the Behavior and Properties of Nanoparticle

There are several challenges in understanding the behavior and properties of nanoparticles, which include:

Complexity of nanoparticles: The behavior and properties of nanoparticles can depend on many factors such as size, shape, surface chemistry, and composition. This complexity can make it difficult to understand how these factors affect the behavior and properties of nanoparticles.

Lack of standardization: There is a lack of standardization in the synthesis and characterization of nanoparticles, which can make it difficult to compare results across different studies and to replicate findings. Additionally, the lack of standardization in the characterization of nanoparticles can make it difficult to accurately measure their properties, such as size and shape (Zahmakıran & Özkar, 2011).

Size and surface effects: The small size of nanoparticles can lead to size and surface effects that are not observed in larger particles. These effects can include changes in physical and chemical properties, such as increased reactivity and increased surface area (Pradhan & Mishra, 2015; Pradhan et al., 2016).

Interaction with the environment: Nanoparticles can interact with their environment in ways that are not well understood, such as by adsorbing onto surfaces or by forming aggregates. These interactions can affect the behavior and properties of nanoparticles (Attar & Kamble, 2022).

Complex behavior in biological systems: The behavior of nanoparticles can be complex in biological systems, such as living organisms and cell cultures. The behavior of nanoparticles can depend on factors such as the size, shape, and surface chemistry of the nanoparticles, as well as the environment in which they are present (Ali et al., 2018).

Long-term effects: The long-term effects of nanoparticles on human health and the environment are not well understood. More research is needed to understand how nanoparticles behave over time and to identify any potential risks associated with their use.

To overcome these challenges, it is important to use advanced characterization techniques to study the properties of nanoparticles, such as transmission electron microscopy, scanning electron microscopy, dynamic light scattering, and small-angle X-ray scattering. Additionally, the use of computational simulations and theoretical models can help to better understand the behavior and properties of nanoparticles.

Scalability and Commercialization Challenges

Scalability and commercialization are significant challenges for nanoparticles and nanofluids research. Some of the main challenges include:

Scale-up of synthesis: The synthesis methods used to produce nanoparticles and nanofluids in laboratory settings may not be easily scalable to industrial production. The methods may be too complex, require expensive equipment, or produce low yields at large scale (Baig et al., 2021).

Cost-effectiveness: The production of nanoparticles and nanofluids can be costly, which can make it difficult to develop cost-effective products and technologies. The cost of raw materials, equipment, and labor can all be significant barriers to commercialization (Bhattad et al., 2018).

Quality control: Ensuring consistent quality of nanoparticles and nanofluids can be a challenge in large-scale production. This can be difficult to control due to variations in the synthesis process, variations in the properties of the base fluid and nanoparticles, and variations in the conditions of the production process (Begum et al., 2022).

Safety and Environmental issues: The safety and environmental risks of nanoparticles and nanofluids are not well understood, which can make it difficult to develop commercial products and technologies that are safe for use and disposal (Zahmakıran & Özkar, 2011).

Regulations and standards: There are currently few regulations and standards in place for the commercial use of nanoparticles and nanofluids, which can make it difficult to develop commercial products and technologies that are compliant with these regulations and standards. Public perception: There is a lack of public awareness and understanding of nanoparticles and nanofluids, which can make it difficult to develop commercial products and technologies that are accepted by the public (A. Ali et al., 2021).

Toxicity Measurement and Remedy

Toxicity measurement of nanoparticles involves determining the potential harmful effects of the particles on living organisms, including humans. This can be done through a variety of methods, such as cell culture studies, animal testing, and in vitro assays.

Remedies for the toxicity of nanoparticles can include using less toxic materials to create the particles, modifying the surface chemistry of the particles to reduce toxicity, and implementing proper safety procedures and protective measures during the handling and disposal of the particles. Additionally, developing and using nanotechnology risk assessment frameworks can also be helpful in identifying and mitigating potential toxicity risks associated with nanoparticles.

ADVANCEMENTS IN NANOFLUID SYNTHESIS

Preparation of Non-Metallic Nanofluids

Carbon Nanotube: Nanofluids

Carbon nanotubes (CNTs) are synthesized through various methods, including chemical vapor deposition (CVD), arc discharge, and laser ablation. CNT-nanofluids are suspensions of CNTs in a liquid medium, typically water or a base fluid such as ethylene glycol. The synthesis of CNT-nanofluids involves dispersing CNTs in the fluid through various methods such as sonication, stirring, or high-pressure homogenization. The properties of CNT-nanofluids, such as thermal conductivity, depend on the size and distribution of CNTs in the fluid, as well as the type of base fluid used. CNT-nanofluids have potential applications in various fields, including heat transfer and thermal management in electronics, as well as energy generation and storage.

Silicon Dioxide-Nanofluids

Silicon dioxide (SiO_2) nanofluids are suspensions of SiO_2 nanoparticles in a liquid medium, typically water or a base fluid such as ethylene glycol. The synthesis of SiO_2-nanofluids involves the preparation of SiO_2 nanoparticles through various methods such as sol-gel synthesis, precipitation, and hydrolysis, followed by their dispersion in the fluid. This can be done through methods such as sonication, stirring, or high-pressure homogenization. The properties of SiO2-nanofluids, such as thermal conductivity and viscosity, depend on the size and distribution of SiO2 nanoparticles in the fluid, as well as the type of base fluid used. SiO2-nanofluids have potential applications in various fields, including heat transfer and thermal management in electronics, as well as energy generation and storage (Ali et al., 2018).

Aluminum Nitride-Nanofluids

Aluminum nitride (AlN) nanofluids are suspensions of AlN nanoparticles in a liquid medium, typically water or a base fluid such as ethylene glycol. The synthesis of AlN-nanofluids involves the preparation of AlN nanoparticles through various methods such as carbothermal reduction and nitridation, followed by their dispersion in the fluid. This can be done through methods such as sonication, stirring, or high-pressure homogenization. The properties of AlN-nanofluids, such as thermal conductivity and viscosity, depend on the size and distribution of AlN nanoparticles in the fluid, as well as the type of base fluid used. AlN-nanofluids have potential applications in various fields, including heat transfer and thermal management in electronics, as well as energy generation and storage (Agarwal et al., 2016).

Metal Oxides-Nanofluids

Metal oxide nanofluids are suspensions of metal oxide nanoparticles in a liquid medium, typically water or a base fluid such as ethylene glycol. The synthesis of metal oxide nanofluids involves the preparation of metal oxide nanoparticles through various methods such as sol-gel synthesis, precipitation, and hydrolysis, followed by their dispersion in the fluid. This can be done through methods such as sonication, stirring, or high-pressure homogenization. The properties of metal oxide nanofluids, such as thermal conductivity and viscosity, depend on the size and distribution of metal oxide nanoparticles in the fluid, as well as the type of base fluid used (Ahamed et al., 2019). Metal oxide nanofluids have potential applications in various fields, including heat transfer and thermal management in electronics, as well as energy generation and storage.

PREPARATION OF METALLIC NANOFLUIDS

Gold and Silver-Nanofluids

Gold (Au) and silver (Ag) nanofluids are suspensions of Au or Ag nanoparticles in a liquid medium, typically water or a base fluid such as ethylene glycol. The synthesis of Au or Ag nanofluids involves the preparation of Au or Ag nanoparticles through various methods such as chemical reduction, followed by their dispersion in the fluid. This can be done through methods such as sonication, stirring, or high-pressure homogenization(Devendiran & Amirtham, 2016).The properties of Au or Ag nanofluids, such as thermal conductivity and viscosity, depend on the size and distribution of Au or Ag nanoparticles in the fluid, as well as the type of base fluid used. Au and Ag nanofluids have potential applications in various fields, including heat transfer and thermal management in electronics, as well as energy generation and storage. Additionally, they have unique optical properties due to their high surface-to-volume ratio, making them useful in various optical applications (Awasthi et al., 2021).

Copper-Nanofluids

Copper (Cu) nanofluids are suspensions of Cu nanoparticles in a liquid medium, typically water or a base fluid such as ethylene glycol. The synthesis of Cu nanofluids involves the preparation of Cu nanoparticles through various methods such as chemical reduction, followed by their dispersion in the fluid. This can be done through methods such as sonication, stirring, or high-pressure homogenization. The properties of Cu nanofluids, such as thermal conductivity and viscosity, depend on the size and distribution of Cu nanoparticles in the fluid, as well as the type of base fluid used. Cu nanofluids have potential applications in various fields, including heat transfer and thermal management in electronics, as well as energy generation and storage. Additionally, they have unique electrical and thermal conductivity properties due to their high surface-to-volume ratio, making them useful in various energy-related applications.

CHALLENGES OF NANOFLUIDS

Long Term Stability of Nanoparticle Dispersion

The long-term stability of nanoparticle dispersions depends on several factors including the type of nanoparticles, the dispersion medium, and the presence of stabilizing agents. It can also be influenced by external factors such as temperature, pH, light exposure, and mechanical stress (Rafati et al., 2012; Huminic & Huminic, 2012).

In general, nanoparticles tend to aggregate and settle over time due to their small size and high surface energy. However, the use of stabilizing agents, such as surfactants or polymers, can help to maintain the dispersion and prevent aggregation (Agarwal et al., 2016).

It is important to note that different types of nanoparticles may have different stability requirements, and that regular monitoring and assessment of the dispersion should be conducted to ensure its stability over time. In addition, measures such as adding stabilizing agents, controlling the temperature, or adjusting the pH may need to be taken to maintain the stability of the dispersion. Challenges of nanofluids due to long term stability of nanoparticle dispersion. The long-term stability of nanoparticle dispersion in nanofluids can pose several challenges, including:

Aggregation and sedimentation: Over time, nanoparticles may aggregate and settle due to their small size and high surface energy. This can lead to decreased thermal conductivity and reduced effectiveness of the nanofluid (Awasthi et al., 2021).

Chemical stability: Nanoparticles may undergo chemical reactions with the fluid or with each other, leading to degradation of the nanoparticles and changes in the properties of the nanofluid. Physical stability: Nanoparticles may be subject to physical forces such as shear and mechanical stress, which can cause them to aggregate or settle.

Compatibility with fluid: The choice of fluid is important for the stability of the nanofluid, as some fluids may be more prone to chemical reactions with the nanoparticles. Cost: Stabilizing agents and other measures to maintain the stability of the nanofluid can add to the cost of production and use (Angayarkanni & Philip, 2015).

To mitigate these challenges, it is important to carefully control the dispersion and formulation of nanofluids, and to regularly assess the stability of the dispersion to ensure that the nanofluid remains effective over time (Agarwal et al., 2016).

The long-term stability of nanoparticle dispersion in nanofluids can pose several challenges, including:

- Aggregation and sedimentation: Over time, nanoparticles may aggregate and settle due to their small size and high surface energy. This can lead to decreased thermal conductivity and reduced effectiveness of the nanofluid.
- Chemical stability: Nanoparticles may undergo chemical reactions with the fluid or with each other, leading to degradation of the nanoparticles and changes in the properties of the nanofluid.

- Physical stability: Nanoparticles may be subject to physical forces such as shear and mechanical stress, which can cause them to aggregate or settle.
- Compatibility with fluid: The choice of fluid is important for the stability of the nanofluid, as some fluids may be more prone to chemical reactions with the nanoparticles.
- Cost: Stabilizing agents and other measures to maintain the stability of the nanofluid can add to the cost of production and use.
- To mitigate these challenges, it is important to carefully control the dispersion and formulation of nanofluids, and to regularly assess the stability of the dispersion to ensure that the nanofluid remains effective over time.

Thermal Performance in Turbulent Flow and Fully Developed Region

Nanofluids are suspensions of nanoparticles in a base fluid and have been shown to have enhanced thermal properties compared to traditional fluids. However, there are several challenges associated with the use of nanofluids in turbulent flow and fully developed regions, including:

Particle aggregation: Nanoparticles in nanofluids tend to aggregate due to Van der Waals forces, resulting in the formation of larger particles that can clog the flow channels and reduce heat transfer performance (Devendiran & Amirtham, 2016; Harilal et al., 2017).

Non-Newtonian behavior: Nanofluids can exhibit non-Newtonian behavior due to the presence of nanoparticles, which can affect their flow characteristics and heat transfer properties (Rafati et al., 2012;Huminic & Huminic, 2012; Harilal et al., 2017).

Viscosity: The addition of nanoparticles to the base fluid can increase the viscosity of nanofluids, which can lead to higher pressure drop and reduced heat transfer performance.

Corrosion: Nanoparticles can be highly reactive and can cause corrosion in the fluid delivery system and heat transfer surfaces, which can limit their practical use.

Particle sedimentation: Nanoparticles in nanofluids tend to settle due to gravitational forces, leading to non-uniform particle distribution and reduced thermal performance.

Measurement challenges: The measurement of thermal properties of nanofluids in turbulent flow and fully developed regions is challenging due to the presence of small and irregularly shaped particles, making it difficult to accurately measure their thermal conductivity, specific heat, and other properties (Devendiran & Amirtham, 2016).

Cost: The production of nanoparticles is often expensive, which can increase the overall cost of using nanofluids in thermal applications.

In conclusion, while nanofluids show promise in enhancing the thermal performance of fluids in turbulent flow and fully developed regions, several challenges related to particle aggregation, non-Newtonian behavior, viscosity, corrosion, particle sedimentation, measurement, and cost must be addressed before their practical use in industrial applications.

Difficulties in Production Process

The production process of nanofluids can be challenging and requires specialized equipment and expertise. Here are some of the difficulties associated with the production process of nanofluids:

Particle size control: The size and distribution of nanoparticles in the nanofluid can have a significant impact on its properties, such as thermal conductivity and viscosity. Controlling the particle size and distribution can be difficult, as nanoparticles tend to agglomerate, which can affect their dispersion in the base fluid (Awasthi et al., 2021).

Stability: Nanoparticles in nanofluids tend to agglomerate and settle over time, which can lead to non-uniform particle distribution and reduced thermal performance. Achieving long-term stability and preventing particle settling can be challenging (Wong & De Leon, 2010).

Synthesis: The synthesis of nanoparticles can be complex and requires specialized equipment and expertise. The choice of synthesis method can affect the size, shape, and surface chemistry of the nanoparticles, which can influence the properties of the nanofluid (Begum et al., 2022).

Cost: The production of nanoparticles can be expensive, which can increase the cost of producing nanofluids. The choice of nanoparticles and the production method can affect the overall cost of the nanofluid (Bhattad et al., 2018).

Contamination: Contamination of the nanofluid can occur during the production process, which can affect its properties and performance. Ensuring the purity of the nanoparticles and the base fluid is critical to the production of high-quality nanofluids.

Scale-up: Scaling up the production of nanofluids from laboratory-scale to industrial-scale can be challenging. The production process must be reproducible and cost-effective, and the quality of the nanofluid must be maintained (Wong & De Leon, 2010).

In conclusion, the production of nanofluids can be challenging due to difficulties in particle size control, stability, synthesis, cost, contamination, and scale-up. Addressing these challenges requires specialized equipment and expertise, and careful consideration of the choice of nanoparticles, synthesis method, and production process.

Challenges of Lower Specific Heat

Nanofluids have been shown to have enhanced thermal conductivity compared to traditional fluids. However, the use of nanofluids can also result in challenges related to lower specific heat. Here are some of the challenges associated with the lower specific heat of nanofluids:

Reduced heat capacity: Nanofluids can have a lower specific heat capacity compared to the base fluid due to the addition of nanoparticles. This can result in a reduced ability to absorb and store heat, which can limit their effectiveness in certain applications.

Temperature gradient: Due to the lower specific heat capacity of nanofluids, they can experience a larger temperature gradient for a given amount of heat input. This can lead to localized overheating and thermal damage in some applications (Rafati et al., 2012; Harilal et al., 2017).

Non-uniformity: The addition of nanoparticles can cause non-uniform heating in the nanofluid, as the nanoparticles tend to agglomerate and settle, leading to a non-uniform distribution of heat in the fluid.

Measurement challenges: Measuring the specific heat of nanofluids can be challenging due to the complex nature of the fluid and the presence of small and irregularly shaped nanoparticles. Accurately measuring specific heat is essential for understanding the thermal properties of the fluid (Devendiran & Amirtham, 2016; Harilal et al., 2017; Bhattad et al., 2018).

Cost: The production of nanofluids can be more expensive compared to traditional fluids due to the higher cost of nanoparticles, which can limit their widespread use in some applications.

In conclusion, while nanofluids can have enhanced thermal conductivity, the lower specific heat of nanofluids can result in challenges related to reduced heat capacity, temperature gradient, non-uniformity, measurement, and cost (Bhattad et al., 2018). Addressing these challenges requires careful consideration of the choice of nanoparticles, the production process, and the application requirements to optimize the performance of nanofluids(Table 5).

CONCLUSION AND RECOMENDATIONS

Recent advancements in nanoparticle and nanofluid research include the development of new synthesis methods for creating nanoparticles with specific size and shape, as well as the discovery of new properties and potential applications for nanoparticles. For example, researchers have developed methods for creating nanoparticles with

Table 5. Comparison data of Nanofluid containing synthesis, shape and size control, recent challenges, and future scope

Nanofluid	Synthesis Method	Nanoparticle Name	Shape	Size	Modern Advancements	Challenges	Future Scope
Al2O3-water	Two-step method, Sol-gel method, Homogeneous precipitation method	Alumina	Spherical	10-60 nm	Improved stability, enhanced thermal conductivity and heat transfer rate	Agglomeration, difficulty in scaling up	Cooling of high power electronic components, advanced manufacturing processes
TiO2-ethylene glycol	Hydrothermal synthesis, Sol-gel method	Titania	Spherical, rod-like, plate-like, flower-like	10-40 nm	Improved heat transfer coefficient, enhanced thermal conductivity, improved flow characteristics	Particle sedimentation, agglomeration, high cost of production	Solar thermal systems, waste heat recovery, energy conversion systems
CuO-water	Chemical precipitation method, Solvothermal method	Copper oxide	Spherical, cubic, rod-like, octahedral	20-80 nm	Enhanced thermal conductivity, improved heat transfer efficiency, anti-corrosion properties	Particle agglomeration, sedimentation, difficulty in long-term stability	Heat transfer fluids, solar thermal systems, automotive and aerospace industries
Fe3O4-kerosene	Co-precipitation method, Hydrothermal synthesis	Magnetite	Spherical, rod-like, octahedral	5-20 nm	High thermal conductivity, improved heat transfer efficiency, magnetic properties	Particle agglomeration, settling, difficulty in controlling particle size and shape	Heat transfer fluids, magnetic refrigeration systems, biomedical applications
SiO2-water	Stober method, Sol-gel method, Chemical vapor deposition	Silica	Spherical, rod-like, plate-like	5-50 nm	Improved thermal conductivity, enhanced heat transfer, stability at high temperatures	Particle agglomeration, sedimentation, difficulty in scaling up	Heat transfer fluids, electronics cooling, advanced manufacturing processes

(Harilal et al., 2017)

tailored physical and chemical properties, such as magnetic or optical properties, which have potential applications in fields such as medicine, electronics, and energy.

In nanofluid research, recent advancements include the development of new types of nanofluids, such as those containing carbon nanotubes and graphene, as well as the discovery of new properties and potential applications for nanofluids, such as enhanced heat transfer and improved lubrication.

Future challenges in nanoparticle and nanofluid research include understanding and mitigating the potential toxicity and environmental risks associated with these materials, as well as scaling up the synthesis and production of nanoparticles and nanofluids for industrial and commercial applications. Additionally, ongoing research efforts are needed to fully understand the properties and behavior of these materials, as well as to develop new methods for characterizing and controlling their properties.

Research directions in nanoparticle and nanofluid research include:

Development of new synthesis methods for creating nanoparticles with specific size, shape, and properties. Investigating the potential applications of nanoparticles in fields such as medicine, electronics, energy, and environmental remediation. Understanding and mitigating the potential toxicity and environmental risks associated with nanoparticles.

Developing new methods for characterizing and controlling the properties of nanoparticles.

Scaling up the synthesis and production of nanoparticles for industrial and commercial applications. Investigating the potential applications of nanofluids in fields such as heat transfer, lubrication, and energy. Implications for future work in nanoparticle and nanofluid research include:

The need for continued research on the properties and behavior of these materials to fully understand their potential uses and risks. The need for further development of synthesis methods and production techniques to create nanoparticles and nanofluids with specific properties for industrial and commercial applications.

The need for ongoing research on the potential toxicity and environmental risks associated with nanoparticles and nanofluids, as well as the development of methods for mitigating these risks.

The potential for these materials to have a significant impact on a wide range of industries and technologies, and therefore the need for continued research and development in this field.

REFERENCES

Agarwal, R., Verma, K., Agrawal, N. K., Duchaniya, R. K., & Singh, R. (2016). Synthesis, characterization, thermal conductivity and sensitivity of CuO nanofluids. *Applied Thermal Engineering*, *102*, 1024–1036. doi:10.1016/j.applthermaleng.2016.04.051

Ahamed, M., Akhtar, M. J., & Alhadlaq, H. A. (2019). Co-Exposure to SiO2 Nanoparticles and Arsenic Induced Augmentation of Oxidative Stress and Mitochondria-Dependent Apoptosis in Human Cells. *International Journal of Environmental Research and Public Health*, *16*(17), 3199. doi:10.3390/ijerph16173199 PMID:31480624

Ali, A., Shah, T., Ullah, R., Zhou, P., Guo, M., Ovais, M., Tan, Z., & Rui, Y. (2021). Review on Recent Progress in Magnetic Nanoparticles: Synthesis, Characterization, and Diverse Applications. *Frontiers in Chemistry*, *9*, 629054. doi:10.3389/fchem.2021.629054 PMID:34327190

Ali, N., Teixeira, J. A., & Addali, A. (2018). A Review on Nanofluids: Fabrication, Stability, and Thermophysical Properties. *Journal of Nanomaterials*, *2018*, 1–33. doi:10.1155/2018/6978130

Angayarkanni, S. A., & Philip, J. (2015). Review on thermal properties of nanofluids: Recent developments. *Advances in Colloid and Interface Science*, *225*, 146–176. doi:10.1016/j.cis.2015.08.014 PMID:26391519

Attar, S. R., & Kamble, S. B. (2022). Recent advances in nanoparticles towards sustainability and their application in organic transformations in aqueous media. *Nanoscale*, *14*(45), 16761–16786. doi:10.1039/D2NR04148K PMID:36341716

Baig, N., Kammakakam, I., & Falath, W. (2021). Nanomaterials: A review of synthesis methods, properties, recent progress, and challenges. *Materials Advances*, *2*(6), 1821–1871. doi:10.1039/D0MA00807A

Begines, B., Ortiz, T., Pérez-Aranda, M., Martínez, G., Merinero, M., Argüelles-Arias, F., & Alcudia, A. (2020). Polymeric Nanoparticles for Drug Delivery: Recent Developments and Future Prospects. *Nanomaterials (Basel, Switzerland)*, *10*(7), 1403. doi:10.3390/nano10071403 PMID:32707641

Begum, S. J. P., Pratibha, S., Rawat, J. M., Venugopal, D., Sahu, P., Gowda, A., Qureshi, K. A., & Jaremko, M. (2022). Recent Advances in Green Synthesis, Characterization, and Applications of Bioactive Metallic Nanoparticles. *Pharmaceuticals (Basel, Switzerland)*, *15*(4), 455. doi:10.3390/ph15040455 PMID:35455452

Bhattad, A., Sarkar, J., & Ghosh, P. (2018). Improving the performance of refrigeration systems by using nanofluids: A comprehensive review. *Renewable & Sustainable Energy Reviews*, *82*, 3656–3669. doi:10.1016/j.rser.2017.10.097

Devendiran, D. K., & Amirtham, V. A. (2016). A review on preparation, characterization, properties and applications of nanofluids. *Renewable & Sustainable Energy Reviews*, *60*, 21–40. doi:10.1016/j.rser.2016.01.055

Ghatnur, S. M., Sonale, R. S., Balaraman, M., & Kadimi, U. S. (2012). Engineering liposomes of leaf extract of seabuckthorn (SBT) by supercritical carbon dioxide (SCCO2)-mediated process. *Journal of Liposome Research*, *22*(3), 215–223. doi:10.3109/08982104.2012.658576 PMID:22397357

Harilal, M., Vidyadharan, B., Misnon, I. I., Anilkumar, G. M., Lowe, A., Ismail, J., Yusoff, M. M., & Jose, R. (2017). One-Dimensional Assembly of Conductive and Capacitive Metal Oxide Electrodes for High-Performance Asymmetric Supercapacitors. *ACS Applied Materials & Interfaces*, *9*(12), 10730–10742. doi:10.1021/acsami.7b00676 PMID:28266837

Huminic, G., & Huminic, A. (2012). Application of nanofluids in heat exchangers: A review. *Renewable & Sustainable Energy Reviews*, *16*(8), 5625–5638. doi:10.1016/j.rser.2012.05.023

Khanafer, K., & Vafai, K. (2018). A review on the applications of nanofluids in solar energy field. *Renewable Energy*, *123*, 398–406. doi:10.1016/j.renene.2018.01.097

Li, L. (2015, July 24). *TEM of Gold Nanoparticles results - FoNS-MAD! Competition 2015*. The Velox Group's Blog. https://blogs.imperial.ac.uk/fonsmad2015-velox/2015/07/18/tem-of-gold-nanoparticles-results/

Li, L. (2013, July 31). *Superparamagnetic Iron Oxide Nanoparticles as MRI contrast agents for Non-invasive Stem Cell Labeling and Tracking*. THNO. https://www.thno.org/v03p0595.html,

Myung, S., Solanki, A., Kim, C., Park, J., Kim, K. S., & Lee, K.-B. (2011). Graphene-Encapsulated Nanoparticle-Based Biosensor for the Selective Detection of Cancer Biomarkers. *Advanced Materials*, *23*(19), 2221–2225. doi:10.1002/adma.201100014 PMID:21469221

Pradhan, S., & Mishra, B. G. (2015). Catalytic application of SO42−/Fe-ZrO2nanoparticles synthesized by a urea hydrolysis method for environmentally benign one pot synthesis of 1,8-dioxodecahydroacridines. *RSC Advances*, *5*(105), 86179–86190. doi:10.1039/C5RA11136F

Pradhan, S., & Mishra, B. G. (2018). CsxH3-xPW12O40 nanoparticles dispersed in the porous network of Zr-pillared α-zirconium phosphate as efficient heterogeneous catalyst for synthesis of spirooxindoles. *Molecular Catalysis*, *446*, 58–71. doi:10.1016/j.mcat.2017.12.013

Pradhan, S., Saha, J., & Mishra, B. G. (2017). Morphology controlled phosphate grafted SnO2–ZrO2 nanocomposite oxides prepared by a urea hydrolysis method as efficient heterogeneous catalysts towards the synthesis of 3-substituted indoles. *New Journal of Chemistry*, *41*(14), 6616–6629. doi:10.1039/C7NJ00249A

Pradhan, S., Sahu, V., & Mishra, B. G. (2016). CaO-ZrO2 nanocomposite oxide prepared by urea hydrolysis method as heterogeneous base catalyst for synthesis of chromene analogues. *Journal of Molecular Catalysis A Chemical*, *425*, 297–309. doi:10.1016/j.molcata.2016.10.031

Rafati, M., Hamidi, A. A., & Shariati Niaser, M. (2012). Application of nanofluids in computer cooling systems (heat transfer performance of nanofluids). *Applied Thermal Engineering*, *45-46*, 9–14. doi:10.1016/j.applthermaleng.2012.03.028

Rajpoot, K. (2019). Solid Lipid Nanoparticles: A Promising Nanomaterial in Drug Delivery. *Current Pharmaceutical Design*, *25*(37), 3943–3959. doi:10.2174/1381612825666190903155321 PMID:31481000

Singh, P., Srivastava, S., & Singh, S. K. (2019). Nanosilica: Recent Progress in Synthesis, Functionalization, Biocompatibility, and Biomedical Applications. *ACS Biomaterials Science & Engineering*, *5*(10), 4882–4898. doi:10.1021/acsbiomaterials.9b00464 PMID:33455238

Wang, Y., & Alocilja, E. C. (2015). *Gold nanoparticle-labeled biosensor for rapid and sensitive detection of bacterial pathogens*. Springer Science and Business Media LLC., doi:10.118613036-015-0014-z

Wong, K. V., & De Leon, O. (2010). Applications of Nanofluids: Current and Future. *Advances in Mechanical Engineering*, *2*, 519659. doi:10.1155/2010/519659

Zahmakıran, M., & Özkar, S. (2011). Metal nanoparticles in liquid phase catalysis; from recent advances to future goals. *Nanoscale*, *3*(9), 3462. doi:10.1039/c1nr10201j PMID:21833406

Compilation of References

Abdelrazek, A. H., Choudhury, I. A., Nukman, Y., & Kazi, S. N. (2020). Metal cutting lubricants and cutting tools: A review on the performance improvement and sustainability assessment. *International Journal of Advanced Manufacturing Technology*, *106*(9-10), 4221–4245. doi:10.100700170-019-04890-w

Abdi, G., Salehi, H., & Khosh-Khui, M. (2008). Nano silver: Anovel nanomaterial for removal of bacterial contamination in valerian (*Valeriana officinalis* L.) tissue culture. *Acta Physiologiae Plantarum*, *30*(5), 709–714. doi:10.100711738-008-0169-z

Abdollahzadeh, A., Shokuhfar, A., Cabrera, J. M., Zhilyaev, A. P., & Omidvar, H. (2019). In-situ nanocomposite in friction stir welding of 6061-T6 aluminum alloy to AZ31 magnesium alloy. *Journal of Materials Processing Technology*, *263*, 296–307. doi:10.1016/j.jmatprotec.2018.08.025

Abdolmohammadi, S., Ghiasi, R., & Ahmadzadeh-Vatani, S. (2016). A highly efficient CuI nanoparticles-catalyzed synthesis of tetrahydrochromenediones and dihydropyrano [c] chromenediones under grinding. *Zeitschrift für Naturforschung. Section B. A Journal of Chemical Sciences*, *71*(7), 777–782. doi:10.1515/znb-2015-0195

Abhilash, P.P, Dheeresh, N., Bhaskar, S., Rajesh, K., & Veerendra, K. (. (2021). Effect of nano-silica in concrete: A review. *Construction & Building Materials*, *278*, 122347. doi:10.1016/j.conbuildmat.2021.122347

Agarwal, R., Verma, K., Agrawal, N. K., Duchaniya, R. K., & Singh, R. (2016). Synthesis, characterization, thermal conductivity and sensitivity of CuO nanofluids. *Applied Thermal Engineering*, *102*, 1024–1036. doi:10.1016/j.applthermaleng.2016.04.051

Ahamed, M., Akhtar, M. J., & Alhadlaq, H. A. (2019). Co-Exposure to SiO2 Nanoparticles and Arsenic Induced Augmentation of Oxidative Stress and Mitochondria-Dependent Apoptosis in Human Cells. *International Journal of Environmental Research and Public Health*, *16*(17), 3199. doi:10.3390/ijerph16173199 PMID:31480624

Ahlawat, J Sehrawat, A. R. (2015). Biological Synthesis of Silver Nanoparticles using Aqueous Leaf Extract of *Capparis decidua* (forsk.) edgew: A better alternative'. *Journal of Pharmacy Research*, *9*(4), 244–249.

Ahlawat, J., & Sehrawat, A. R. (2017). Nano Dimensional (1-20nm) Silver Nanoparticles: Stem Extract of *Capparis decidua* (FORSK.) EDGEW Mediated Synthesis and its Characterization-A Lab to Land Approach'. *International Journal of Current Microbiology and Applied Sciences*, *6*(10), 1874–1883. doi:10.20546/ijcmas.2017.610.226

Ahlawat, J., Sehrawat, A. R., & Choudhary, R. (2020). Biologically synthesized silver nanoparticles eclipse fungal and bacterial contamination in micropropagation of Capparis decidua (FORSK.) Edgew: A substitute to toxic substances. *Indian Journal of Experimental Biology*, *50*, 336–343.

Ahmada, A., Asghara, S., & Alsaedi, A. (2014). Flow and heat transfer of a nanofluid over a hyperbolically stretching sheet. *Chinese Physics B*, *23*(7), 074401. doi:10.1088/1674-1056/23/7/074401

Akter, M., Sikder, M.T., & Rahman, M.M. (2018). A systematic review on silver nanoparticles-induced cytotoxicity: physicochemical properties and perspectives. *J Adv Res*, *9*, 1–16. doi:. jare.2017.10.008 doi:10.1016/j

Akturk, O., Kismet, K., Yasti, A. C., Kuru, S., Duymus, M. E., Kaya, F., Caydere, M., Hucumenoglu, S., & Keskin, D. (2016). Collagen/gold nanoparticle nanocomposites: A potential skin wound healing biomaterial. *Journal of Biomaterials Applications*, *31*(2), 283–301. doi:10.1177/0885328216644536 PMID:27095659

Akyildiz, F. T., Bellout, H., Vajravelu, K., & Van Gorder, R. A. (2011). Existence results for third order nonlinear boundary value problems arising in nano boundary layer fluid flows over stretching surfaces. *Nonlinear Analysis Real World Applications*, *12*(6), 2919–2930. doi:10.1016/j.nonrwa.2011.02.017

Ali, A., Shah, T., Ullah, R., Zhou, P., Guo, M., Ovais, M., Tan, Z., & Rui, Y. (2021). Review on Recent Progress in Magnetic Nanoparticles: Synthesis, Characterization, and Diverse Applications. *Frontiers in Chemistry*, *9*, 629054. doi:10.3389/fchem.2021.629054 PMID:34327190

Ali, N., Teixeira, J. A., & Addali, A. (2018). A Review on Nanofluids: Fabrication, Stability, and Thermophysical Properties. *Journal of Nanomaterials*, *2018*, 1–33. doi:10.1155/2018/6978130

Anand, R., & Sridhar, V. G. (2021). Effects of SiC and Al2O3 Reinforcement of Varied Volume Fractions on Mechanical and Micro Structure Properties of Interlock FSW Dissimilar Joints AA7075-T6-AA7475-T7. *Silicon*, *13*(9), 3017–3029. doi:10.100712633-020-00630-y

Angayarkanni, S. A., & Philip, J. (2015). Review on thermal properties of nanofluids: Recent developments. *Advances in Colloid and Interface Science*, *225*, 146–176. doi:10.1016/j.cis.2015.08.014 PMID:26391519

Aranaz, I., Mengíbar, M., Harris, R., Paños, I., Miralles, B., Acosta, N., Galed, G., & Heras, Á. (2018). Cosmetics and cosmeceutical applications of chitin, chitosan and their derivatives. *Polymers*, *10*(2), 213. doi:10.3390/polym10020213 PMID:30966249

Arifuddin, A., Redhwan, A. A. M., Azmi, W. H., & Zawawi, N. N. M. (2022). Performance of Al2O3/TiO2 Hybrid Nano-Cutting Fluid in MQL Turning Operation via RSM Approach. *Lubricants (Basel, Switzerland)*, *10*(12), 366. doi:10.3390/lubricants10120366

Arora, N., Agarwal, S., & Murthy, R. S. R. (2012). Latest technology advances in cosmaceuticals. *International Journal of Pharmaceutical Sciences and Drug Research*, *4*(3), 168–182.

Arunprasad, R. V. (2019). Alternate Refrigerants for Minimization Environmental Impacts: A Review. In *Advances in Engineering Technology* (p. 69). AkiNik Publications New Delhi. https://www.researchgate.net/publication/340939609

Ashok, B. & Mahender, B. (2017). Effect of Nano Silica on the Compressive Strength of Concrete. *International Journal of Professional Engineering Studies.*, *8*(2), 222–226.

Attar, S. R., & Kamble, S. B. (2022). Recent advances in nanoparticles towards sustainability and their application in organic transformations in aqueous media. *Nanoscale*, *14*(45), 16761–16786. doi:10.1039/D2NR04148K PMID:36341716

Avio, C. G., Gorbi, S., & Regoli, F. (2017). Plastics and microplastics in the oceans: From emerging pollutants to emerged threat. *Marine Environmental Research*, *128*, 2–11. doi:10.1016/j.marenvres.2016.05.012 PMID:27233985

Awoke, N., Pandey, D., & Habtemariam, A. B. (2021). Synthesis of Tin(IV) Oxide Nanoparticles Using Plant Leaf Extracts of Vernonia amygdalina and Mentha spicata. *Regenerative Engineering and Translational Medicine*, 1–6.

Babu, B. S., Kamalakannan, J., Meenatchi, N., M, S. K. S., S, K., & Boopathi, S. (2023). Economic impacts and reliability evaluation of battery by adopting Electric Vehicle. *IEEE Explore*, (pp. 1–6). IEEE. doi:10.1109/ICPECTS56089.2022.10046786

Bafakeeh, O. T., Shewakh, W. M., Abu-Oqail, A., Abd-Elaziem, W., Abdel Ghafaar, M., & Abu-Okail, M. (2021). Synthesis and Characterization of Hybrid Fiber-Reinforced Polymer by Adding Ceramic Nanoparticles for Aeronautical Structural Applications. *Polymers*, *13*(23), 4116. doi:10.3390/polym13234116 PMID:34883619

Bahrami, M., Dehghani, K., & Besharati Givi, M. K. (2014). A novel approach to develop aluminum matrix nano-composite employing friction stir welding technique. *Materials & Design*, *53*, 217–225. doi:10.1016/j.matdes.2013.07.006

Bahrami, M., Farahmand Nikoo, M., & Besharati Givi, M. K. (2015). Microstructural and mechanical behaviors of nano-SiC-reinforced AA7075-O FSW joints prepared through two passes. *Materials Science and Engineering A*, *626*, 220–228. doi:10.1016/j.msea.2014.12.009

Bahrami, M., Helmi, N., Dehghani, K., & Givi, M. K. B. (2014). Exploring the effects of SiC reinforcement incorporation on mechanical properties of friction stir welded 7075 aluminum alloy: Fatigue life, impact energy, tensile strength. *Materials Science and Engineering A*, *595*, 173–178. doi:10.1016/j.msea.2013.11.068

Baig, N., Kammakakam, I., & Falath, W. (2021). Nanomaterials: A review of synthesis methods, properties, recent progress, and challenges. *Materials Advances*, *2*(6), 1821–1871. doi:10.1039/D0MA00807A

Barbosa, T. C., Nascimento, L. É. D., Bani, C., Almeida, T., Nery, M., Santos, R. S., Menezes, L. R. O., Zielińska, A., Fernandes, A. R., Cardoso, J. C., Jäger, A., Jäger, E., Sanchez-Lopez, E., Nalone, L., Souto, E. B., & Severino, P. (2019). Development, cytotoxicity and eye irritation profile of a new sunscreen formulation based on benzophenone-3-poly (ε-caprolactone) nanocapsules. *Toxics*, *7*(4), 51. doi:10.3390/toxics7040051 PMID:31546707

Bárcena, C., Sra, A. K., & Gao, J. (2009). Applications of Magnetic Nanoparticles in Biomedicine. In *Nanoscale Magnetic Materials and Applications* (pp. 591–626). Springer. doi:10.1007/978-0-387-85600-1_20

Baroli, B., Ennas, M. G., Loffredo, F., Isola, M., Pinna, R., & López-Quintela, M. A. (2007). Penetration of metallic nanoparticles in human full-thickness skin. *The Journal of Investigative Dermatology*, *127*(7), 1701–1712. doi:10.1038j.jid.5700733 PMID:17380118

Batyuk, L. V., & Kizilova, N. M. (2019). Maghetohydrodynamic flows of micro/nano fluids through thin capillaries. *Bulletin of Taras Shevchenko National University of Kyiv. Series. Physics and Mathematics*, *3*(3), 32–36. doi:10.17721/1812-5409.2019/3.4

Begines, B., Ortiz, T., Pérez-Aranda, M., Martínez, G., Merinero, M., Argüelles-Arias, F., & Alcudia, A. (2020). Polymeric Nanoparticles for Drug Delivery: Recent Developments and Future Prospects. *Nanomaterials (Basel, Switzerland)*, *10*(7), 1403. doi:10.3390/nano10071403 PMID:32707641

Begum, S. J. P., Pratibha, S., Rawat, J. M., Venugopal, D., Sahu, P., Gowda, A., Qureshi, K. A., & Jaremko, M. (2022). Recent Advances in Green Synthesis, Characterization, and Applications of Bioactive Metallic Nanoparticles. *Pharmaceuticals (Basel, Switzerland)*, *15*(4), 455. doi:10.3390/ph15040455 PMID:35455452

Behera, S., & Nayak, P. L. (2013). *In vitro* Antibacterial Activity of Green Synthesized Silver Nanoparticles Using Jamun Extract Against Multiple Drug Resistant Bacteria. *World J. Nano Sci. Tech.*, *2*(1), 62–65.

Bhattad, A., Sarkar, J., & Ghosh, P. (2018). Improving the performance of refrigeration systems by using nanofluids: A comprehensive review. *Renewable & Sustainable Energy Reviews*, *82*, 3656–3669. doi:10.1016/j.rser.2017.10.097

Bhupendra, G., Prajapati Niklesh, K., Manan, M., Rakesh, P. P. (2013). Topical Liposomes in Drug Delivery. *International Journal of Pavement Research and Technology*, *4*(1), 39–44.

Biswas, P., & Wu, C.-Y. (2005). Nanoparticles and the environment. *Journal of the Air \& Waste Management Association*, *55*(6), 708–746.

Boopathi, S. (2021). Improving of Green Sand-Mould Quality using Taguchi Technique. *Journal of Engineering Research*. doi:10.36909/jer.14079

Boopathi, S., Arigela, S. H., Raman, R., Indhumathi, C., Kavitha, V., & Bhatt, B. C. (2023). Prominent Rule Control-based Internet of Things: Poultry Farm Management System. *IEEE Explore*, (pp. 1–6). IEEE. doi:10.1109/ICPECTS56089.2022.10047039

Boopathi, S., Khare, R., Jaya Christiyan, K. G., Muni, T. V., & Khare, S. (2023). Additive manufacturing developments in the medical engineering field. In Development, Properties, and Industrial Applications of 3D Printed Polymer Composites (pp. 86–106). IGI Global. doi:10.4018/978-1-6684-6009-2.ch006

Boopathi, S., Siva Kumar, P. K., & Meena, R. S. J., S. I., P., S. K., & Sudhakar, M. (2023). Sustainable Developments of Modern Soil-Less Agro-Cultivation Systems. In Human Agro-Energy Optimization for Business and Industry (pp. 69–87). IGI Global. doi:10.4018/978-1-6684-4118-3.ch004

Boopathi, S. (2022a). An experimental investigation of Quench Polish Quench (QPQ) coating on AISI 4150 steel. *Engineering Research Express*, *4*(4), 45009. doi:10.1088/2631-8695/ac9ddd

Boopathi, S. (2022a). An Extensive Review on Sustainable Developments of Dry and Near-Dry Electrical Discharge Machining Processes. *Journal of Manufacturing Science and Engineering*, *144*(5), 50801. doi:10.1115/1.4052527

Boopathi, S. (2022b). An investigation on gas emission concentration and relative emission rate of the near-dry wire-cut electrical discharge machining process. *Environmental Science and Pollution Research International*, *29*(57), 86237–86246. doi:10.100711356-021-17658-1 PMID:34837614

Boopathi, S. (2022c). Cryogenically treated and untreated stainless steel grade 317 in sustainable wire electrical discharge machining process: A comparative study. *Environmental Science and Pollution Research International*, 1–10. doi:10.100711356-022-22843-x PMID:36057706

Boopathi, S. (2022d). Experimental investigation and multi-objective optimization of cryogenic Friction-stir-welding of AA2014 and AZ31B alloys using MOORA technique. *Materials Today. Communications*, *33*, 104937. doi:10.1016/j.mtcomm.2022.104937

Boopathi, S. (2022e). Performance Improvement of Eco-Friendly Near-Dry Wire-Cut Electrical Discharge Machining Process Using Coconut Oil-Mist Dielectric Fluid. *Journal of Advanced Manufacturing Systems*, 1–20. doi:10.1142/S0219686723500178

Boopathi, S., Balasubramani, V., Kumar, R. S., & Singh, G. R. (2021). The influence of human hair on kenaf and Grewia fiber-based hybrid natural composite material: An experimental study. *Functional Composites and Structures*, *3*(4), 45011. doi:10.1088/2631-6331/ac3afc

Boopathi, S., Balasubramani, V., & Sanjeev Kumar, R. (2023). Influences of various natural fibers on the mechanical and drilling characteristics of coir-fiber-based hybrid epoxy composites. *Engineering Research Express*, *5*(1), 15002. doi:10.1088/2631-8695/acb132

Boopathi, S., Haribalaji, V., Mageswari, M., & Asif, M. M. (2022). Influences of Boron Carbide Particles on the Wear Rate and Tensile Strength of Aa2014 Surface Composite Fabricated By Friction-Stir Processing. *Materiali in Tehnologije*, *56*(3), 263–270. doi:10.17222/mit.2022.409

Boopathi, S., Jeyakumar, M., Singh, G. R., King, F. L., Pandian, M., Subbiah, R., & Haribalaji, V. (2022). An experimental study on friction stir processing of aluminium alloy (AA-2024) and boron nitride (BNp) surface composite. *Materials Today: Proceedings*, *59*, 1094–1099. doi:10.1016/j.matpr.2022.02.435

Boopathi, S., Kumaresan, A., Manohar, N., & Krishna Moorthi, R. (2017). Review on Effect of Process Parameters-Friction Stir Welding Process. [IRJET]. *International Research Journal of Engineering and Technology*, *4*(7), 272–278.

Boopathi, S., Lewise, K. A. S., Subbiah, R., & Sivaraman, G. (2021). Near-dry wire-cut electrical discharge machining process using water-air-mist dielectric fluid: An experimental study. *Materials Today: Proceedings*, *49*(5), 1885–1890. doi:10.1016/j.matpr.2021.08.077

Boopathi, S., & Myilsamy, S. (2021). Material removal rate and surface roughness study on Near-dry wire electrical discharge Machining process. *Materials Today: Proceedings*, *45*(9), 8149–8156. doi:10.1016/j.matpr.2021.02.267

Boopathi, S., & Sivakumar, K. (2012a). Experimental Analysis of Eco-friendly Near-dry Wire Electrical Discharge Machining Process. *Archives des Sciences*, *65*(10), 334–346.

Boopathi, S., & Sivakumar, K. (2012b). Experimental comparative study of near-dry wire-cut electrical discharge machining (WEDM). *European Journal of Scientific Research*, *75*(4), 472–481.

Boopathi, S., & Sivakumar, K. (2012c). Parametric Study of Dry WEDM Using Taguchi Method. *International Journal of Engineering Research and Development*, *2*(4), 63–68. www.ijerd.com

Boopathi, S., & Sivakumar, K. (2013). Experimental investigation and parameter optimization of near-dry wire-cut electrical discharge machining using multi-objective evolutionary algorithm. *International Journal of Advanced Manufacturing Technology*, *67*(9–12), 2639–2655. doi:10.100700170-012-4680-4

Boopathi, S., & Sivakumar, K. (2014). Study of water assisted dry wire-cut electrical discharge machining. *Indian Journal of Engineering and Materials Sciences*, *21*(1), 75–82.

Boopathi, S., & Sivakumar, K. (2016). Optimal parameter prediction of oxygen-mist near-dry Wire-cut EDM. *International Journal of Manufacturing Technology and Management*, *30*(3–4), 164–178. doi:10.1504/IJMTM.2016.077812

Boopathi, S., Sureskumar, M., Jeyakumar, M., Sanjeev Kumar, R., & Subbiah, R. (2022). Influences of Fabrication Parameters on Natural Fiber Reinforced Polymer Composite (NFRPC) Material: A Review. *Materials Science Forum*, *1075*, 115–124. doi:10.4028/p-095f0t

Boopathi, S., Thillaivanan, A., Azeem, M. A., Shanmugam, P., & Pramod, V. R. (2022). Experimental investigation on abrasive water jet machining of neem wood plastic composite. *Functional Composites and Structures*, *4*(2), 25001. doi:10.1088/2631-6331/ac6152

Boopathi, S., Venkatesan, G., & Anton Savio Lewise, K. (2023). Mechanical Properties Analysis of Kenaf–Grewia–Hair Fiber-Reinforced Composite. In *Lecture Notes in Mechanical Engineering* (pp. 101–110). Springer., doi:10.1007/978-981-16-9057-0_11

Borowska, S., & Brzóska, M. M. (2015). Metals in cosmetics: Implications for human health. *Journal of Applied Toxicology*, *35*(6), 551–572. doi:10.1002/jat.3129 PMID:25809475

Brown, J. S., Zeman, K. L., & Bennett, W. D. (2002). Ultrafine particle deposition and clearance in the healthy and obstructed lung. *American Journal of Respiratory and Critical Care Medicine*, *166*(9), 1240–1247. doi:10.1164/rccm.200205-399OC PMID:12403694

Buongiorno, J. (2006). *Convective transport in nanofluids*. Academic Press.

Casati, R., & Vedani, M. (2014). Metal matrix composites reinforced by nano-particles—A review. *Metals*, *4*(1), 65–83. doi:10.3390/met4010065

Castranova, V. (2011). Overview of Current Toxicological Knowledge of Engineered Nanoparticles. *Journal of Occupational and Environmental Medicine*, *53*(Supplement 6S), S14–S17. doi:10.1097/JOM.0b013e31821b1e5a PMID:21606847

Cevc, G., & Blume, G. (1992). Lipid vesicles penetrate into intact skin owing to the transdermal osmotic gradients and hydration force. *Biochimica et Biophysica Acta (BBA)-. Biomembranes*, *1104*(1), 226–232. doi:10.1016/0005-2736(92)90154-E PMID:1550849

Chaudhri, S. K., & Jain, N. K. (2009). History of cosmetics. [AJP]. *Asian Journal of Pharmaceutics*, *3*(3).

Chaudhry, Q.Scientific Committee on Consumer. (2016). Chaudhry Q. Opinion of the scientific committee on consumer safety (SCCS)—Second revision of the opinion on carbon black, nano-form, in cosmetic products. *Regulatory Toxicology and Pharmacology*, *79*, 103–104. doi:10.1016/j.yrtph.2016.02.021 PMID:26946407

Chen, M., Ma, C. Y., Mahmud, T., Darr, J. A., & Wang, X. Z. (2011). Modelling and simulation of continuous hydrothermal flow synthesis process for nano-materials manufacture. *The Journal of Supercritical Fluids*, *59*, 131–139. doi:10.1016/j.supflu.2011.07.002

Cheraghian, G. (2021). Nanoparticles in drilling fluid: A review of the state-of-the-art. *Journal of Materials Research and Technology*, *13*, 737–753. doi:10.1016/j.jmrt.2021.04.089

Choi, S. (1998). Nanofluid technology: current status and future research. In *Energy*. http://www.ncbi.nlm.nih.gov/pubmed/11048%5Cnhttp://www.osti.gov/energycitations/product.biblio.jsp?osti_id=11048

Choi, S. U., & Eastman, J. A. (1995). Enhancing thermal conductivity of fluids with nanoparticles (No. ANL/MSD/CP-84938; CONF-951135-29). Argonne National Lab.

Choi, S. U. S. (1995). *Enhancing Thermal Conductivity of Fluids with Nanoparticles, Developments and Applications of Non-Newtonian Flows*. ASME.

Chopkar, M., Das, P. K., & Manna, I. (2006). Synthesis and characterization of nanofluid for advanced heat transfer applications. *Scripta Materialia*, *55*(6), 549–552. doi:10.1016/j.scriptamat.2006.05.030

Coelho, C. C., Grenho, L., Gomes, P. S., Quadros, P. A., & Fernandes, M. H. (2019). Nano-hydroxyapatite in oral care cosmetics: Characterization and cytotoxicity assessment. *Scientific Reports*, *9*(1), 11050. doi:10.103841598-019-47491-z PMID:31363145

Cornier, J., Keck, C., & Voorde, M. (2019). *Nanocosmetics from Ideas to Products: From Ideas to Products*. Springer. doi:10.1007/978-3-030-16573-4

Cosmetics Info. (n.d.). History of Cosmetics from Ancient Times. *Cosmetics Info*. https://www.cosmeticsinfo.org/Ancient-history-cosmetics

Courville, I. G., Y. B., & A. (2016). Deep learning. In *Nature, 29*. MIT press. http://deeplearning.net/

Couteau, C., Paparis, E., Chauvet, C., & Coiffard, L. (2015). Tris-biphenyl triazine, a new ultraviolet filter studied in terms of photoprotective efficacy. *International Journal of Pharmaceutics*, *487*(1-2), 120–123. doi:10.1016/j.ijpharm.2015.03.077 PMID:25843762

Daehn, G. S. (2014). 13 - Sustainable design and manufacture of lightweight vehicle structures. In R. Folkson (Ed.), *Alternative Fuels and Advanced Vehicle Technologies for Improved Environmental Performance* (pp. 433–461). Woodhead Publishing. doi:10.1533/9780857097422.2.433

Danish, M., Ginta, T., Yasir, M., & Abdul-Rani, A.-M. (2018). Light alloys and their machinability. *RE:view*, 1–34. doi:10.1201/b22153-1

Das, A., Patel, S. K., Arakha, M., Dey, A., & Biswal, B. B. (2021). Processing of hardened steel by MQL technique using nano cutting fluids. *Materials and Manufacturing Processes*, *36*(3), 316–328. doi:10.1080/10426914.2020.1832688

Das, M., Saxena, N., & Dwivedi, P. D. (2009). Emerging trends of nanoparticles application in food technology: Safety paradigms. *Nanotoxicology*, *3*(1), 10–18. doi:10.1080/17435390802504237

Dass james, A., & Boopathi, S. (2016). Experimental Study of Eco-friendly Wire-Cut Electrical Discharge Machining Processes. *International Journal of Innovative Research in Science, Engineering and Technology, 5*(1), 666–675.

Daveiga, J., & Ferreira, P. (2005). Smart and Nano materials in architecture. *ACADIA 2005 Conference: Smart Architecture*, 58–67. 10.52842/conf.acadia.2005.058

Dayanand, S., Boppana, S. B., Auradi, V., Nagaral, M., Udaya Ravi, M., & Bharath. (2021). Evaluation of Wear Properties of Heat-Treated Al-AlB2 in-situ Metal Matrix Composites. *Journal of Bio- and Tribo-Corrosion*, *7*(2), 40. doi:10.100740735-021-00476-w

Devendiran, D. K., & Amirtham, V. A. (2016). A review on preparation, characterization, properties and applications of nanofluids. *Renewable & Sustainable Energy Reviews*, *60*, 21–40. doi:10.1016/j.rser.2016.01.055

Devi, S. P. A., & Devi, S. S. U. (2016). Numerical investigation of hydromagnetic hybrid Cu-Al2O3/water nanofluid flow over a permeable stretching sheet with suction. *International Journal of Nonlinear Sciences and Numerical Simulation*, *17*(5), 249–257. doi:10.1515/ijnsns-2016-0037

Dhapte-Pawar, V., Kadam, S., Saptarsi, S., & Kenjale, P. P. (2020). Nanocosmeceuticals: facets and aspects. Fut Sci OA 6: FSO613.

Dhapte-Pawar, V.; Kadam, S.; Saptarsi, S.; Kenjale, P.P.(2020), Nanocosmeceuticals: Facets and aspects. *Futur. Sci. OA*.

Dingler, A., & Gohla, S. (2002). Production of solid lipid nanoparticles (SLN): Scaling up feasibilities. *Journal of Microencapsulation*, *19*(1), 11–16. doi:10.1080/02652040010018056 PMID:11811752

Doak, S. (2016). Opinion of the Scientific Committee on consumer safety (SCCS)-Opinion on the use of 2, 2′-methylene-bis-(6-(2H-benzotriazol-2-yl)-4-(1, 1, 3, 3-tetramethylbutyl) phenol) (nano)–S79–In cosmetic products. *Regulatory Toxicology and Pharmacology*, *76*, 215–216. doi:10.1016/j.yrtph.2016.01.002 PMID:26776753

Domakonda, V. K., Farooq, S., Chinthamreddy, S., Puviarasi, R., Sudhakar, M., & Boopathi, S. (2023). Sustainable Developments of Hybrid Floating Solar Power Plants. In *Human Agro-Energy Optimization for Business and Industry* (pp. 148–167). IGI Global. doi:10.4018/978-1-6684-4118-3.ch008

Dong, C., Cairney, J., Sun, Q., Maddan, O. L., He, G., & Deng, Y. (2010). Investigation of Mg(OH)2 nanoparticles as an antibacterial agent. *Journal of Nanoparticle Research*, *12*(6), 2101–2109. doi:10.100711051-009-9769-9

Dorri Moghadam, A., Omrani, E., Menezes, P. L., & Rohatgi, P. K. (2015). Mechanical and tribological properties of self-lubricating metal matrix nanocomposites reinforced by carbon nanotubes (CNTs) and graphene – A review. *Composites. Part B, Engineering*, *77*, 402–420. doi:10.1016/j.compositesb.2015.03.014

Duan, X. P., & Li, Y. P. (2013). Physicochemicalcharacteristicsofnanoparticlesaffectcirculation,biodistribution,cellular internalization, and trafficking. *Small*, *9*(9-10), 1521–1532. doi:10.1002mll.201201390 PMID:23019091

Duarah, S., Pujari, K., Durai, R. D., & Narayanan, V. H. B. (2016). Nanotechnology-based cosmeceuticals: A review. *Int J Appl Pharm*, *8*(1), 8–12.

Dureja, H., Kaushik, D., Gupta, M., Kumar, V., & Lather, V. (2005). Cosmeceuticals: An emerging concept. *Indian Journal of Pharmacology*, *37*(3), 155–159. doi:10.4103/0253-7613.16211

Dushyanth, V., Babu, R., Mukesh, P., Menka, P., Harikaran, M., Balaji, G., & Monisha, K.M. (2022). Experimental Study of Concrete By Replacement of Cement With Polymer Resin. *IOP Conf. Series: Earth and Environmental Science*. IOP. . doi:10.1088/1755-1315/1125/1/012014

Dwivedi, R. (2013). Silver nanoparticles ecofriendly green synthesis by using two medicinal plant extract. *International Journal of Bio-Technology and Research*, *3*(4), 61–68.

Ebaid, A., & Aly, E. H. (2013). Exact analytical solution of the peristaltic nanofluids flow in an asymmetric channel with flexible walls and slip condition: Application to the cancer treatment. *Computational and Mathematical Methods in Medicine*, *2013*, 1–8. doi:10.1155/2013/825376 PMID:24151526

Ehsan, E.-B., Charjouei, M. M., Hamid, N., Weerapun, D., & Somchai, W. (2016). Experimental and numerical investigation of nanofluids heat transfer characteristics for application in solar heat exchangers. *International Journal of Heat and Mass Transfer*, *92*, 1041–1052. doi:10.1016/j.ijheatmasstransfer.2015.08.107

Elavazhagan, T., & Arunachalam, K. D. (2011). Memecylon edule leaf extract mediated green synthesis of silver and gold nanoparticles. *International Journal of Nanomedicine*, *6*, 1265–1278. doi:10.2147/IJN.S18347 PMID:21753878

El-Ramady, H., Alshaal, T., Abowaly, M., Abdalla, N., Taha, H. S., Al-Saeedi, A. H., Shalaby, T., Amer, M., Fári, M., & Domokos-Szabolcsy, É. (2017). Nanoremediation for Sustainable Crop Production. In S. Ranjan, N. Dasgupta, & E. Lichtfouse (Eds.), *Nanoscience in Food and agriculture* (1st ed., Vol. 1, pp. 335–363). Springer. doi:10.1007/978-3-319-58496-6_12

Emami, S., Dibaei Bonab, M. H., Mohammadiun, M., Mohammadiun, H., & Sadi, M. (2020). Effect of nano-fluid types on the improvement of heat transfer in a micro-channel with regular hexagonal pattern. *Proceedings of the Institution of Mechanical Engineers. Part E, Journal of Process Mechanical Engineering*, *234*(6), 657–664. doi:10.1177/0954408920932349

Enami, M., Farahani, M., & Farhang, M. (2019). Novel study on keyhole less friction stir spot welding of Al 2024 reinforced with alumina nanopowder. *International Journal of Advanced Manufacturing Technology*, *101*(9), 3093–3106. doi:10.100700170-018-3142-z

Esfe, M. H., Bahiraei, M., & Mir, A. (2020). Application of conventional and hybrid nanofluids in different machining processes: A critical review. *Advances in Colloid and Interface Science*, *282*, 102199. doi:10.1016/j.cis.2020.102199 PMID:32679398

Faivre, D., & Bennet, M. (2016). Materials science: Magnetic nanoparticles line up. *Nature*, *535*(7611), 235–236. doi:10.1038/535235a PMID:27411627

Fang, K., Song, L., Gu, Z., Yang, F., Zhang, Y., & Gu, N. (2015). Magnetic field activated drug release system based on magnetic PLGA microspheres for chemo-thermal therapy. *Colloids and Surfaces. B, Biointerfaces*, *136*, 712–720. doi:10.1016/j.colsurfb.2015.10.014 PMID:26513754

Faraji, G., & Asadi, P. (2011). Characterization of AZ91/alumina nanocomposite produced by FSP. *Materials Science and Engineering A*, *528*(6), 2431–2440. doi:10.1016/j.msea.2010.11.065

Fatima, M., Monawwar, S., Mohapatra, S., Alex, T. S., Ahmed, A., Taleuzzaman, M., Ali, A., Ansari, M. J., Mirza, M. A., & Iqbal, Z. (2021). In silico drug screening based development of novel formulations for onychomycosis management. *Gels (Basel, Switzerland)*, *2021*(7), 221. doi:10.3390/gels7040221 PMID:34842710

Feltin, C., & Brun, G. (2015). Cosmetic Composition Based on Supramolecular Polymer and An Absorbent Filler. U.S. Patent US9000051B2.

Filipe, P., Silva, J. N., Silva, R., De Castro, J. C., Gomes, M. M., Alves, L. C., & Pinheiro, T. (2009). Stratum corneum is an effective barrier to TiO2 and ZnO nanoparticle percutaneous absorption. *Skin Pharmacology and Physiology*, *22*(5), 266–275. doi:10.1159/000235554 PMID:19690452

Florence, A. T. (2012). Targeting Nanoparticles: The constraints of physical laws and physical barriers. *Journal of Controlled Release*, *164*(2), 115–124. doi:10.1016/j.jconrel.2012.03.022 PMID:22484196

Fox, C. (1998). *Cosmetic and Pharmaceutical Vehicles: Skin Care, Hair Care, Makeup and Sunscreens* (Vol. 113). Cosmetics & Toiletries.

Frazier, N., & Ghandehari, H. (2015). Hyperthermia approaches for enhanced delivery of nanomedicines to solid tumors. *Biotechnology and Bioengineering*, *112*(10), 1967–1983. doi:10.1002/bit.25653 PMID:25995079

Fruchon, S., & Poupot, R. (2017). Pro-inflammatory versus anti-inflammatory effects of dendrimers: The two faces of immunomodulatory nanoparticles. *Nanomaterials (Basel, Switzerland)*, *7*(9), 251. doi:10.3390/nano7090251 PMID:28862693

Fuskele, V., & Sarviya, R. M. (2017). Recent developments in Nanoparticles Synthesis, Preparation and Stability of Nanofluids. *Materials Today: Proceedings*, *4*(2), 4049–4060. doi:10.1016/j.matpr.2017.02.307

Fytianos, G., Rahdar, A., & Kyzas, G. Z. (2020). Nanomaterials in cosmetics: Recent updates. *Nanomaterials (Basel, Switzerland)*, *10*(5), 979. doi:10.3390/nano10050979 PMID:32443655

Gajbhiye, S., & Sakharwade, S. (2016). Silver nanoparticles in cosmetics. *Journal of Cosmetics. Dermatological Sciences and Applications*, *6*(1), 48–53.

Gardašević, G., Veletić, M., Maletić, N., Vasiljević, D., Radusinović, I., Tomović, S., & Radonjić, M. (2017). The IoT Architectural Framework, Design Issues and Application Domains. *Wireless Personal Communications*, *92*(1), 127–148. doi:10.100711277-016-3842-3

Gautam, A., Singh, D., & Vijayaraghavan, R. (2011). Dermal exposure of nanoparticles: An understanding. *Journal of Cell and Tissue Research*, *11*(1), 2703–2708.

Gavhane, A J., & Padmanabhan, P, Suresh P Kamble and Suresh N Jangle. (2012). Synthesis of silver nanoparticles using extract of Neem leaf and Triphala and evaluation of their antimicrobial activities. *International Journal of Pharma and Bio Sciences*, *3*(3), 88–100.

Geethalakshmi, R., & Sarada, D. V. L. (2012). Gold and Silver nanoparticles from *Trianthema decandra*: Characterization and antimicrobial properties. *International Journal of Nanomedicine*, *7*, 5375–5384. doi:10.2147/IJN.S36516 PMID:23091381

Ge, X., Wei, M., He, S., & Yuan, W. E. (2019). Advances of non-ionic surfactant vesicles (niosomes) and their application in drug delivery. *Pharmaceutics, 11*(2), 55. doi:10.3390/pharmaceutics11020055 PMID:30700021

Ghatnur, S. M., Sonale, R. S., Balaraman, M., & Kadimi, U. S. (2012). Engineering liposomes of leaf extract of seabuckthorn (SBT) by supercritical carbon dioxide (SCCO2)-mediated process. *Journal of Liposome Research, 22*(3), 215–223. doi:10.3109/08982104.2012.658576 PMID:22397357

Glaser, D. A. (2004). Anti-aging products and cosmeceuticals. *Facial Plastic Surgery Clinics of North America, 12*(3), 363–372. doi:10.1016/j.fsc.2004.03.004 PMID:15261173

Gnana, D., Gnana, J., Gurusamy, A., & Chellapandian, K. (2012). Green synthesis of silver nanoparticle using *Elettaria Cardamomom* and assessment of its antimicrobial activity. *International Journal of Pharmaceutical Sciences and Research, 3*, 323–330.

Gokulakannan, S., Harikaran, M., Mounesh, K., Sankaranarayanan, S., Sathishkumar, R., & Sivakumar, R. (2023). Experimental studies on durability and deflection behavior of concrete in metakaolin cement concrete. *Materials Today: Proceedings*. doi:10.1016/j.matpr.2023.05.236

Gowramma, B., Keerthi, U., Rafi, M., & Muralidhara Rao, D. (2015). Biogenic silver nanoparticles production and characterization from native stain of *Corynebacterium* species and its antimicrobial activity. *Biotech., 5*, 195–201. PMID:28324578

Gowri, N. V., Dwivedi, J. N., Krishnaveni, K., Boopathi, S., Palaniappan, M., & Medikondu, N. R. (2023). Experimental investigation and multi-objective optimization of eco-friendly near-dry electrical discharge machining of shape memory alloy using Cu/SiC/Gr composite electrode. *Environmental Science and Pollution Research International*, 1–19. doi:10.100711356-023-26983-6 PMID:37126160

Goyal, M., & Bhargava, R. (2014). Numerical study of thermodiffusion effects on boundary layer flow. *Microfluidics and Nanofluidics, 17*(3), 591–604. doi:10.100710404-013-1326-2

Grong, Ø., Sandnes, L., & Berto, F. (2019). Progress in Solid State Joining of Metals and Alloys. *Procedia Structural Integrity, 17*, 788–798. doi:10.1016/j.prostr.2019.08.105

Gunasekaran, K., Suthanthira Vanitha, N., Radhika, K., & Suresh, P. (2022). Nanomedicine in healthcare: Impact and challenges for future generation. Research Gate.

Gunasekaran, K., Suthanthira Vanitha, N., Radhika, K., & Suresh, P. (2022). Nanomedicine in healthcare: Impact and challenges for future generation. Taylor & Francis. doi:10.1201/9781003239888-12

Gunasekaran, K., Boopathi, S., & Sureshkumar, M. (2022). Analysis of a Cryogenically Cooled Near-Dry Wedm Process Using Different Dielectrics. *Materiali in Tehnologije, 56*(2), 179–186. doi:10.17222/mit.2022.397

Guo, J., Gougeon, P., Nadeau, F., & Chen, X.-G. (2012). Joining of AA1100–16 vol.-%B 4 C metal matrix composite using laser welding and friction stir welding. *Canadian Metallurgical Quarterly, 51*(3), 277–283. doi:10.1179/1879139512Y.0000000003

Gupta, M. K., Khan, A. M., Song, Q., Liu, Z., Khalid, Q. S., Jamil, M., Kuntoğlu, M., Usca, Ü. A., Sarıkaya, M., & Pimenov, D. Y. (2021). A review on conventional and advanced minimum quantity lubrication approaches on performance measures of grinding process. *International Journal of Advanced Manufacturing Technology*, *117*(3-4), 729–750. doi:10.100700170-021-07785-x

Gupta, V., Mohapatra, S., Mishra, H., Farooq, U., Kumar, K., Ansari, M. J., Aldawsari, M., Alalaiwe, A., Mirza, M., & Iqbal, Z. (2022). Nanotechnology in cosmetics and cosmeceuticals—A review of latest advancements. *Gels (Basel, Switzerland)*, *8*(3), 173. doi:10.3390/gels8030173 PMID:35323286

Hachani, R., Lowdell, M., Birchall, M., & Thanh, N. T. (2013). Tracking stem cells in tissue-engineered organs using magnetic nanoparticles. *Nanoscale*, *5*(23), 11362–11372. doi:10.1039/c3nr03861k PMID:24108444

Hamad, E. M., Khaffaf, A., Yasin, O., Abu El-Rub, Z., Al-Gharabli, S., Al-Kouz, W., & Chamkha, A. J. (2021). Review of Nanofluids and Their Biomedical Applications. *Journal of Nanofluids*, *10*(4), 463–477. doi:10.1166/jon.2021.1806

Harikaran, M. & Balasundaram, N. (2019). Evaluation of Concrete Using Reshaped Waste Tyre Rubber as Partial Replacement of Coarse Aggregate. *International Journal of Recent Technology and Engineering (IJRTE)*, *7*(6s5), 2277-3878.

Harikaran, M., Gokulakannan, S., Loganathan, A., Chandiran, R. D., Ajith, M., & Dhanasekar, V. (2023). Metakaolin cement concrete evaluation using industrial by-products as fine aggregate. *Materials Today: Proceedings*. doi:10.1016/j.matpr.2023.04.586

Harikaran, M., Keerthanadevi, P., Inbarasan, S., Tamilan, N., Manimaran, R., & Jayavikram, S. (2023). Experimental examination of concrete using hazardous plasti-fiber and rub – fiber. *Materials Today: Proceedings*. doi:10.1016/j.matpr.2023.04.598

Harikaran, M., Kulanthaivel, P., Krishnaraja, A. R., & Murugan, P. C. (2023). Performance enhancement of concrete by using ceramic waste as a partial replacement for coarse aggregate. *Journal of Ceramic Processing Research*, *24*(2), 266–273.

Harikaran, M., Mukesh, P., Suganth, R. S., Nithishkumar, R., & Vijay, V. (2022). Study of structural bonding strength with tyre rubber instead of coarse aggregate and water hyacinth ash instead of cement. *Materials Today: Proceedings*, *68*, 1330–1334. doi:10.1016/j.matpr.2022.06.359

Harikaran, M., & Sampath Boopathi, S. (2023). *Gokulakannan & M. Poonguzhali*. Study on the Source of E-Waste Management and Disposal Methods. Sustainable Approaches and Strategies for E-Waste Management and Utilization. doi:10.4018/978-1-6684-7573-7.ch003

Harilal, M., Vidyadharan, B., Misnon, I. I., Anilkumar, G. M., Lowe, A., Ismail, J., Yusoff, M. M., & Jose, R. (2017). One-Dimensional Assembly of Conductive and Capacitive Metal Oxide Electrodes for High-Performance Asymmetric Supercapacitors. *ACS Applied Materials & Interfaces*, *9*(12), 10730–10742. doi:10.1021/acsami.7b00676 PMID:28266837

Harish, S., Ishikawa, K., Einarsson, E., Aikawa, S., Chiashi, S., Shiomi, J., & Maruyama, S. (2012). Enhanced thermal conductivity of ethylene glycol with single-walled carbon nanotube inclusions. *International Journal of Heat and Mass Transfer*, *55*(13-14), 3885–3890. doi:10.1016/j.ijheatmasstransfer.2012.03.001

Hasan, B., & Nihal, S. (2014). Comparative Study of the Characteristics of Nano Silica, Silica Fumeand Fly Ash Incorporated Cement Mortars. *IJESE*, *17*(3), 570–582. doi:10.1590/S1516-14392014005000054

Hasan, S. (2015). A review on nanoparticles: Their synthesis and types. *Research Journal of Recent Sciences*, *2277*, 2502.

Hashemabadi, D. (2014). The role of silver nano-particles and silver thiosulfate on the longevity of cut carnation (*Dianthus caryophyllus*) flowers. *Journal of Environmental Biology*, *35*, 661–666. PMID:25004750

Hegab, H., Darras, B., & Kishawy, H. A. (2018). Sustainability Assessment of Machining with Nano-Cutting Fluids. *Procedia Manufacturing*, *26*, 245–254. doi:10.1016/j.promfg.2018.07.033

He, X., Park, E. Y. H., Fowler, A., Yarmush, M. L., & Toner, M. (2008). Vitrification by ultra-fast cooling at a low concentration of cryoprotectants in a quartz micro-capillary: A study using murine embryonic stem cells. *Cryobiology*, *56*(3), 223–232. doi:10.1016/j.cryobiol.2008.03.005 PMID:18462712

Hilton, L. (n.d.). Cosmetic Dermatology, Aesthetic Technology. *Dermatology Times*. http://www.dermatologytimes.com

Hind, A. R., Bhargava, S. K., & McKinnon, A. (2001). At the solid/liquid interface: FTIR/ATR—The tool of choice. *Advances in Colloid and Interface Science*, *93*(1-3), 91–114. doi:10.1016/S0001-8686(00)00079-8 PMID:11591110

Hong, W. X., Che Sidik, N., & Saidur, R. (2020). Impact of different surfactants and ultrasonication time on the stability and thermophysical properties of hybrid nanofluids. *ICHMT, 110*, 104389.

Hong, S.-T., Das, H., Oh, H.-S., Al Nasim, M. N. E. A., & Chun, D.-M. (2017). Combination of nano-particle deposition system and friction stir spot welding for fabrication of carbon/aluminum metal matrix composite joints of dissimilar aluminum alloys. *CIRP Annals*, *66*(1), 261–264. doi:10.1016/j.cirp.2017.04.115

Hope, M. J., & Kitson, C. N. (1993). Liposomes: A perspective for dermatologists. *Dermatologic Clinics*, *11*(1), 143–154. doi:10.1016/S0733-8635(18)30291-2 PMID:8435909

Hosseinkhani, B., Callewaert, C., Vanbeveren, N., & Boon, N. (2015). Novel biocompatible nanocapsules for slow release of fragrances on the human skin. *New Biotechnology*, *32*(1), 40–46. doi:10.1016/j.nbt.2014.09.001 PMID:25224920

Hougeir, F. G., & Kircik, L. (2012). A review of delivery systems in cosmetics. *Dermatologic Therapy*, *25*(3), 234–237. doi:10.1111/j.1529-8019.2012.01501.x PMID:22913440

Huang, H., & Yang, X. (2004). synthesis of polysaccharide-stabilized gold and silver nanoparticles: A green method. *Caries Research, 339*(15), 2627–2631. doi:10.1016/j.carres.2004.08.005 PMID:15476726

Huminic, G., & Huminic, A. (2012). Application of nanofluids in heat exchangers: A review. *Renewable & Sustainable Energy Reviews, 16*(8), 5625–5638. doi:10.1016/j.rser.2012.05.023

Hund-Rinke, K., & Simon, M. (2006). Ecotoxic effect of photocatalytic active nanoparticles (TiO2) on algae and daphnids (8 pp). *Environmental Science and Pollution Research International, 13*(4), 225–232. doi:10.1065/espr2006.06.311 PMID:16910119

Huseien, G. F., Khalid, N. H. A., & Mirza, J. (2022). *Nanotechnology for smart concrete*. CRC Press.

Hu, Z., Liao, M., & Chen, Y. (2012). A novel preparation method for silicone oil nanoemulsions and its application for coating hair with silicone. *International Journal of Nanomedicine, 7*, 5719–5724. PMID:23166436

Indira, T. K., & Lakshmi, P. K. (2010). Magnetic nanoparticles—A review. [IJPSN]. *International Journal of Pharmaceutical Sciences and Nanotechnology, 3*(3), 1035–1042. doi:10.37285/ijpsn.2010.3.3.1

Irshad, A., Zahid, M., Husnain, T., Rao, A. Q., Sarwar, N., & Hussain, I. (2020). A proactive model on innovative biomedical applications of gold nanoparticles. *Applied Nanoscience, 10*(8), 2453–2465. doi:10.100713204-019-01165-4

Is It Safe to Use Cosmetics Containing Silica in Nanoform? (n.d.). Europea. https://ec.europa.eu/health/sites/health/files/scientific_committees/docs/citizens_nanosilica_en.pdf

Ishfaq, K., Maqsood, M. A., Anwar, S., Harris, M., Alfaify, A., & Zia, A. W. (2022). EDM of Ti6Al4V under nano-graphene mixed dielectric: A detailed roughness analysis. *International Journal of Advanced Manufacturing Technology, 120*(11–12), 7375–7388. doi:10.100700170-022-09207-y

Jafari, H., Mansouri, H., & Honarpisheh, M. (2019). Investigation of residual stress distribution of dissimilar Al-7075-T6 and Al-6061-T6 in the friction stir welding process strengthened with SiO2 nanoparticles. *Journal of Manufacturing Processes, 43*, 145–153. doi:10.1016/j.jmapro.2019.05.023

Jailani, S., Franks, G. V., & Healy, T. W. (2008). The potential of nanoparticle suspensions: Effect of electrolyte concentration, particle size and volume fraction. *Journal of the American Ceramic Society, 91*(4), 1141–1147. doi:10.1111/j.1551-2916.2008.02277.x

Janardhana, K., Singh, V., Singh, S. N., Babu, T. S. R., Bano, S., & Boopathi, S. (2023). Utilization Process for Electronic Waste in Eco-Friendly Concrete: Experimental Study. In Sustainable Approaches and Strategies for E-Waste Management and Utilization (pp. 204–223). IGI Global.

Janardhana, K., Anushkannan, N. K., Dinakaran, K. P., Puse, R. K., & Boopathi, S. (2023). *Experimental Investigation on Microhardness, Surface Roughness, and White Layer Thickness of Dry EDM*. Engineering Research Express. doi:10.1088/2631-8695/acce8f

Jaya, R. P. (2016). A Review of Microstructure Properties of Porous Concrete Pavement Incorporating Nano Silica. *Journal of Engineering and Applied Sciences (Asian Research Publishing Network)*, *11*(20), 1819–6608.

Jeevanantham, Y. A., A, S., V, V., J, S. I., Boopathi, S., & Kumar, D. P. (2023). Implementation of Internet-of Things (IoT) in Soil Irrigation System. *IEEE Explore*, 1–5. doi:10.1109/ICPECTS56089.2022.10047185

Jones, D. E., Ghandehari, H., & Facelli, J. C. (2016). A review of the applications of data mining and machine learning for the prediction of biomedical properties of nanoparticles. *Computer Methods and Programs in Biomedicine*, *132*, 93–103. doi:10.1016/j.cmpb.2016.04.025 PMID:27282231

Joo, H. H., Lee, H. Y., & Kim, J. C. (2009). Stability, release property and skin penetration of stearic acid nanoparticles. *Molecular Crystals and Liquid Crystals (Philadelphia, Pa.)*, *508*(1), 137–149. doi:10.1080/15421400903060466

Joseph, J., & BN, V. H. (2018). Experimental optimization of Lornoxicam liposomes for sustained topical delivery. *European Journal of Pharmaceutical Sciences*, *112*, 38–51. doi:10.1016/j.ejps.2017.10.032 PMID:29111151

Kah, P., Rajan, R., Martikainen, J., & Suoranta, R. (2015). Investigation of weld defects in friction-stir welding and fusion welding of aluminium alloys. *International Journal of Mechanical and Materials Engineering*, *10*(1), 26. doi:10.118640712-015-0053-8

Kamyar, A., Saidur, R., & Hasanuzzaman, M. (2012). Application of computational fluid dynamics (CFD) for nanofluids. *International Journal of Heat and Mass Transfer*, *55*(15-16), 4104–4115. doi:10.1016/j.ijheatmasstransfer.2012.03.052

Kannan, E., Trabelsi, Y., Boopathi, S., & Alagesan, S. (2022). Influences of cryogenically treated work material on near-dry wire-cut electrical discharge machining process. *Surface Topography : Metrology and Properties*, *10*(1), 15027. doi:10.1088/2051-672X/ac53e1

Karakizis, P. N., Pantelis, D. I., Fourlaris, G., & Tsakiridis, P. (2018a). Effect of SiC and TiC nanoparticle reinforcement on the microstructure, microhardness, and tensile performance of AA6082-T6 friction stir welds. *International Journal of Advanced Manufacturing Technology*, *95*(9), 3823–3837. doi:10.100700170-017-1446-z

Karakizis, P. N., Pantelis, D. I., Fourlaris, G., & Tsakiridis, P. (2018b). The role of SiC and TiC nanoparticle reinforcement on AA5083-H111 friction stir welds studied by electron microscopy and mechanical testing. *International Journal of Advanced Manufacturing Technology*, *94*(9), 4159–4176. doi:10.100700170-017-1147-7

Karim, K. M. (2010). A future of targeted drug delivery systems. *Journal of Advanced Pharmaceutical Technology & Research*, *1*(4), 374–380. doi:10.4103/0110-5558.76435 PMID:22247876

Karthikeyan, P., & Mahadevan, K. (2015). Investigation on the effects of SiC particle addition in the weld zone during friction stir welding of Al 6351 alloy. *International Journal of Advanced Manufacturing Technology*, *80*(9–12), 1919–1926. doi:10.100700170-015-7160-9

Kartsonakis, I. A., Dragatogiannis, D. A., Koumoulos, E. P., Karantonis, A., & Charitidis, C. A. (2016). Corrosion behaviour of dissimilar friction stir welded aluminium alloys reinforced with nanoadditives. *Materials & Design*, *102*, 56–67. doi:10.1016/j.matdes.2016.04.027

Katna, R., Agrawal, N., & Suhaib, M. (2020). A Short Review on Machining with Ultrasonic MQL Method. *IOP Conference Series. Materials Science and Engineering*, *802*(1), 12011. doi:10.1088/1757-899X/802/1/012011

Katz, L. M., Dewan, K., & Bronaugh, R. L. (2015). Nanotechnology in cosmetics. *Food and Chemical Toxicology*, *85*, 127–137. doi:10.1016/j.fct.2015.06.020 PMID:26159063

Kaul, S., Gulati, N., Verma, D., Mukherjee, S., & Nagaich, U. (2018). Role of nanotechnology in cosmeceuticals: A review of recent advances. *Journal of Pharmaceutics*, *2018*(34202043), 1–19. doi:10.1155/2018/3420204 PMID:29785318

Keck, C. M., & Müller, R. H. (2006). Drug nanocrystals of poorly soluble drugs produced by high pressure homogenisation. *European Journal of Pharmaceutics and Biopharmaceutics*, *62*(1), 3–16. doi:10.1016/j.ejpb.2005.05.009 PMID:16129588

Khanafer, K., Eltaggaz, A., Deiab, I., Agarwal, H., & Abdul-Latif, A. (2020). Toward sustainable micro-drilling of Inconel 718 superalloy using MQL-Nanofluid. *International Journal of Advanced Manufacturing Technology*, *107*(7-8), 3459–3469. doi:10.100700170-020-05112-4

Khanafer, K., & Vafai, K. (2018). A review on the applications of nanofluids in solar energy field. *Renewable Energy*, *123*, 398–406. doi:10.1016/j.renene.2018.01.097

Khanam, P. N., & Hasan, A. (2019). Biosynthesis and characterization of graphene by using non-toxic reducing agent from Allium Cepa extract: Anti-bacterial properties. *International Journal of Biological Macromolecules*, *126*, 151–158. doi:10.1016/j.ijbiomac.2018.12.213 PMID:30584937

Khandekar, S., Sankar, M. R., Agnihotri, V., & Ramkumar, J. (2012). Nano-cutting fluid for enhancement of metal cutting performance. *Materials and Manufacturing Processes*, *27*(9), 963–967. doi:10.1080/10426914.2011.610078

Khan, F., Soota, T., & ... (2020). Optimization of EDM process parameter for stainless steel D3. *Materials Today: Proceedings*, *25*, 635–638. doi:10.1016/j.matpr.2019.07.529

Khan, R., & Irchhaiya, R. (2016). Niosomes: A Potential Tool for Novel Drug Delivery. *Journal of Pharmaceutical Investigation*, *46*(3), 195–204. doi:10.100740005-016-0249-9

Khdher, A. M., Sidik, N. A. C., Hamzah, W. A. W., & Mamat, R. (2016). An experimental determination of thermal conductivity and electrical conductivity of bio glycol based Al2O3 nanofluids and development of new correlation. *International Communications in Heat and Mass Transfer*, *73*, 75–83. doi:10.1016/j.icheatmasstransfer.2016.02.006

Kim, T. H., Lee, S., & Chen, X. (2013). Nanotheranostics for personalized medicine. *Expert Review of Molecular Diagnostics*, *13*(3), 257–269. doi:10.1586/erm.13.15 PMID:23570404

Kittelson, D. B. (1998). Engines and nanoparticles: A review. *Journal of Aerosol Science*, *29*(5–6), 575–588. doi:10.1016/S0021-8502(97)10037-4

Klajnert, B., & Bryszewska, M. (2001). Dendrimers: Properties and Applications. *Acta Biochimica Polonica*, *48*(1), 199–208. doi:10.18388/abp.2001_5127 PMID:11440170

Kocic, H., Stankovic, M., Tirant, M., Lotti, T., & Arsic, I. (2020). Favorable effect of creams with skimmed donkey milk encapsulated in nanoliposomes on skin physiology. *Dermatology and Therapy*, *2020*(33), e13511. PMID:32372458

Kokura, S., Handa, O., Takagi, T., Ishikawa, T., Naito, Y., & Yoshikawa, T. (2010). Silver nanoparticles as a safe preservative for use in cosmetics. *Nanomedicine; Nanotechnology, Biology, and Medicine*, *6*(4), 570–574. doi:10.1016/j.nano.2009.12.002 PMID:20060498

Köpke, D., & Pyo, S. M. (2020). Symurban nanocrystals for advanced anti-pollution skincare. *Cosmetics*, *2020*(7), 17. doi:10.3390/cosmetics7010017

Kos, L., Sójka-Ledakowicz, J., Michalska, K., & Perkowski, J. (2014). .Decomposition of azodye C.I. Direct Yellow 86by the fenton process in the presence of nanoparticles of iron oxides. *Fibres Text. East. Eur*, *5*, 114–120.

Kothandapani, M., & Prakash, J. (2015). The peristaltic transport of Carreau nanofluids under effect of a magnetic field in a tapered asymmetric channel: Application of the cancer therapy. *Journal of Mechanics in Medicine and Biology*, *15*(03), 1550030. doi:10.1142/S021951941550030X

Koul, B., & Taak, P. (2018). Chemical methods for soil remediation. In B. Koul & P. Taak (Eds.), *Biotechnological Strategies for Effective Remediation of Polluted Soils* (1st ed., Vol. 1, pp. 77–84). Springer. doi:10.1007/978-981-13-2420-8_4

Kumara, V., Mohanaprakash, T. A., Fairooz, S., Jamal, K., Babu, T., & B., S. (2023). Experimental Study on a Reliable Smart Hydroponics System. In *Human Agro-Energy Optimization for Business and Industry* (pp. 27–45). IGI Global. doi:10.4018/978-1-6684-4118-3.ch002

Kumar, A. D. (2012). Rapid and green synthesis of Silver Nanoparticles using the leaf extracts of *Parthenium Hysterophorus*: A Novel Biological Approach. *Int. Res. J.Pharm.*, *3*(2), 169–173.

Kuppusamy, S., Thavamani, P., Venkateswarlu, K., Lee, Y. B., Naidu, R., & Megharaj, M. (2017). Remediationapproaches forpolycyclicaromatichydrocarbons(PAHs)contaminatedsoils: Technologicalconstraints,emergingtrends and future directions. *Chemosphere*, *168*, 944–968. doi:10.1016/j.chemosphere.2016.10.115 PMID:27823779

Labhasetwar, V., & Leslie-Pelecky, D. L. (Eds.). (2007). *Biomedical applications of nanotechnology*. doi:10.1002/9780470152928

Lai, F. C., & Kulacki, F. A. (1990). The effect of variable viscosity on convective heat transfer along a vertical surface in a saturated porous medium. *International Journal of Heat and Mass Transfer*, *33*(5), 1028–1031. doi:10.1016/0017-9310(90)90084-8

Latief, F. H., Sherif, E.-S. M., Almajid, A. A., & Junaedi, H. (2011). Fabrication of exfoliated graphite nanoplatelets-reinforced aluminum composites and evaluating their mechanical properties and corrosion behavior. *Journal of Analytical and Applied Pyrolysis*, *92*(2), 485–492. doi:10.1016/j.jaap.2011.09.003

Lee, P. H., Nam, T. S., Li, C., & Lee, S. W. (2010). Environmentally-friendly nano-fluid minimum quantity lubrication (MQL) meso-scale grinding process using nano-diamond particles. *Proceedings - 2010 International Conference on Manufacturing Automation, ICMA 2010*, (pp. 44–49). IEEE. 10.1109/ICMA.2010.27

Lee, R., Shenoy, D., & Sheel, R. (2010). Micellar Nanoparticles: Applications for Topical and Passive Transdermal Drug Delivery. In Handbook of Non-Invasive Drug Delivery Systems. William Andrew.

Lee, J., & Mudawar, I. (2007). Assessment of the effectiveness of nanofluids for single-phase and two-phase heat transfer in micro-channels. *International Journal of Heat and Mass Transfer*, *50*(3-4), 452–463. doi:10.1016/j.ijheatmasstransfer.2006.08.001

Lee, R., Shenoy, D., & Sheel, R. (2010). Micellar Nanoparticles: Applications for Topical and Passive Transdermal Drug Delivery. In *Handbook of Non-Invasive Drug Delivery Systems*. William Andrew. doi:10.1016/B978-0-8155-2025-2.10002-2

Lee, Y. J., & Nam, G. W. (2020). Sunscreen boosting effect by solid lipid nanoparticles-loaded fucoxanthin formulation. *Cosmetics*, *2020*(7), 14. doi:10.3390/cosmetics7010014

Li, L. (2013, July 31). *Superparamagnetic Iron Oxide Nanoparticles as MRI contrast agents for Non-invasive Stem Cell Labeling and Tracking*. THNO. https://www.thno.org/v03p0595.html,

Li, L. (2015, July 24). *TEM of Gold Nanoparticles results - FoNS-MAD! Competition 2015*. The Velox Group's Blog. https://blogs.imperial.ac.uk/fonsmad2015-velox/2015/07/18/tem-of-gold-nanoparticles-results/

Lim, E.-K., Jang, E., Lee, K., Haam, S., & Huh, Y.-M. (2013). Delivery of cancer therapeutics using nanotechnology. *Pharmaceutics*, *5*(2), 294–317. doi:10.3390/pharmaceutics5020294 PMID:24300452

Lim, S. M., Shin, B. S., & Kim, K. (2017). Characterization of products using additive manufacturing with graphene/photopolymer - Resin nano-fluid. *Journal of Nanoscience and Nanotechnology*, *17*(8), 5492–5495. doi:10.1166/jnn.2017.14159

Li, N., Sioutas, C., Cho, A., Schmitz, D., Misra, C., Sempf, J., Wang, M., Oberley, T., Froines, J., & Nel, A. (2003). Ultrafine particulate pollutants induce oxidative stress and mitochondrial damage. *Environmental Health Perspectives*, *111*(4), 455–460. doi:10.1289/ehp.6000 PMID:12676598

Lin, P. C., Lin, S., Wang, P. C., & Sridhar, R. (2014). Techniques for physicochemical characterization of nanomaterials. *Biotechnology Advances*, *32*(4), 711–726. doi:10.1016/j.biotechadv.2013.11.006 PMID:24252561

Liu, J., Morales-Narváez, E., Vicent, T., Merkoçi, A., & Zhong, G. H. (2018). Microorganism-decoratednanocellulosefor effcient diuron removal. *Chemical Engineering Journal*, *354*, 1083–1091. doi:10.1016/j.cej.2018.08.035

Liu, Z. Y., Xiao, B. L., Wang, W. G., & Ma, Z. Y. (2014). Tensile Strength and Electrical Conductivity of Carbon Nanotube Reinforced Aluminum Matrix Composites Fabricated by Powder Metallurgy Combined with Friction Stir Processing. *Journal of Materials Science and Technology*, *30*(7), 649–655. doi:10.1016/j.jmst.2014.04.016

Logothethetidis, S. (2012). *Nanoscience and Technology* (1st ed.). Springer.

Lohani, A., Verma, A., Joshi, H., Yadav, N., & Karki, N. (2014). Nanotechnology-based cosmeceuticals. *International Scholarly Research Notices*, *2014*, 1–14. doi:10.1155/2014/843687 PMID:24963412

Luna-Sánchez, J. L., Jiménez-Pérez, J. L., Carbajal-Valdez, R., Lopez-Gamboa, G., Pérez-González, M., & Correa-Pacheco, Z. N. (2019). Green synthesis of silver nanoparticles using Jalapeño Chili extract and thermal lens study of acrylic resin nanocomposites. *Thermochimica Acta*, *678*, 178314. doi:10.1016/j.tca.2019.178314

Lu, P. J., Huang, S. C., Chen, Y. P., Chiueh, L. C., & Shih, D. Y. C. (2015). Analysis of titanium dioxide and zinc oxide nanoparticles in cosmetics. *Yao Wu Shi Pin Fen Xi*, *23*(3), 587–594. PMID:28911719

M, S., S, N., & Davidson, B. S. (2021). Friction stir welding parameters and their influence on mechanical properties of welded AA6061 and AA5052 aluminium plates. *Materials Research Express*, *8*(10), 106525. doi:10.1088/2053-1591/ac2daf

M., A., M., M., & S., B. (2021). Tribological characterization of friction stir welded dissimilar aluminum alloy AA6061–AA5083 reinforced with CeO and LaO nanoparticles. *Industrial Lubrication and Tribology*, *73*(5), 783–788. doi:10.1108/ILT-01-2021-0009

Mabood, F., Shateyi, S., Rashidi, M. M., Momoniat, E., & Freidoonimehr, N. (2016). MHD stagnation point flow heat and mass transfer of nanofluids in porous medium with radiation, viscous dissipation and chemical reaction. *Advanced Powder Technology*, *27*(2), 742–749. doi:10.1016/j.apt.2016.02.033

Madesh, S., Arul, T. G., & Perumal, V. (2022). *Advances in Welding Techniques for Similar and Dissimilar Materials.*, doi:10.4018/978-1-7998-9574-9.ch011

Mahapatra, O., Bhagat, M., Gopalakrishnan, C., & Arunachalam, K. D. (2008). Ultrafine dispersed CuO nanoparticles and their antibacterial activity. *Journal of Experimental Nanoscience, 3*(3), 185–193. doi:10.1080/17458080802395460

Makinde, O., & Aziz, A. (2011). Boundary layer flow of a nanofluid past a stretching sheet with a convective boundary condition. *International Journal of Thermal Sciences, 50*(7), 1326–1332. doi:10.1016/j.ijthermalsci.2011.02.019

Malabadi,, R B., & Raju,, K., Chalannavar, N., Meti, T., Gangadhar, S., Mulgund, K., Nataraja, S., & Vijaya, K. (2012). Synthesis of antimicrobial silver nanoparticles by callus cultures and *in vitro* derived plants of *Catharanthus roseus. Research in Pharmacy, 2*(6), 18–31.

Malakar, P., Thakur, M. S. H., Nahid, S. M., & Islam, M. M. (2022). Data-Driven Machine Learning to Predict Mechanical Properties of Monolayer Transition-Metal Dichalcogenides for Applications in Flexible Electronics. *ACS Applied Nano Materials, 5*(11), 16489–16499. doi:10.1021/acsanm.2c03564

Mansourian-Tabaei, M., Asiaee, A., Hutton-Prager, B., & Nouranian, S. (2022). Thermal barrier coatings for cellulosic substrates: A statistically designed molecular dynamics study of the coating formulation effects on thermal conductivity. *Applied Surface Science, 587*, 152879. doi:10.1016/j.apsusc.2022.152879

Mayyas, A. T., Mayyas, A. R., & Omar, M. (2016). 11 - Sustainable lightweight vehicle design: A case study in eco-material selection for body-in-white. In J. Njuguna (Ed.), *Lightweight Composite Structures in Transport* (pp. 267–302). Woodhead Publishing. doi:10.1016/B978-1-78242-325-6.00011-6

Mebarek-Oudina, F., & Chabani, I. (2022). Review on Nano-Fluids Applications and Heat Transfer Enhancement Techniques in Different Enclosures. *Journal of Nanofluids, 11*(2), 155–168. doi:10.1166/jon.2022.1834

Mebert, A. M., Baglole, C. J., Desimone, M. F., & Maysinger, D. (2017). Nanoengineered silica: Properties, applications and toxicity. *Food and Chemical Toxicology, 109*, 753–770. doi:10.1016/j.fct.2017.05.054 PMID:28578101

Mehdi, H., & Mishra, R. S. (2021). Effect of multi-pass friction stir processing and SiC nanoparticles on microstructure and mechanical properties of AA6082-T6. *Advances in Industrial and Manufacturing Engineering, 3*, 100062. doi:10.1016/j.aime.2021.100062

Mehrali, M., Sadeghinezhad, E., Akhiani, A. R., Latibari, S. T., Talebian, S., Dolatshahi-Pirouz, A., & Mehrali, M. (2016). An ecofriendly graphene-based nanofluid for heat transfer applications. *Journal of Cleaner Production, 137*, 555–566. doi:10.1016/j.jclepro.2016.07.136

Micha,. D., Marcin, L., Grzegorz, D., & Andrzej, G.(2012). Preparation of metal oxide-water nanofluids by two-step method. *Prosimy Cytowa Jako: Inżynieria I Aparatura Chemiczna, 51*, 213-215.

Mirjavadi, S. S., Alipour, M., Emamian, S., Kord, S., Hamouda, A. M. S., Koppad, P. G., & Keshavamurthy, R. (2017). Influence of TiO2 nanoparticles incorporation to friction stir welded 5083 aluminum alloy on the microstructure, mechanical properties and wear resistance. *Journal of Alloys and Compounds*, *712*, 795–803. doi:10.1016/j.jallcom.2017.04.114

Mittal, J., Batra, A., Singh, A., & Sharma, M. M. (2014). Phytofabrication of nanoparticles through plant as nanofactories. *Adv. Nat. Sci: Nanosci. Nanotechnol.*, *5*(4), 043002. doi:10.1088/2043-6262/5/4/043002

Mohammed, Y. H., Holmes, A., Haridass, I. N., Sanchez, W. Y., Studier, H., Grice, J. E., Benson, H. A. E., & Roberts, M. S. (2019). Support for the safe use of zinc oxide nanoparticle sunscreens: Lack of skin penetration or cellular toxicity after repeated application in volunteers. *The Journal of Investigative Dermatology*, *139*(2), 308–315. doi:10.1016/j.jid.2018.08.024 PMID:30448212

Mohanraj, V. J., & Chen, Y. (2006). Nanoparticles-a review. *Tropical Journal of Pharmaceutical Research*, *5*(1), 561–573.

Mohanty, A., Venkateswaran, N., Ranjit, P. S., Tripathi, M. A., & Boopathi, S. (2023). Innovative Strategy for Profitable Automobile Industries: Working Capital Management. In Handbook of Research on Designing Sustainable Supply Chains to Achieve a Circular Economy (pp. 412–428). IGI Global.

Moita, A., Moreira, A., & Pereira, J. (2021). Nanofluids for the next generation thermal management of electronics: A review. *Symmetry*, *13*(8), 1362. doi:10.3390ym13081362

Monsé, C., Hagemeyer, O., Raulf, M., Jettkant, B., van Kampen, V., Kendzia, B., Gering, V., Kappert, G., Weiss, T., Ulrich, N., Marek, E.-M., Bünger, J., Brüning, T., & Merget, R. (2018). Concentration-dependent systemic response after inhalation of nano-sized zinc oxide particles in human volunteers. *Particle and Fibre Toxicology*, *15*(1), 1–11. doi:10.118612989-018-0246-4 PMID:29429408

Montazerian, M. H., Movahedi, M., & Jondi, M. R. (2019). Effect of graphene and process parameters on mechanical performance and electrical resistance of aluminum to copper friction stir joint. *Materials Research Express*, *6*(4), 046561. doi:10.1088/2053-1591/aafe1a

Montgomery, J., Abu-Lebdeh, T., Hamoush, S. A., & Picornell, M. (2016). Effect of nano-silicaon the compressive strength of hardening cement paste at different stages of hydration. *American Journal of Engineering and Applied Sciences*, *9*(1), 166–177. doi:10.3844/ajeassp.2016.166.177

Moradi, M. M., Aval, H. J., & Jamaati, R. (2018). Microstructure and mechanical properties in nano and microscale SiC-included dissimilar friction stir welding of AA6061-AA2024. *Materials Science and Technology*, *34*(4), 388–401. doi:10.1080/02670836.2017.1393976

Moradi, M. M., Jamshidi Aval, H., & Jamaati, R. (2017). Effect of pre and post welding heat treatment in SiC-fortified dissimilar AA6061-AA2024 FSW butt joint. *Journal of Manufacturing Processes*, *30*, 97–105. doi:10.1016/j.jmapro.2017.08.014

Morales, M. E., Gallardo, V., Clares, B., García, M. B., & Ruiz, M. A. (2009). Study and description of hydrogels and organogels as vehicles for cosmetic active ingredients. *Journal of Cosmetic Science*, *60*(6), 627–636. PMID:20038351

Morganti, P., & Morganti, G. (2008). Chitin nanofibrils for advanced cosmeceuticals. *Clinics in Dermatology*, *26*(4), 334–340. doi:10.1016/j.clindermatol.2008.01.003 PMID:18691512

Mu, L., & Sprando, R. L. (2010). Application of nanotechnology in cosmetics. *Pharmaceutical Research*, *27*(8), 1746–1749. doi:10.100711095-010-0139-1 PMID:20407919

Muller, H. R., Shegokar, R., & Keck, M. C. (2011). 20 years of lipid nanoparticles (SLN & NLC): Present state of development & industrial applications. *Current Drug Discovery Technologies*, *8*, 207–227. doi:10.2174/157016311796799062 PMID:21291409

Müller, R. H., Alexiev, U., Sinambela, P., & Keck, C. M. (2016). *Nanostructured lipid carriers (NLC): The second generation of solid lipid nanoparticles Percutaneous Penetration Enhancers chemical Methods in Penetration Enhancement*. Springer., doi:10.1007/978-3-662-47862-2_11

Müller, R. H., Petersen, R. D., Hommoss, A., & Pardeike, J. (2007). Nanostructured lipid carriers (NLC) in cosmetic dermal products. *Advanced Drug Delivery Reviews*, *59*(6), 522–530. doi:10.1016/j.addr.2007.04.012 PMID:17602783

Murari, G., Nahak, B., & Pratap, T. (2023). Hybrid surface modification for improved tribological performance of IC engine components--a review. *Proceedings of the Institution of Mechanical Engineers, Part E: Journal of Process Mechanical Engineering*, 09544089221150718.

Murdock, R. C., Braydich-Stolle, L., Schrand, A. M., Schlager, J. J., & Hussain, S. M. (2008). Characterization of nanomaterial dispersion in solution prior to in vitro exposure using dynamic light scattering technique. *Toxicological Sciences*, *101*(2), 239–253. doi:10.1093/toxsci/kfm240 PMID:17872897

Musazzi, U. M., Franzè, S., Minghetti, P., & Casiraghi, A. (2018). Emulsion versus nanoemulsion: How much is the formulative shift critical for a cosmetic product? *Drug Delivery and Translational Research*, *8*(2), 414–421. doi:10.100713346-017-0390-7 PMID:28508377

Myilsamy, S., & Boopathi, S. (2017). Grey Relational Optimization of Powder Mixed Near-Dry Wire Cut Electrical Discharge Machining of Inconel 718 Alloy. *Asian Journal of Research in Social Sciences and Humanities*, *7*(3), 18. doi:10.5958/2249-7315.2017.00157.5

Myilsamy, S., Boopathi, S., & Yuvaraj, D. (2021). A study on cryogenically treated molybdenum wire electrode. *Materials Today: Proceedings*, *45*(9), 8130–8135. doi:10.1016/j.matpr.2021.02.049

Myilsamy, S., & Sampath, B. (2021). Experimental comparison of near-dry and cryogenically cooled near-dry machining in wire-cut electrical discharge machining processes. *Surface Topography : Metrology and Properties*, *9*(3), 35015. doi:10.1088/2051-672X/ac15e0

Myung, S., Solanki, A., Kim, C., Park, J., Kim, K. S., & Lee, K.-B. (2011). Graphene-Encapsulated Nanoparticle-Based Biosensor for the Selective Detection of Cancer Biomarkers. *Advanced Materials*, *23*(19), 2221–2225. doi:10.1002/adma.201100014 PMID:21469221

Nafisi, S., & Maibach, H. I. (2017). Nanotechnology in cosmetics. *Cosmetic science and technology: theoretical principles and applications, 337.*

Nargundkar, A., Gulia, V., & Khan, A. (2023). Nano-Abrasives Assisted Abrasive Water Jet Machining of Bio-Composites—An Experimental and Optimization Approach. *Journal of Advanced Manufacturing Systems*, 1–24. doi:10.1142/S0219686723500373

Nasir, A. (2010). Nanotechnology and dermatology: Part II-risks of nanotechnology. *Clinics in Dermatology*, *28*(5), 581–588. doi:10.1016/j.clindermatol.2009.06.006 PMID:20797523

Neto, E. T., & Filho, E. P. B. (2010). Preparation methods of nano-fluids to obtain stable dispersions. *In: 13th Brazilian Congress of Thermal Sciences and Engineering*. IEEE.

Nicole, F. (2019) A Brief History of Makeup. *Her Campus*. https://www.hercampus.com

Nigam, P. K. (2009). Adverse reactions to cosmetics and methods of testing. *Indian Journal of Dermatology, Venereology and Leprology*, *75*(1), 10–19. doi:10.4103/0378-6323.45214 PMID:19172025

Niu, S., Ji, S., Yan, D., Meng, X., & Xiong, X. (2019). AZ31B/7075-T6 alloys friction stir lap welding with a zinc interlayer. *Journal of Materials Processing Technology*, *263*, 82–90. doi:10.1016/j.jmatprotec.2018.08.009

Nohynek, G. J., Lademann, J., Ribaud, C., & Roberts, M. S. (2007). Grey Goo on the skin? Nanotechnology, cosmetic and sunscreen safety. *Critical Reviews in Toxicology*, *37*(3), 251–277. doi:10.1080/10408440601177780 PMID:17453934

Nosko, M., Štepánek, M., Zifčák, P., Orovčík, L., Nagy, Š., Dvorák, T., Oslanec, P., Khodabakhshi, F., & Gerlich, A. P. (2019). Solid-state joining of powder metallurgy Al-Al2O3 nanocomposites via friction-stir welding: Effects of powder particle size on the weldability, microstructure, and mechanical property. *Materials Science and Engineering A*, *754*, 190–204. doi:10.1016/j.msea.2019.03.074

Nowack, B., Schulin, R., & Robinson, B. H. (2006). Critical assessment of chelant-enhanced metal phytoextraction. *Environmental Science & Technology*, *40*(17), 5225–5232. doi:10.1021/es0604919 PMID:16999093

Okonkwo, E. C., Essien, E. A., Abid, M., Kavaz, D., & Ratlamwala, T. A. (2018). Thermal performance analysis of a parabolic trough collector using water-based green-synthesized nanofluids. *Solar Energy*, *170*, 658–670. doi:10.1016/j.solener.2018.06.012

Okonkwo, E. C., Wole-Osho, I., Kavaz, D., Abid, M., & Al-Ansari, T. (2020). Thermodynamic evaluation and optimization of a flat plate collector operating with alumina and iron mono and hybrid nanofluids. *Sustainable Energy Technologies and Assessments*, *37*, 100636. doi:10.1016/j.seta.2020.100636

Omiddezyani, S., Gharehkhani, S., Yousefi-Asli, V., Khazaee, I., Ashjaee, M., Nayebi, R., Shemirani, F., & Houshfar, E. (2021). Experimental investigation on thermo-physical properties and heat transfer characteristics of green synthesized highly stable CoFe2O4/rGO nanofluid. *Colloids and Surfaces. A, Physicochemical and Engineering Aspects*, *610*, 125923. doi:10.1016/j.colsurfa.2020.125923

Ostiguy, C., Roberge, B., Woods, C., & Soucy, B. (2010), Engineered Nanoparticles: Current Knowledge about OHS Risks and Prevention Measures. Studies and Research Projects. Robert-Sauvé Research Institute for Occupational Health and Safety (IRSST).

Otanicar, T. P., Phelan, P., Prasher, R., Rosengarten, G., & Taylor, A. (2010). Nanofluid-based direct absorption solar collector. *Journal of Renewable and Sustainable Energy, 2*(3).

Otanicar, T. P., Phelan, P. E., & Golden, J. S. (2009). Optical properties of liquids for direct absorption solar thermal energy systems. *Solar Energy, 2009*(83), 969–977. doi:10.1016/j.solener.2008.12.009

Ovsianikov, A., Chichkov, B., Mente, P., Monteiro-Riviere, N. A., Doraiswamy, A., & Narayan, R. J. (2007). Two photon polymerization of polymer–ceramic hybrid materials for transdermal drug delivery. *International Journal of Applied Ceramic Technology*, *4*(1), 22–29. doi:10.1111/j.1744-7402.2007.02115.x

Owens, D. E. III, & Peppas, N. A. (2006). Opsonization, biodistribution, and pharmacokinetics of polymeric nanoparticles. *International Journal of Pharmaceutics*, *307*(1), 93–102. doi:10.1016/j.ijpharm.2005.10.010 PMID:16303268

Paidar, M., & Sarab, M. L. (2016). Friction stir spot welding of 2024-T3 aluminum alloy with SiC nanoparticles. *Journal of Mechanical Science and Technology*, *30*(1), 365–370. doi:10.100712206-015-1241-4

Palaniappan, M., Tirlangi, S., Mohamed, M. J. S., Moorthy, R. M. S., Valeti, S. V., & Boopathi, S. (2023). Fused deposition modelling of polylactic acid (PLA)-based polymer composites: A case study. In Development, Properties, and Industrial Applications of 3D Printed Polymer Composites (pp. 66–85). IGI Global. doi:10.4018/978-1-6684-6009-2.ch005

Pankhurst, Q. A., Connolly, J., Jones, S. K., & Dobson, J. (2003). Applications of magnetic nanoparticles in biomedicine. *Journal of Physics. D, Applied Physics*, *36*(13), R167–R181. doi:10.1088/0022-3727/36/13/201

Pardeike, J., Hommoss, A., & Müller, R. H. (2009). Lipid nanoparticles (SLN, NLC) in cosmetic and pharmaceutical dermal products. *International Journal of Pharmaceutics*, *366*(1-2), 170–184. doi:10.1016/j.ijpharm.2008.10.003 PMID:18992314

Park, K. (2014). Controlled drug delivery systems: Past forward and future back. *Journal of Controlled Release*, *190*, 3–8. doi:10.1016/j.jconrel.2014.03.054 PMID:24794901

Patel, M., Chaudhary, B., Murugesan, J., Jain, N. K., & Patel, V. (2022). Enhancement of tensile and fatigue properties of hybrid aluminium matrix composite via multipass friction stir processing. *Journal of Materials Research and Technology*, *21*, 4811–4823. doi:10.1016/j.jmrt.2022.11.073

Patil, P., Kardekar, N., & Kamble, D. (2023). Performance analysis and optimization of nano particle base Libr-H2o vapor absorption system by using Taguchi, Grey relational analysis and ANOVA. *Materials Today: Proceedings*, *72*, 1413–1419. doi:10.1016/j.matpr.2022.09.340

Patil-Tekale, R. A., Gadekar, A., Gadhade, Y., Parakh, L., Balaji, R., & Selokar, A. (2021). Recent Investigation on Ultrasonic Machining of Aluminum Metal Matrix Composite. *Intelligent Manufacturing and Energy Sustainability: Proceedings of ICIMES 2020*, (pp. 619–636). IEEE.

Peña-Parás, L., Maldonado-Cortés, D., Rodríguez-Villalobos, M., Sauceda, A., Aguilar, I. J., Villanueva, J. A., & Cárdenas, P. A. (2023). Nanoparticle lubricant additives applied in a CNC lathe ball screw component for improving the quality of machined workpieces: A case study. *Wear*, *523*, 204752. doi:10.1016/j.wear.2023.204752

Pentek, T., Newenhouse, E., O'Brien, B., & Chauhan, A. S. (2017). Development of a topical resveratrol formulation for commercial applications using dendrimer nanotechnology. *Molecules (Basel, Switzerland)*, *22*(1), 137. doi:10.3390/molecules22010137 PMID:28098828

Pereira, L., Dias, N., Carvalho, J., Fernandes, S., Santos, C., & Lima, N. (2014). Synthesis, characterization and antifungal activity of chemically and fungal-produced silver nanoparticles against Trichophyton rubrum. *Journal of Applied Microbiology*, *117*(6), 1601–1613. doi:10.1111/jam.12652 PMID:25234047

Petersen, R. (2010). *Nanocrystals for Use in Topical Cosmetic Formulations and Method of Production Thereof.* US Patent US 20100047297A1.

Poland, C. A., Larsen, P. B., Read, S. A. K., Varet, J., Hankin, S. M., & Lam, H. R. (2016). *Assessment if Nano-Enabled Technologies in Cosmetics*. The Danish Environmental Protection Agency.

Poletto, F. S., Beck, R. C. R., Guterres, S. S., & Pohlmann, A. R. (2011). Polymeric nanocapsules: Concepts and applications. In R. Beck, S. Guterres, & A. Pohlmann (Eds.), *Nanocosmetics and Nanomedicines*. Springer. doi:10.1007/978-3-642-19792-5_3

Pradhan, S., & Mishra, B. G. (2015). Catalytic application of SO42−/Fe–ZrO2nanoparticles synthesized by a urea hydrolysis method for environmentally benign one pot synthesis of 1,8-dioxodecahydroacridines. *RSC Advances*, *5*(105), 86179–86190. doi:10.1039/C5RA11136F

Pradhan, S., & Mishra, B. G. (2018). CsxH3-xPW12O40 nanoparticles dispersed in the porous network of Zr-pillared α-zirconium phosphate as efficient heterogeneous catalyst for synthesis of spirooxindoles. *Molecular Catalysis*, *446*, 58–71. doi:10.1016/j.mcat.2017.12.013

Pradhan, S., Saha, J., & Mishra, B. G. (2017). Morphology controlled phosphate grafted SnO2–ZrO2 nanocomposite oxides prepared by a urea hydrolysis method as efficient heterogeneous catalysts towards the synthesis of 3-substituted indoles. *New Journal of Chemistry*, *41*(14), 6616–6629. doi:10.1039/C7NJ00249A

Pradhan, S., Sahu, V., & Mishra, B. G. (2016). CaO-ZrO2 nanocomposite oxide prepared by urea hydrolysis method as heterogeneous base catalyst for synthesis of chromene analogues. *Journal of Molecular Catalysis A Chemical*, *425*, 297–309. doi:10.1016/j.molcata.2016.10.031

Purohit, D. K., Nandgude, T. D., & Poddar, S. S. (2016). Nano-lipid carriers for topical application: Current scenario. *Asian Journal of Pharmaceutics*, *9*(5), 544–553.

Pyo, S. M., Meinke, M., Keck, C. M., & Müller, R. H. (2016). Rutin—Increased antioxidant activity and skin penetration by nanocrystal technology (smartCrystals). *Cosmetics*, *3*(1), 9. doi:10.3390/cosmetics3010009

Quddoos, A., Anand, A., Mishra, G. K., & Nag, P. (2014). Nano-fluids: Introduction, preparation, stability analysis and stability enhancement techniques. *IJPAS*, *1*, 31–36.

Rafati, M., Hamidi, A. A., & Shariati Niaser, M. (2012). Application of nanofluids in computer cooling systems (heat transfer performance of nanofluids). *Applied Thermal Engineering*, *45-46*, 9–14. doi:10.1016/j.applthermaleng.2012.03.028

Rahman, M., Woon, K. S., & Neo, W. K. (2020). Tool-Based Micro/Nano Machining: Development of Innovative Machine and Machining Processes. *International Manufacturing Science and Engineering Conference, 84263*. IEEE. 10.1115/MSEC2020-8580

Rahmanzadeh, B., Rahmani, K., & Piroti, S. (2018). Experimental study of the effect of water-cement ratio on compressive strength, abrasion resistance, porosity and permeability of Nanosilica concrete. *Fratturaed Integrit Struttural.*, *12*(44), 16–24. doi:10.3221/IGF-ESIS.44.02

Raja, S., Muhamad, M. R., Jamaludin, M. F., & Yusof, F. (2020). A review on nanomaterials reinforcement in friction stir welding. *Journal of Materials Research and Technology*, *9*(6), 16459–16487. doi:10.1016/j.jmrt.2020.11.072

Rajkumar, R., Akkineni, S., & Ramya, S. (2016). Experimental Study on the Strength and Durability of Nano Concrete. *International Journal of Applied Engineering Research, 11*(4), 2854-2858.

Rajpoot, K. (2019). Solid Lipid Nanoparticles: A Promising Nanomaterial in Drug Delivery. *Current Pharmaceutical Design*, *25*(37), 3943–3959. doi:10.2174/1381612825666190903155321 PMID:31481000

Rajput, N. (2015). Methods of preparation of nanoparticles-a review. *International Journal of Advances in Engineering \& Technology, 7*(6), 1806.

Ramírez-García, R., Gohil, N., & Singh, V. (2019). Recent Advances, Challenges, and Opportunities in Bioremediation of Hazardous Materials. In P. V. Chandra & K. Bauddh (Eds.), *Phyto management of Polluted Sites* (1st ed., Vol. 1, pp. 517–568). Elsevier. doi:10.1016/B978-0-12-813912-7.00021-1

Ramis, J. M., Coelho, C. C., Córdoba, A., Quadros, P. A., & Monjo, M. (2018). Safety assessment of nano-hydroxyapatite as an oral care ingredient according to the EU Cosmetics Regulation. *Cosmetics*, *5*(3), 53. doi:10.3390/cosmetics5030053

Rana, P., & Bhargava, R. (2012). Flow and heat transfer of a nanofluid over a nonlinearly stretching sheet. *Communications in Nonlinear Science and Numerical Simulation*, *17*(1), 212–226. doi:10.1016/j.cnsns.2011.05.009

Ranjbarzadeh, R., Moradikazerouni, A., Bakhtiari, R., Asadi, A., & Afrand, M. (2019). An experimental study on stability and thermal conductivity of water/silica nanofluid: Eco-friendly production of nanoparticles. *Journal of Cleaner Production*, *206*, 1089–1100. doi:10.1016/j.jclepro.2018.09.205

Rashidi, M.M., Freidoonimehr, N., Momoniat, E., & Rostami, B. (2015). Study of nonlinear MHD tribological squeeze film at generalized magnetic reynolds numbers using DTM. *PLoS One, 10*(8). http://dx.doi.org/.pone.0135004. e0135004 doi:10.1371/journal

Rashidi, M. M., Freidoonimehr, N., Hosseini, A., Anwar Bég, O., & Hung, T. K. (2014). Homotopy simulation of nanofluid dynamics from a non-linearly stretching isothermal permeable sheet with transpiration of nanofluids over a power law stretching sheet. *Meccanica*, *49*(2), 469–482. doi:10.100711012-013-9805-9

Ravikumar, M., Sivashankar, A., & Suresh, S. (2020). Materials Today : Proceedings Effects of rice bran oil blends and engine load on emission characteristics of a variable compression ratio diesel engine. *Materials Today: Proceedings*. Science Direct. doi:10.1016/j.matpr.2020.03.680

Reddy, P. M., Kumar, T. G., & Reddy, C. V. R. G. (2020). *Nano-fluids and its Applications.*

Reijnders, L. (2006). Cleaner nanotechnology and hazard reduction of manufactured nanoparticles. *Journal of Cleaner Production*, *14*(2), 124–133. doi:10.1016/j.jclepro.2005.03.018

Ribeiro, R. C. D. A., Barreto, S. M. D. A. G., Ostrosky, E. A., Rocha-Filho, P. A., Veríssimo, L. M., & Ferrari, M. (2015). Production and characterization of cosmetic nanoemulsions containing Opuntia ficus-indica (L.) Mill extract as moisturizing agent. *Molecules (Basel, Switzerland)*, *20*(2), 2492–2509. doi:10.3390/molecules20022492 PMID:25648593

Rohrer, G. S. (2010). "Introduction to Grains, Phases, and Interfaces—An Interpretation of Microstructure," Trans. AIME, 1948, vol. 175, pp. 15–51, by C.S. Smith. *Metallurgical and Materials Transactions. A, Physical Metallurgy and Materials Science*, *41*(5), 1063–1100. doi:10.100711661-010-0215-5

Rosen, J., Landriscina, A., & Friedman, A. (2015). NanotechnologyBased Cosmetics for Hair Care. *Cosmetics*, *2*(4), 211–224. doi:10.3390/cosmetics2030211

Rubbi, F., Habib, K., Saidur, R., Aslfattahi, N., Yahya, S. M., & Das, L. (2020). Performance optimization of a hybrid PV/T solar system using Soybean oil/MXene nanofluids as A new class of heat transfer fluids. *Solar Energy*, *208*, 124–138. doi:10.1016/j.solener.2020.07.060

Rutuja, S., Rajguru, N., Madhuri, P., Suraj, S., Rohan, W., Kishor, Z. & Ghogare, R. (2017). Review Paper on Multi-functional use of Nano Silica in Concrete. *International Journal of Engineering Science and Computing*. 10780-10782.

Ryman-Rasmussen, J. P., Riviere, J. E., & Monteiro-Riviere, N. A. (2006). Penetration of intact skin by quantum dots with diverse physicochemical properties. *Toxicological Sciences*, *91*(1), 159–165. doi:10.1093/toxsci/kfj122 PMID:16443688

S., P. K., Sampath, B., R., S. K., Babu, B. H., & N., A. (2022). Hydroponics, Aeroponics, and Aquaponics Technologies in Modern Agricultural Cultivation. In *Trends, Paradigms, and Advances in Mechatronics Engineering* (pp. 223–241). IGI Global. doi:10.4018/978-1-6684-5887-7.ch012

Sadegh, K., Hossein, T., Arezoo, K., & Mehdi, R. M. (2015). Unsteady convective heat and mass transfer in pseudoplastic nanofluid over a stretching wall. *Advanced Powder Technology*, *26*(5), 1319–1326. doi:10.1016/j.apt.2015.07.006

Sadri, R., Hosseini, M., Kazi, S. N., Bagheri, S., Ahmed, S. M., Ahmadi, G., Zubir, N., Sayuti, M., & Dahari, M. (2017). Study of environmentally friendly and facile functionalization of graphene nanoplatelet and its application in convective heat transfer. *Energy Conversion and Management*, *150*, 26–36. doi:10.1016/j.enconman.2017.07.036

Sadri, R., Hosseini, M., Kazi, S. N., Bagheri, S., Zubir, N., Ahmadi, G., Dahari, M., & Zaharinie, T. (2017). A novel, eco-friendly technique for covalent functionalization of graphene nanoplatelets and the potential of their nanofluids for heat transfer applications. *Chemical Physics Letters*, *675*, 92–97. doi:10.1016/j.cplett.2017.02.077

Sadri, R., Hosseini, M., Kazi, S. N., Bagheri, S., Zubir, N., Solangi, K. H., Zaharinie, T., & Badarudin, A. (2017). A bio-based, facile approach for the preparation of covalently functionalized carbon nanotubes aqueous suspensions and their potential as heat transfer fluids. *Journal of Colloid and Interface Science*, *504*, 115–123. doi:10.1016/j.jcis.2017.03.051 PMID:28531649

Safavi, K., Esfahanizadeh, M., Mortazaeinezahad, F., & Dastjerd, H. (2011) The study of nano silver anti microbiaactivity and evaluation of using NS in tissue culture media. *International Conference on Life and Technology*. Research Gate.

Sahal, B. C., R, D., A, A., Thrinath, B. V. S., Boopathi, S., J. R., & Sudhakar, M. (n.d.). *IoT Based Smart Energy Meter For Smart Grid.*

Sahud, K. (2014). *Carbon black particle exhibits size dependent toxicity in human monocytes*, 827019.

Saidur, R., Leong, K. Y., & Mohammed, H. A. (2011). A review on applications and challenges of nanofluids. *Renewable & Sustainable Energy Reviews*, *15*(3), 1646–1668. doi:10.1016/j.rser.2010.11.035

Said, Z., Gupta, M., Hegab, H., Arora, N., Khan, A. M., Jamil, M., & Bellos, E. (2019). A comprehensive review on minimum quantity lubrication (MQL) in machining processes using nano-cutting fluids. *International Journal of Advanced Manufacturing Technology, 105*(5-6), 2057–2086. doi:10.100700170-019-04382-x

Saifuddin, N., Wong, C. W., & Yasumira, A. A. N. (2009). Rapid biosynthesis of silver nanoparticles using culture supernatant of bacteria with microwave irradiation. *E-Journal of Chemistry, 6*(1), 61–70. doi:10.1155/2009/734264

Saleh, H., Alali, E., & Ebaid, A. (2017). Medical applications for the flow of carbon-nanotubes suspended nanofluids in the presence of convective condition using Laplace transform. *J Assoc Arab Univ Basic Appl Sci., 24*(1), 206–212. doi:10.1016/j.jaubas.2016.12.001

Salloum, M., Ma, R., & Zhu, L. (2008). An in-vivo experimental study of temperature elevations in animal tissue during magnetic nanoparticle hyperthermia. *International Journal of Hyperthermia, 24*(7), 589–601. doi:10.1080/02656730802203377 PMID:18979310

Samikannu, R., Koshariya, A. K., Poornima, E., Ramesh, S., Kumar, A., & Boopathi, S. (2023). Sustainable Development in Modern Aquaponics Cultivation Systems Using IoT Technologies. In *Human Agro-Energy Optimization for Business and Industry* (pp. 105–127). IGI Global. doi:10.4018/978-1-6684-4118-3.ch006

Sampath, B. C. S., & Myilsamy, S. (2022). Application of TOPSIS Optimization Technique in the Micro-Machining Process. In Trends, Paradigms, and Advances in Mechatronics Engineering (pp. 162–187). IGI Global. doi:10.4018/978-1-6684-5887-7.ch009

Sampath, B., & Myilsamy, S. (2021). Experimental investigation of a cryogenically cooled oxygen-mist near-dry wire-cut electrical discharge machining process. *Strojniski Vestnik. Jixie Gongcheng Xuebao, 67*(6), 322–330. doi:10.5545v-jme.2021.7161

Sampath, B., Sureshkumar, M., Yuvaraj, T., & Velmurugan, D. (2021). Experimental investigations on eco-friendly helium-mist near-dry wire-cut edm of m2-hss material. *Materials Research Proceedings, 19*, 175–180. doi:10.21741/9781644901618-22

Sanchaya, M., Harikaran, M., Cheran, K., Varoon, M., Jayaprakash, P., & Navin, G. (2022). Experimental investigation on rice husk ash and coconut shell used in building materials. *Materials Today: Proceedings, 68*(1), 1697–1702. doi:10.1016/j.matpr.2022.08.264

Sankar. P, Suresh, T., Aravind, V., Buvanesvari, K., Harikaran, M., & R. Raja. (2023). Analysis and Optimization of Supply Chain Travelling Salesman Problem Using K-Means Cluster based Genetic Algorithm. *IJIRSET, 12*(1), doi:. doi:10.15680/IJIRSET.2023.1201095

Sankar, P., Suresh, P., Aravind, T., Buvanesvari, V. K., Harikaran, M., & Raja, R. (2023). Investigation on Supply Chain Distribution Network Problem using RFID Technique [IJIRSET]. *International Journal of Innovative Research in Science, Engineering and Technology, 12*(1). doi:10.15680/IJIRSET.2023.1201094

Sano, N., Akazawa, H., Kikuchi, T., & Kanki, T. (2003). Separated synthesis of iron-included carbon nanocapsules and nanotubes by pyrolysis of ferrocene in pure hydrogen. *Carbon, 41*. 2159-2162. doi: 00215-X. doi:10.1016/S0008-6223(03)

Santos, A. C., Morais, F., Simões, A., Pereira, I., Sequeira, J. A., Pereira-Silva, M., Veiga, F., & Ribeiro, A. (2019). Nanotechnology for the development of new cosmetic formulations. *Expert Opinion on Drug Delivery, 16*(4), 313–330. doi:10.1080/17425247.2019.1585426 PMID:30793641

Sapsford, K. E., Tyner, K. M., Dair, B. J., Deschamps, J. R., & Medintz, I. L. (2011). Analyzing nanomaterial bioconjugates: A review of current and emerging purification and characterization techniques. *Analytical Chemistry, 83*(12), 4453–4488. doi:10.1021/ac200853a PMID:21545140

Saravanan, A., Venkatasubramanian, R., Khare, R., Surakasi, R., Boopathi, S., Ray, S., & Sudhakar, M. (2022). *Policy Trends of Renewable Energy and Non Renewable Energy.* Research Gate.

Sasikala, G., Jothiprakash, V. M., Pant, B., Subalakshmi, R., Thirumal Azhagan, M., Arul, K., Alonazi, W. B., Karnan, M., & Praveen Kumar, S. (2022). Optimization of Process Parameters for Friction Stir Welding of Different Aluminum Alloys AA2618 to AA5086 by Taguchi Method. *Advances in Materials Science and Engineering, 2022*, 1–9. doi:10.1155/2022/3808605

Sasivarman, B., Nagalakshmi, R., & Harikaran, M. (2022). Treatment of wastewater by slow sand process with polystyrene. *Materials Today: Proceedings, 68*(1), 1648–1653. doi:10.1016/j.matpr.2022.08.082

Satyavani, K., Gurudeeban, S., Ramanathan, T., & Balasubramanian, T. (2011). Biomedical potential of silver nanoparticles from calli cells of *Citrullus colocynthis* (L.) Schrad. *Journal of Nanobiotechnology, 43*, 1–8. PMID:21943321

Savithramma, N. (2011). M Linga Rao, K Rukmini and P Suvarnalatha devi (2011). Antimicrobial activity of silver nanoparticles synthesized by using medicinal plants. African. *Australian Journal of Basic and Applied Sciences, 3*, 265.

SCCP (Scientific Committee on Consumer Products), Preliminary Opinion On Safety of nanomaterials in cosmetic products.

SCENIHR. (n.d.). *Opinion on The Appropriateness of The Risk Assessment Methodology in Accordance with The Technical Guidance Documents for New and Existing Substances for Assessing The Risks Of Nanomaterials.* SCENIHR. http://www.ec.europa.eu/health/ph_risk/committees/04_scenihr/docs/scenihr_o_023.pdf

Schulte, P., Geraci, C., Zumwalde, R., Hoover, M., & Kuempe, E. (2008). Occupational risk management of engineered nanoparticles. *Journal of Occupational and Environmental Hygiene, 5*(4), 239–249. doi:10.1080/15459620801907840 PMID:18260001

Scientific Committee on Consumer Safety. (2019). *Scientific Committee on Consumer Safety Melatonin.* Scientific Committee on Consumer Safety.

Senthil, T. S. R. Ohmsakthi vel, Puviyarasan, M., Babu, S. R., Surakasi, R., & Sampath, B. (2023). Industrial Robot-Integrated Fused Deposition Modelling for the 3D Printing Process. In Development, Properties, and Industrial Applications of 3D Printed Polymer Composites (pp. 188–210). IGI Global. doi:10.4018/978-1-6684-6009-2.ch011

Sermakkani, M. & Thangapandian, V. (2012). Biological synthesis of silver nanoparticles using medicinal plant(*cassia italica*) leaves . *Int. J. Curr. Res. 4*(10), 053-058.

Seth, G. S., Bhattacharyya, A., Kumar, R., & Mishra, M. K. (2019). Modelling and numerical simulation of hydromagnetic natural convection Casson fluid flow with n-th order chemical reaction and Newtonian heating in porous medium. *Journal of Porous Media*, *22*(9), 1141–1157. doi:10.1615/JPorMedia.2019025699

Seth, G. S., Bhattacharyya, A., & Mishra, M. K. (2019). Study of partial slip mechanism on free convection flow of viscoelastic fluid past a non-linearly stretching surface. Comput Thermal Sci. *International Journal (Toronto, Ont.)*, *11*(1–2), 105–117. doi:10.1615/ComputThermalScien.2018024728

Seth, G. S., Kumar, R., Tripathi, R., & Bhattacharyya, A. (2018). Double diffusive MHD Casson fluid flow in a non-Darcy porous medium with Newtonian heating and thermo-diffusion effects. *Int J Heat Tech*, *36*(4), 1517–1527. doi:10.18280/ijht.360446

Shafiq, T., Uzair, M., Iqbal, M. J., Zafar, M., Hussain, S. J., & Shah, S. A. A. (2021). Green Synthesis of Metallic Nanoparticles and Their Potential in Bio-Medical Applications. *Nano Biomedicine \& Engineering, 13*(2).

Shah, M., & Shah, D. Kaushal, V., & Niragi, D. (2019). Effect of GGBS and Nano Silica on the Strength Properties of Ternary Concrete. *4th world congress on civil, structural and environmental engineering, 522*. IEEE. 10.1088/1757-899X/522/1/012011

Shailesh, L., Patwekar Snehal, R., Ashwini, P., Manoj, S., & Arvind, B. (2016). Nanostructured lipid carriers in stability improvement forcosmetic nanoparticles. *International Journal of Pharmacy and Pharmaceutical Research*, *6*(1), 168–180.

Sharifi Asl, N., Mirsalehi, S. E., & Dehghani, K. (2019). Effect of TiO2 nanoparticles addition on microstructure and mechanical properties of dissimilar friction stir welded AA6063-T4 aluminum alloy and AZ31B-O magnesium alloy. *Journal of Manufacturing Processes*, *38*, 338–354. doi:10.1016/j.jmapro.2019.01.023

Sharma, A. K., Tiwari, A. K., & Dixit, A. R. (2015). Mechanism of nanoparticles functioning and effects in machining processes: A review. *Materials Today: Proceedings*, *2*(4–5), 3539–3544. doi:10.1016/j.matpr.2015.07.331

Sharma, A. K., Tiwari, A. K., & Dixit, A. R. (2016). Effects of Minimum Quantity Lubrication (MQL) in machining processes using conventional and nanofluid based cutting fluids: A comprehensive review. *Journal of Cleaner Production*, *127*, 1–18. doi:10.1016/j.jclepro.2016.03.146

Sharma, A., Sharma, V. M., Gugaliya, A., Rai, P., Pal, S. K., & Paul, J. (2020). Friction stir lap welding of AA6061 aluminium alloy with a graphene interlayer. *Materials and Manufacturing Processes*, *35*(3), 258–269. doi:10.1080/10426914.2020.1718694

Sharma, N., Khan, Z. A., Siddiquee, A. N., Shihab, S. K., & Wahid, M. A. (2018). Effect of process parameters on microstructure and electrical conductivity during FSW of Al-6101 and Pure Copper. *Materials Research Express*, *5*(4), 046519. doi:10.1088/2053-1591/aabbe0

Sharma, P. (2008). *Cosmetic Formulation Manufacture and Quality Control* (4th ed.). Vandanna Publication.

Sharma, S. (2012). Cosmeceuticals and herbal drugs: Practical uses. *International Journal of Pharmaceutical Sciences and Research*, *3*(1), 59–65.

Sharma, V. K., Yngard, R. A., & Lin, Y. (2009). Silver Nanoparticles: Green Synthesis and Their Antimicrobial Activities. *Advances in Colloid and Interface Science*, *145*(1-2), 83–96. doi:10.1016/j.cis.2008.09.002 PMID:18945421

Sharma, V., Prakash, U., & Kumar, B. V. M. (2015). Surface composites by friction stir processing: A review. *Journal of Materials Processing Technology*, *224*, 117–134. doi:10.1016/j.jmatprotec.2015.04.019

Sharmila, G., Muthukumaran, C., Sangeetha, E., Saraswathi, H., Soundarya, S., & Kumar, N. M. (2019). Green fabrication, characterization of Pisonia alba leaf extract derived MgO nanoparticles and its biological applications. *Nano-Structures & Nano-Objects*, *20*, 100380. doi:10.1016/j.nanoso.2019.100380

Sharmin, I., Gafur, M. A., & Dhar, N. R. (2020). Preparation and evaluation of a stable CNT-water based nano cutting fluid for machining hard-to-cut material. *SN Applied Sciences*, *2*(4), 1–18. doi:10.100742452-020-2416-x

Sheikhpour, M., Arabi, M., Kasaeian, A., Rokn Rabei, A., & Taherian, Z. (2020). Role of Nanofluids in Drug Delivery and Biomedical Technology: Methods and Applications, *Dove Press journal. Nanotechnology, Science and Applications*, *13*, 47–59. doi:10.2147/NSA.S260374 PMID:32801669

Shirazi, S. F. S., Gharehkhani, S., Yarmand, H., Badarudin, A., Metselaar, H. S. C., & Kazi, S. N. (2015). Nitrogen doped activated carbon/graphene with high nitrogen level: Green synthesis and thermo-electrical properties of its nanofluid. *Materials Letters*, *152*, 192–195. doi:10.1016/j.matlet.2015.03.110

Shokrollahi, H. (2013). Contrast agents for MRI. *Materials Science and Engineering C*, *33*(8), 4485–4497. doi:10.1016/j.msec.2013.07.012 PMID:24094150

Simonetti, N., Simonetti, G., Bougnol, F., & Scalzo, M. (1992). Electrochemical Ag+ for preservative use. *Applied and Environmental Microbiology*, *58*(12), 3834–3836. doi:10.1128/aem.58.12.3834-3836.1992 PMID:1476427

Singh Sidhu, H., Singh, B., & Kumar, P. (2020). To study the corrosion behavior of friction stir processed magnesium alloy AZ91. *Materials Today: Proceedings, 44*, 4633–4639. doi:10.1016/j.matpr.2020.10.920

Singh, T. (2021a). Processing of friction stir welded AA6061-T6 joints reinforced with nanoparticles. *Results in Materials, 12*, 100210. doi.org/10.1016/j.rinma.2021.100210

Singh, A. V., Rosenkranz, D., Ansari, M. H. D., Singh, R., Kanase, A., Singh, S. P., Johnston, B., Tentschert, J., Laux, P., & Luch, A. (2020). Artificial Intelligence and Machine Learning Empower Advanced Biomedical Material Design to Toxicity Prediction. *Advanced Intelligent Systems, 2*(12), 2070125. doi:10.1002/aisy.202070125

Singh, P., Srivastava, S., & Singh, S. K. (2019). Nanosilica: Recent Progress in Synthesis, Functionalization, Biocompatibility, and Biomedical Applications. *ACS Biomaterials Science & Engineering, 5*(10), 4882–4898. doi:10.1021/acsbiomaterials.9b00464 PMID:33455238

Singh, R., Tiwari, S., & Tawaniya, J. (2013). Review on nanotechnology with several aspects. *Int. J. Res. Computer Eng. Electron, 2*, 1–8.

Singh, S., & Bhargava, R. (2014). Simulation of phase transition during cryosurgical treatment of a tumor tissue loaded with nanoparticles using meshfree approach. *Journal of Heat Transfer, 136*(12), 121101. doi:10.1115/1.4028730

Singh, T. (2021b). Processing of friction stir welded AA6061-T6 joints reinforced with nanoparticles. *Results in Materials, 12*, 100210. doi:10.1016/j.rinma.2021.100210

Singh, T. G., & Sharma, N. (2016). Nanobiomaterials in cosmetics: Current status and future prospects. In A. M. Grumezescu (Ed.), *Nanobiomaterials in Galenic Formulations and Cosmetics* (pp. 149–174). William Andrew Publishing. doi:10.1016/B978-0-323-42868-2.00007-3

Singh, T., Tiwari, S. K., & Shukla, D. K. (2019). Friction-stir welding of AA6061-T6: The effects of Al2O3 nano-particles addition. *Results in Materials, 1*, 100005. doi:10.1016/j.rinma.2019.100005

Singh, T., Tiwari, S. K., & Shukla, D. K. (2020). Mechanical and microstructural characterization of friction stir welded AA6061-T6 joints reinforced with nano-sized particles. *Materials Characterization, 159*, 110047. doi:10.1016/j.matchar.2019.110047

Sivanskar, M., & Kumar, B. P. (2010). Role of Nanoparticle in Drug Delivery System. *International Journal of Research in Pharmaceutical and Biomedical Sciences, 1*, 41–66.

Sivaranjani, K., & Meenakshisundaram, M. (2013). Biological synthesis of silver nanoparticles by using *Ocimum basillicum* leaf extract and their antibacterial activity. *IJPR, 4*(1), 225–229.

Sizochenko, N., Syzochenko, M., Gajewicz, A., Leszczynski, J., & Puzyn, T. (2017). Predicting physical properties of nanofluids by computational modeling. *The Journal of Physical Chemistry C, 121*(3), 1910–1917. doi:10.1021/acs.jpcc.6b08850

Smaisim, G. F., Mohammed, D. B., Abdulhadi, A. M., Uktamov, K. F., Alsultany, F. H., Izzat, S. E., Ansari, M. J., Kzar, H. H., Al-Gazally, M. E., & Kianfar, E. (2022). Nanofluids: Properties and applications. *Journal of Sol-Gel Science and Technology*, *104*(1), 1–35. doi:10.100710971-022-05859-0

Smijs, T. G., & Pavel, S. (2011). Titanium dioxide and zinc oxide nanoparticles in sunscreens: Focus on their safety and efectiveness. *Nanotechnology, Science and Applications*, *4*(1), 95–112. doi:10.2147/NSA.S19419 PMID:24198489

Soltanloo, S., Alimohammadi, M., Ramezanpour, S. S., & Najar, M. B. (2010). Nanosilver colloid: A novel antimicrobial candidate applicable in plant tissue culturemedium. *Australian Journal of Basic and Applied Sciences*, *4*, 5338–5345.

Sone, B. T., Diallo, A., Fuku, X. G., Gurib-Fakim, A., & Maaza, M. (2020). Biosynthesized CuO nano-platelets: Physical properties & enhanced thermal conductivity nanofluidics. *Arabian Journal of Chemistry*, *13*(1), 160–170. doi:10.1016/j.arabjc.2017.03.004

Song, J. Y., & Kim, B. S. (2009). Rapid Biological Synthesis of Silver Nanoparticles Using Plant Leaf Extracts. *Bioprocess and Biosystems Engineering*, *32*(1), 79–84. doi:10.100700449-008-0224-6 PMID:18438688

Sonneville-Aubrun, O., Simonnet, J. T., & L'alloret, F. (2004). Nanoemulsions: A new vehicle for skincare products. *Advances in Colloid and Interface Science*, *108*, 145–149. doi:10.1016/j.cis.2003.10.026 PMID:15072937

Sonneville-Aubrun, O., Yukuyama, M. N., & Pizzino, A. (2018). Application of nanoemulsions in cosmetics. In *Nanoemulsions* (pp. 435–475). Academic Press. doi:10.1016/B978-0-12-811838-2.00014-X

Spielvogel, J., Guo, X., Pesch, M., Keck, L., & Hagler, R. (2009). A New Real-time Exposure Monitor for Measuring Airborne Nanoparticles. *GRIMM Aerosol Techni GmbH & Co, Bayern, Germany*. Stephanie, S.G. History of Makeup. https://www.webmd.com/beauty/history-makeup

Sreelakshmy, K. R., Nair, A. S., Vidhya, K., Saranya, T., & Nair, S. C. (2014). An overview of recent nanofluid research. *International Research Journal of Pharmacy, 5*(4), 239-243.

Sridher, J., Vivek, D., & Jegatheeswaran, D. (2019). Mechanical and Flexural Behavior of High-performance Concrete Containing Nano Silica. [IJEAT]. *International Journal of Engineering and Advanced Technology*, *9*(2), 2249–8958. doi:10.35940/ijeat.B3164.129219

Srinivas, K. (2016). The current role of nanomaterials in cosmetics. *Journal of Chemical and Pharmaceutical Research*, *8*(5), 906–914.

Stankic, S., Suman, S., Haque, F., & Vidic, J. (2016). Pure and multi metal oxide nanoparticles: Synthesis, antibacterial and cytotoxic properties. *Journal of Nanobiotechnology*, *14*(1), 1–20. doi:10.118612951-016-0225-6 PMID:27776555

Stephen, U. S. (1998). Nanofluid Technology. Current Status and Future Research, 1, 1-3.

Subramani, V., Jayavel, B., Sengottuvelu, R., & Lazar, P. (2019). Assessment of Microstructure and Mechanical Properties of Stir Zone Seam of Friction Stir Welded Magnesium AZ31B through Nano-SiC. *Materials (Basel), 12*(7), 1044. doi:10.3390/ma12071044 PMID:30934939

Sudhagar, S., & Gopal, P. M. (2021). Investigation on Mechanical and Tribological Characteristics Cu/Si3N4 Surface Composite Developed Through Friction Stir Processing. *Silicon*, 1–10.

Sudhagar, S., Sakthivel, M., & Ganeshkumar, P. (2019). Monitoring of friction stir welding based on vision system coupled with Machine learning algorithm. *Measurement, 144*, 135–143. doi:10.1016/j.measurement.2019.05.018

Su, H. Y., Chang-Qiang, W., & Dan-Yang, L. (2015). Self-assembled superparamagnetic nanoparticles as MRI contrast agents - A review. *Chinese Physics B, 24*(12), 127506. doi:10.1088/1674-1056/24/12/127506

Sukkasamy, S., & Gopal, P. M. (2022). Monitoring of the Friction Stir Welding Process. In Advanced Manufacturing Techniques for Engineering and Engineered Materials (pp. 173–182). doi:10.4018/978-1-7998-9574-9.ch010

Suresh, S., Elango, N., Venkatesan, K., Lim, W. H., Palanikumar, K., & Rajesh, S. (2020). Sustainable friction stir spot welding of 6061-T6 aluminium alloy using improved non-dominated sorting teaching learning algorithm. *Journal of Materials Research and Technology, 9*(5), 11650–11674. doi:10.1016/j.jmrt.2020.08.043

Suresh, S., Natarajan, E., Franz, G., & Rajesh, S. (2022). Differentiation in the SiC Filler Size Effect in the Mechanical and Tribological Properties of Friction-Spot-Welded AA5083-H116 Alloy. *Fibers (Basel, Switzerland), 10*(12), 109. doi:10.3390/fib10120109

Suresh, S., Natarajan, E., & Shanmugam, R. (2022). Strategized friction stir welded AA6061-T6/SiC composite lap joint suitable for sheet metal applications. *Journal of Materials Research and Technology, 21*, 30–39. doi:10.1016/j.jmrt.2022.09.022

Suresh, S., Natarajan, E., Vinayagamurthi, P., Venkatesan, K., Viswanathan, R., & Rajesh, S. (2023). Optimum Tool Traverse Speed Resulting Equiaxed Recrystallized Grains and High Mechanical Strength at Swept Friction Stir Spot Welded AA7075-T6 Lap Joints. In E. Natarajan, S. Vinodh, & V. Rajkumar (Eds.), *Materials, Design and Manufacturing for Sustainable Environment* (pp. 547–555). Springer Nature Singapore. doi:10.1007/978-981-19-3053-9_41

Suresh, S., Venkatesan, K., & Natarajan, E. (2018). Influence of SiC Nanoparticle Reinforcement on FSS Welded 6061-T6 Aluminum Alloy. *Journal of Nanomaterials, 2018*, 1–11. doi:10.1155/2018/7031867

Suresh, S., Venkatesan, K., Natarajan, E., & Rajesh, S. (2020a). Influence of tool rotational speed on the properties of friction stir spot welded AA7075-T6/Al2O3 composite joint. *Materials Today: Proceedings, 27*, 62–67. doi:10.1016/j.matpr.2019.08.220

Suresh, S., Venkatesan, K., Natarajan, E., & Rajesh, S. (2021). Performance Analysis of Nano Silicon Carbide Reinforced Swept Friction Stir Spot Weld Joint in AA6061-T6 Alloy. *Silicon, 13*(10), 3399–3412. doi:10.100712633-020-00751-4

Suresh, S., Venkatesan, K., Natarajan, E., Rajesh, S., & Lim, W. H. (2019). Evaluating weld properties of conventional and swept friction stir spot welded 6061-t6 aluminium alloy. *Materials Express, 9*(8), 851–860. doi:10.1166/mex.2019.1584

Suresh, S., Venkatesan, K., & Rajesh, S. (2019). Optimization of process parameters for friction stir spot welding of AA6061/Al2O3 by Taguchi method. *AIP Conference Proceedings, 2128*(July), 1–10. doi:10.1063/1.5117961

Suresh, S., Venkitaraj, K. P., Selvakumar, P., & Chandrasekar, M. (2011). Synthesis of Al2O3-Cu/water hybrid nanofluids using two step method and its thermo physical properties. *Colloids and Surfaces. A, Physicochemical and Engineering Aspects, 388*(1-3), 41–48. doi:10.1016/j.colsurfa.2011.08.005

Svarc, F., Ranocchia, R. P., Perullini, A. M., Jobbagy, M., & Aldabe, S. A. (2018). *A New Route to Obtain Perfluorodecalin Nanocapsules as An Oxygen Carrier in Cosmetic Formulations.*

Tadros, T. (2004). Application of rheology for assessment and prediction of the long-term physical stability of emulsions. *Advances in Colloid and Interface Science, 108*, 227–258. doi:10.1016/j.cis.2003.10.025 PMID:15072944

Tahmassbi, D., Zharghami, R., Vatanpour Azghandi, A., & Chaeich, M. (2011). Effects of nanosilver and nitrogen biofertilizer on yield and yield components of potato microtubes. *International Journal of Agriculture and Biology, 13*, 986–990.

Thornfeldt, C. (2005). Cosmeceuticals containing herbs: Fact, fiction, and future. *Dermatologic Surgery, 31*, 873–881. doi:10.1111/j.1524-4725.2005.31734 PMID:16029681

Thorny, C., Wirth, A., Kreimeyer, M., Wagner, F., & Vollertsen, F. (2007). Joining of dissimilar materials - New perspectives for lightweight design in the transportation industries. *Welding in the World, 51*, 311–326.

Tiwari, R. K., & Das, M. K. (2007). Heat transfer augmentation in a twosided lid-driven differentially heated square cavity utilizing nanofluids. *International Journal of Heat and Mass Transfer, 50*(9-10), 2002–2018. doi:10.1016/j.ijheatmasstransfer.2006.09.034

Tran, N., & Webster, T. J. (2010). Magnetic nanoparticles: Biomedical applications and challenges. *Journal of Materials Chemistry, 20*(40), 8760–8767. doi:10.1039/c0jm00994f

Triazine, B. (n.d.). TINOSORB® A 2 B. *Career Creations.* https://www.carecreations.basf.com/product-formulations/prod
uct-highlights/product-highlights-detail/TINOSORB-A2B/304781
25

Tripathy, S., & Dureja, H. (2015). Cosmetics: Regulatory Scenario in USA, EU and India. *J. Pharm. Technol. Res. Manag, 3*(2), 127–139. doi:10.15415/jptrm.2015.32010

Tripura, P., & Anushree, H. (2017). Novel delivery systems: Current trend in cosmetic industry. *European Journal Pharmaceutical and Medical Research*, *4*(8), 617–627.

Trisakti, V., & Wongwises, S. (2007). Critical review of heat transfer characteristics of nanofluids. *Renewable & Sustainable Energy Reviews*, *11*(3), 512–523. doi:10.1016/j.rser.2005.01.010

Trojovský, P., Dhasarathan, V., & Boopathi, S. (2023). Experimental investigations on cryogenic friction-stir welding of similar ZE42 magnesium alloys. *Alexandria Engineering Journal*, *66*(1), 1–14. doi:10.1016/j.aej.2022.12.007

U. S. Food Drug Administration. (2018). *Is It a Cosmetic, a Drug, or Both? (Or is it soap?)*. FDA. https://www.fda.gov/cosmetics/guidanceregulation/lawsregulations/ucm074201.htm

Ulutaş, A. (2022). *Joining Techniques Like Welding in Lightweight Material Structures.*, doi:10.4018/978-1-7998-7864-3.ch006

US Food and Drug Administration. (n.d.). *Cosmeceuticals*. FDA. https://www.fda.gov

Usman, M. S., El Zowalaty, M. E., & Shameli, K. (2013). Synthesis, characterization, and antimicrobial properties of copper nanoparticles. *International Journal of Nanomedicine*, *8*, 4467–4479. doi:10.2147/IJN.S50837 PMID:24293998

Vajravelu, K., Prasad, K. V., Lee, J., Lee, C., Pop, I., & Van Gorder, R. A. (2011). Convective heat transfer in the flow of viscous Ag-water and Cu-water nanofluids over a stretching surface. *International Journal of Thermal Sciences*, *50*(5), 843–851. doi:10.1016/j.ijthermalsci.2011.01.008

Vallejo, J., Sani, E., Żyła, G., & Lugo, L. (2019). Tailored silver/graphene nanoplatelet hybrid nanofluids for solar applications. *Journal of Molecular Liquids*, *296*, 112007. doi:10.1016/j.molliq.2019.112007

Van Gorder, R. A., Sweet, E., & Vajravelu, K. (2010). Nano boundary layers over stretching surfaces. *Communications in Nonlinear Science and Numerical Simulation*, *15*(6), 1494–1500. doi:10.1016/j.cnsns.2009.06.004

Vanitha, S. K. R., & Boopathi, S. (2023). Artificial Intelligence Techniques in Water Purification and Utilization. In *Human Agro-Energy Optimization for Business and Industry* (pp. 202–218). IGI Global., doi:10.4018/978-1-6684-4118-3.ch010

Vasanthi, P, Senthil selvan S, Benita Hexton D, Mohamed Salman Farooq I & Suresh P. (2017). Durability Study on Nano Concrete. *International Journal of Chemtech Research*, *10*(3), 693–698.

Vasu, K., Chelladurai, H., Ramaswamy, A., Malarvizhi, S., & Balasubramanian, V. (2019). Effect of fusion welding processes on tensile properties of armor grade, high thickness, non-heat treatable aluminium alloy joints. *Defence Technology*, *15*(3), 353–362. https://doi.org/10.1016/j.dt.2018.11.004

Vázquez-Núñez, E., Molina-Guerrero, C., Peña-Castro, J., Fernández-Luqueño, F., & Guadalupe de la Rosa-Álvarez, M. (2020). Use of Nanotechnology for the Bioremediation of Contaminants. *Processes (Basel, Switzerland)*, *8*, 826. doi:10.3390/pr8070826

Vennila, T., Karuna, M. S., Srivastava, B. K., Venugopal, J., Surakasi, R., & B., S. (2023). New Strategies in Treatment and Enzymatic Processes. In *Human Agro-Energy Optimization for Business and Industry* (pp. 219–240). IGI Global. doi:10.4018/978-1-6684-4118-3.ch011

Vidhu, V. K., & Philip, D. (2015). Phytosynthesis and applications of bioactive SnO2 nanoparticles. *Materials Characterization*, *101*, 97–105. doi:10.1016/j.matchar.2014.12.027

Vijaya Kumar, P., Madhusudhan Reddy, G., & Srinivasa Rao, K. (2015). Microstructure and pitting corrosion of armor grade AA7075 aluminum alloy friction stir weld nugget zone – Effect of post weld heat treatment and addition of boron carbide. *Defence Technology*, *11*(2), 166–173. doi:10.1016/j.dt.2015.01.002

Vimalraj, C., & Kah, P. (2021). Experimental review on friction stir welding of aluminium alloys with nanoparticles. *Metals*, *11*(3), 390. doi:10.3390/met11030390

Wang, H., & Chu, P. K. (2013). Chapter 4 — surface characterization of biomaterials. In: Amit B, Susmita B, (eds.) Characterization of biomaterials. Oxford: Academic Press.

Wang, Y., & Alocilja, E. C. (2015). *Gold nanoparticle-labeled biosensor for rapid and sensitive detection of bacterial pathogens*. Springer Science and Business Media LLC., doi:10.118613036-015-0014-z

Watchtower Bible and Tract Society of Pennsylvania (1995) Fashion—Ancient Greek Style. Awake (March 8), 21-24.

Watchtower Bible and Tract Society of Pennsylvania. (2012). *The Use of Cosmetics in Bible Times*. The Watchtower Public Edition (December 1).

Welch, C. M., & Compton, R. G. (2006). The use of nanoparticles in electroanalysis: A review. *Analytical and Bioanalytical Chemistry*, *384*(3), 601–619. doi:10.100700216-005-0230-3 PMID:16402180

Westesen, K., Bunjes, H., & Koch, M. H. J. (1997). Physicochemical characterization of lipid nanoparticles and evaluation of their drug loading capacity and sustained release potential. *Journal of Controlled Release*, *48*(2-3), 223–236. doi:10.1016/S0168-3659(97)00046-1

Wiedeman, M.P. (1963). Dimensions of blood vessels from distributing artery to collecting vein. *Circ Res.*, *12*(4), 375–378. doi:. RES.12.4.375 doi:10.1161/01

Wissing, S. A., & Müller, R. H. (2003). Cosmetic applications for solid lipid nanoparticles (SLN). *International Journal of Pharmaceutics*, *254*(1), 65–68. doi:10.1016/S0378-5173(02)00684-1 PMID:12615411

Wissing, S. A., & Müller, R. H. (2003). The influence of solid lipid nanoparticles on skin hydration and viscoelasticity–in vivo study. *European Journal of Pharmaceutics and Biopharmaceutics*, *56*(1), 67–72. doi:10.1016/S0939-6411(03)00040-7 PMID:12837483

Wong, K. V., & De Leon, O. (2017). Applications of Nanofluids: Current and Future, in Nanotechnology and Energy. Jenny Stanford Publishing.

Wong, K. V., & De Leon, O. (2009). Applications of Nanofluids. *Current and Future*, *2*, 5.

Wong, K. V., & De Leon, O. (2010). Applications of Nanofluids: Current and Future. *Advances in Mechanical Engineering*, *2*, 519659. doi:10.1155/2010/519659

Wu, D., Shen, J., Lv, L., Wen, L., & Xie, X. (2017). Effects of nano-SiC particles on the FSSW welded AZ31 magnesium alloy joints. *Materials Science and Technology*, *33*(8), 998–1003. do i:10.1080/02670836.2016.1254891

Wu, Q., Miao, W. S., Du Zhang, Y., Gao, H. J., & Hui, D. (2020). Mechanical properties of nanomaterials: A review. *Nanotechnology Reviews*, *9*(1), 259–273. doi:10.1515/ntrev-2020-0021

Wu, X., & Guy, R. H. (2009). Applications of nanoparticles in topical drug delivery and in cosmetics. *Journal of Drug Delivery Science and Technology*, *19*(6), 371–384. doi:10.1016/S1773-2247(09)50080-9

Wu, X., Li, C., Zhou, Z., Nie, X., Chen, Y., Zhang, Y., Cao, H., Liu, B., Zhang, N., Said, Z., Debnath, S., Jamil, M., Ali, H. M., & Sharma, S. (2021). Circulating purification of cutting fluid: An overview. *International Journal of Advanced Manufacturing Technology*, *117*(9–10), 2565–2600. doi:10.100700170-021-07854-1 PMID:34465936

Xu, X., Costa, A. P., Khan, M. A., & Burgess, D. J. (2012). Application of quality by design to formulation and processing of protein liposomes. *International Journal of Pharmaceutics*, *434*(1-2), 349–359. doi:10.1016/j.ijpharm.2012.06.002 PMID:22683453

Yang, B., Gao, L., Xue, M., Wang, H., Hou, Y., Luo, Y., Xiao, H., Hu, H., Cui, C., Wang, H., Zhang, J., Li, Y. F., Xie, G., Tong, X., & Xie, Y. (2021). Experimental and simulation research on the preparation of carbon nano-materials by chemical vapor deposition. *Materials (Basel)*, *14*(23), 7356. doi:10.3390/ma14237356 PMID:34885507

Yang, J., Oliveira, J. P., Li, Y., Tan, C., Gao, C., Zhao, Y., & Yu, Z. (2022). Laser techniques for dissimilar joining of aluminum alloys to steels: A critical review. *Journal of Materials Processing Technology*, *301*, 117443. doi:10.1016/j.jmatprotec.2021.117443

Yang, L., Ji, W., Huang, J., & Xu, G. (2019). An updated review on the influential parameters on thermal conductivity of nano-fluids. *Journal of Molecular Liquids*, *296*, 111780. doi:10.1016/j.molliq.2019.111780

Yang, L., & Watts, D. J. (2005). Particle surface characteristics may play an important role in phytotoxicity of alumina nanoparticles. *Toxicology Letters*, *158*(2), 122–132. doi:10.1016/j.toxlet.2005.03.003 PMID:16039401

Yang, R., Zhang, Z., Zhao, Y., Chen, G., Guo, Y., Liu, M., & Zhang, J. (2015). Effect of multi-pass friction stir processing on microstructure and mechanical properties of Al3Ti/A356 composites. *Materials Characterization, 106*, 62–69. doi:10.1016/j.matchar.2015.05.019

Yapar, E. A., & Inal, O. (2012). Nanomatrials and cosmetics. *Journal of Pharmacy of Istanbul University, 42*(1), 43–70.

Yeo, P. L., Lim, C. L., Chye, S. M., Ling, A. P., & Kand Koh, R. Y. (2018). Niosomes: A review of their structure, properties, methods of preparation,and medical applications. *Asian Biomedicine, 11*(4), 301–314. doi:10.1515/abm-2018-0002

Ye, W., Li, X., Luo, J., Wang, X., & Sun, R. (2017). Lignin as a green reductant and morphology directing agent in the fabrication of 3D graphene-based composites for high-performance supercapacitors. *Industrial Crops and Products, 109*, 410–419. doi:10.1016/j.indcrop.2017.08.047

Yoo, D. H., Hong, K. S., & Yang, H. S. (2007). Study of thermal conductivity of nanofluids for the application of heat transfer fluids. *Thermochimica Acta, 455*(1–2), 66–69. doi:10.1016/j.tca.2006.12.006

You, G. L., Ho, N. J., & Kao, P. W. (2013). In-situ formation of Al2O3 nanoparticles during friction stir processing of AlSiO2 composite. *Materials Characterization, 80*, 1–8. doi:10.1016/j.matchar.2013.03.004

Yu, W., & Huaqing, X. A. (2012). Review on nano-fluids: Preparation, stability mechanisms, and applications. Journal of Nanomaterials.

Yupapin, P., Trabelsi, Y., Nattappan, A., & Boopathi, S. (2023). Performance Improvement of Wire-Cut Electrical Discharge Machining Process Using Cryogenically Treated Super-Conductive State of Monel-K500 Alloy. *Iranian Journal of Science and Technology. Transaction of Mechanical Engineering, 47*(1), 267–283. doi:10.100740997-022-00513-0

Yurtkuran, H. (2021). An Evaluation on Machinability Characteristics of Titanium and Nickel Based Superalloys Used in Aerospace Industry. *İmalat Teknolojileri ve Uygulamaları, 2*(2), 10–29. doi:10.52795/mateca.940261

Yu, W., France, D. M., Routbort, J. L., & Choi, S. U. S. (2008). Review and comparison of nanofluid thermal conductivity and heat transfer enhancements. *Heat Transfer Engineering, 29*(5), 432–460. doi:10.1080/01457630701850851

Yu, W., & Huaqing, X. A. (2012). Review on nano-fluids: Preparation, stability mechanisms, and applications. *Journal of Nanomaterials, 2012*(435873), 17p.

Zahmakıran, M., & Özkar, S. (2011). Metal nanoparticles in liquid phase catalysis; from recent advances to future goals. *Nanoscale, 3*(9), 3462. doi:10.1039/c1nr10201j PMID:21833406

Zhang, X.-F., Liu, Z.-G., Wei, S., & Sangiliyandi, G. (2016). Silver Nanoparticles: Synthesis, Characterization, Properties,Applications,andTherapeuticApproaches. *International Journal of Molecular Sciences, 17*(9), 1534. doi:10.3390/ijms17091534 PMID:27649147

Zhu, Y., Zhao, Q., Li, Y., Cai, X., & Li, W. (2006). The interaction and toxicity of multi-walled carbon nanotubes with Stylonychia mytilus. *Journal of Nanoscience and Nanotechnology*, *6*(5), 1357–1364. doi:10.1166/jnn.2006.194 PMID:16792365

About the Contributors

Sampath Boopathi is an accomplished individual with a strong academic background and extensive research experience. He completed his undergraduate studies in Mechanical Engineering and pursued his postgraduate studies in the field of Computer-Aided Design. Dr. Boopathi obtained his Ph.D. from Anna University, focusing his research on Manufacturing and optimization. Throughout his career, Dr. Boopathi has made significant contributions to the field of engineering. He has authored and published over 100 research articles in internationally peer-reviewed journals, highlighting his expertise and dedication to advancing knowledge in his area of specialization. His research output demonstrates his commitment to conducting rigorous and impactful research. In addition to his research publications, Dr. Boopathi has also been granted one patent and has three published patents to his name. This indicates his innovative thinking and ability to develop practical solutions to real-world engineering challenges. With 16 years of academic and research experience, Dr. Boopathi has enriched the engineering community through his teaching and mentorship roles. He has served in various engineering colleges in Tamilnadu, India, where he has imparted knowledge, guided students, and contributed to the overall academic development of the institutions. Dr. Sampath Boopathi's diverse background, ranging from mechanical engineering to computer-aided design, along with his specialization in manufacturing and optimization, positions him as a valuable asset in the field of engineering. His research contributions, patents, and extensive teaching experience exemplify his expertise and dedication to advancing engineering knowledge and fostering innovation.

J. Paulo Davim is a Full Professor at the University of Aveiro, Portugal. He is also distinguished as honorary professor in several universities/colleges/institutes in China, India and Spain. He received his Ph.D. degree in Mechanical Engineering in 1997, M.Sc. degree in Mechanical Engineering (materials and manufacturing processes) in 1991, Mechanical Engineering degree (5 years) in 1986, from the University of Porto (FEUP), the Aggregate title (Full Habilitation) from the University of Coimbra in 2005 and the D.Sc. (Higher Doctorate) from London Metropolitan

University in 2013. He is Senior Chartered Engineer by the Portuguese Institution of Engineers with an MBA and Specialist titles in Engineering and Industrial Management as well as in Metrology. He is also Eur Ing by Engineers Europe FEANI-Brussels and Fellow (FIET) of IET-London. He has more than 35 years of teaching and research experience in Manufacturing, Materials, Mechanical and Industrial Engineering, with special emphasis in Machining & Tribology. He has also interest in Management, Engineering Education and Higher Education for Sustainability. He has guided large numbers of postdoc, Ph.D. and master's students as well as has coordinated and participated in several financed research projects. He has received several scientific awards and honors. He has worked as evaluator of projects for ERC-European Research Council and other international research agencies as well as examiner of Ph.D. thesis for many universities in different countries. He is the Editor in Chief of several international journals, Guest Editor of journals, books Editor, book Series Editor and Scientific Advisory for many international journals and conferences.

***

A. Kalaiyarasan is working as Assistant Professor, Muthayammal Engineering College, Namakkal, Tamilnadu India.

Mimi Akash Pateria, Professor in the Department of Applied Physics at Shri Shankaracharya Technical Campus, located in Bhilai, Chhattisgarh 490020, India.

M Anusuya is a faculty member in the Department of Physics at Indra Ganesan College of Engineering, situated in Manikandam, Tamil Nadu 620012, India.

Manohar. B.S. is a faculty member in the Department of Mechanical Engineering at Aditya College of Engineering, located in East Godavari, Andhra Pradesh 533437, India.

Anitha D. received her Ph.D degree from Anna University, Chennai, India. She is having 17 years of experience in research and teaching. She is the life member of Indian Society for Technical Education, Institute for Engineering Research and Publication, International Association of Engineers and National Institute for Technical Training and Skill Development. So far she has published 10 research papers in various national, International conferences and journals. Her area of interest is Image Processing, Data Mining, Machine Learning, Cloud Computing and Neural Network. She always engaged to fill the gap between the students and industry with latest technologies. Currently she is working as an Associate Professor in Department of Information Technology, Muthayammal Engineering College, Rasipuram, Tamilnadu, India.

Pankaj Dadheech received his Ph.D. degree in Computer Science & Engineering from Suresh Gyan Vihar University (Accredited by NAAC with 'A' Grade), Jaipur, Rajasthan, India. He received his M.Tech. degree in Computer Science & Engineering from Rajasthan Technical University, Kota and he has received his B.E. in Computer Science & Engineering from University of Rajasthan, Jaipur. He has more than 18 years of experience in teaching. He is currently working as an Associate Professor & Dy. HOD in the Department of Computer Science & Engineering (NBA Accredited), Swami Keshvanand Institute of Technology, Management & Gramothan (SKIT), Jaipur, Rajasthan, India. He has published 23 Patents at Intellectual Property India, Office of the Controller General of Patents, Design and Trade Marks, Department of Industrial Policy and Promotion, Ministry of Commerce and Industry, Government of India. He has published 5 Australian Patents at Commissioner of Patents, Intellectual Property Australia, Australian Government. He has 60 publications in various International & National Journals, 55 papers in various National & International conferences. He has published 8 Books & 18 Book Chapters.

Yagya Dutta Dwivedi is a faculty member in the Department of Aeronautical Engineering at the Institute of Aeronautical Engineering, located in Hyderabad, Telangana 500043, India.

M. Elangovan holds the esteemed position of Professor in the Department of Aerospace Engineering at SNS College of Technology, situated at Saravanampatti post, Coimbatore district, Tamil Nadu, with the pin code 641035.

Kavitha G. is a Professor in the Department of Computer Science & Engineering, Muthayammal Engineering College,TamilNadu, India. She has a good Academics & Research experience in various areas of Computer Science. She is completed Ph.D. from Anna University, Chennai. Her area of research includes Networking, Machine Learning, Data Science and Big Data. She has filed successfully seven Patents and received award called Distinguished Professor, Research Excellence Award, Global Eminent Teacher Award,Academic Excellence Awards. She has published number of research papers in various International/National Journals (including Elsevier, Springer etc.). She is an active member of various Professional International Societies(IE(I),ACM,ISTE,CSI). She has also attended seminars, workshops and short-term courses in Various University. She has published four text book and three book Chapter content in reputed publications.

Sudarmozhi G. is working as a Professor in Science and Humanities with an experience of more than 30 years in academic

Parth Gaud is a faculty member in the Department of Mechanical Engineering at Parul Institute of Engineering and Technology, which is a part of Parul University, located in Vadodara, Gujarat 391760, India.

Kode Jaya Prakash is Assistant Professor Department of Mechanical Engineering VNR Vignana Jyothi Institute of Engineering & Technology Hyderabad.

Bashkaran K. is currently working as a Assistant Professor in Kongunadu College of Engineering and Technology in the department of Biomedical Engineering. I am willing to Research in the area of BioMEMS.

Radhika K. is working as a Professor in the Department of Electronics and Communication Engineering at Muthayammal Engineering College, Rasipuram.. She has 15 years of Teaching and Research experience. She has published papers in reputed International Journals and Conferences,07 Books/Chapters and published 06 Patents.She is the Life Member of ISTE, IETE. Her research interest includes VLSI, Embedded Systems and Instrumentation.

Sankar K. is working as Assistant Professor, Muthayammal Engineering College, Namakkal, Tamilnadu India

Shemalaa K. is expertise in Theory of computation, Software Engineering and programming in C for computer science Engineering.

L. Karthick is a faculty member in the Department of Mechanical Engineering at Hindusthan College of Engineering and Technology, located in Coimbatore 641032, India.

Francis Luther King M is currently an Associate Professor at Swarnandhra College of Engineering and Technology in the Department of Mechanical Engineering, He obtained his PhD in Manufacturing Engineering from FEAT, Annamalai University, Chidambaram, Tamilnadu, India and Published research articles in various reputed journals and participated in several high profile Conferences, Seminars and Workshops.

Harikaran M. is working as Assistant Professor in Muthayammmal Engineering College, India.

Sanchaya M. is Assistant Professor in Muthayammmal Engineering College, India

Shenbagapriya M. received "Mentor Appreciation Award" by Texas University Program and Department of Science and Technology (DST) Govt of India. Received MTLC "UTLP Champion Award" from Wipro Technologies, Bangalore. Areas of interests include Biomedical Image Processing and Applications of soft computing techniques in image analysis using Artificial intelligence, Machine and deep learning. Also interested in content writing, Documentation, creative writing, Technical writing, proof reading and revising. As a Coordinator, received Sponsors from various funding Agencies also coordinated AICTE sponsored SDP and STTP, IEEE MAS Sponsored Symposium / Workshops and Seminars . Published / presented 41 papers in the International / National Conferences and International Journals. Invited for guest lecturers in CSI Polytechnic Sponsored by IEI – Institute of Engineers India. Delivered lectures in IEEE MAS Sponsored National Level Seminars and also chaired the sessions in International/ National conferences and symposiums.

Elango Natarajan (Senior Member, IEEE) received the Ph.D. degree in mechanical engineering from Anna University, Chennai, India, in 2010. He worked as a Postdoctoral Research Fellow at UTM, Skudai, Malaysia, in 2013. He has served for engineering colleges/university for about 20 years in various academic positions. He is now associated with the Department of Mechanical and Mechanical Engineering, UCSI University, Malaysia, as an Associate Professor. He has completed many funded projects and published 105research articles in Scopus/Wos as of today. He is currently the Chairman of the IEEE/Robotics and Automation Society, Malaysia. He is also a Chartered Mechanical Engineer (C.Eng.), who specialized in mechanical engineering design, composite materials, CAE, and soft robotics.

Keerthanadevi P. is an Assistant Professor in Muthayammmal Engineering College, India

Kulandhaivel P. is working as Assistant Professor in Kongu Engineering College, India

P. Suresh completed his PhD in manufacturing in 2010 and having 18 years of Teaching and research experience.He has published 106 paper in national and international journal.He is a Fellow Member Institution of Engineers (FIE), Senior Member IEEE, Life Member of ISTE, Life Member of ISNT, Life Member of IIIE, Life Member of Indian Society of Ergonomics (MISE), Life Member of IACSIT, Life Member of IAENG.He is a recipient of Best Engineer Award.

Binay Kumar Pandey currently working as an Assistant Professor in Department of Information Technology of Govind Ballabh Pant University of Agriculture and Technology Pantnagar Uttrakhand, India. He obtained his M. Tech with Specialization in Bioinformatics from Maulana Azad National Institute of Technology Bhopal M. P. India, in 2008 . He obtained his First Degree B. Tech at the IET Lucknow (Uttar Pradesh Technical University, Uttar Pradesh and Lucknow) India, in 2005. In 2010, he joined Department of Information Technology of College of Technology in Govind Ballabh Pant University of Agriculture and Technology Pantnagar as an Assistant Professor and worked for various UG and PG projects till date. He has more than ten years of experience in the field of teaching and research.He has more than 40 publications in reputed peer journal reputed journal Springer,Inderscience,(sci and socopus indexed journal and others) and 3 patent.He has many awards such PM Scholarship etc . He session chair in IEEE International Conference on Advent Trends in Multidisciplinary Research and Innovation (ICATMRI-2020) on December 30, 2020 organized by Pankaj Laddhad Institute of Technology and Management Studies; Buldhana, Maharashtra, India

Digvijay Pandey is a Lecturer in the Department of technical Education, Research Scholar IET Lucknow, India.

G. Rathinasabapathi is a faculty member in the Department of Mechanical Engineering at Panimalar Engineering College, situated in Chennai, Tamil Nadu 600123, India.

Khandekar Reza is Assistant professor at MAKAUT, West Bengal. Department of Pharmaceutical Technology.

Rajesh S. has areas of the research including thermal power system, bio diesel and manufacturing.

Sherin Nithya S. is currently an Associate Professor at Kasturba Gandhi Nursing College, in the Department of Child Health Nursing, Currently pursuing her PhD from Sri Balaji Vidyapeeth University, Pondicherry, India and Published research articles in reputed journals and participated in several Conferences, Seminars and Workshops.

Sundaram S. is working as Professor and Dean, Muthayammal Engineering College, Namakkal, Tamilnadu India.

Suresh S. has interest in the research areas of friction stir welding, friction stir processing and 3D printing.

C. Sailaja is a faculty member in the Department of Mechanical Engineering at Bangalore College of Engineering and Technology, situated in Chandapura, Bangalore 560099, India.

M. Sudhakar is in the department of Mechanical Engineering, Sri Sairam Engineering College, Chennai, Tamil Nadu.

Meka Umareddy is a renowned lecturer in the field of Chemical Engineering at the University of Technology and Applied Sciences, located in Salalah, Dhofar Region, Oman.

N. Suthanthira Vanitha is a Professor in the Department of Electrical and Electronics Engineering and Dean - Quality Assurance at Muthayammal Engineering College, Rasipuram. She has 21 years of experience in Teaching and Research. She has published 166 papers in International Journals and Conferences,12 Books/ Chapters and 13 Patents. She is the Life Member in IEEE, ISTE and Fellow Member in IE(I). Her research area of specialization includes Embedded Systems, Biomedical Instrumentation and Renewable Energy.

Index

F

G

I

L

M

N

P

S

T

U

www.ingramcontent.com/pod-product-compliance
Lightning Source LLC
LaVergne TN
LVHW080605300926
849162LV00005B/1051